HOLOGRAPHY HANDBOOK

Making Holograms the Easy Way

Hologram Included

Fred Unterseher   Jeannene Hansen   Bob Schlesinger

Holography Handbook

HOLOGRAPHY HANDBOOK

Making Holograms the Easy Way

Fred Unterseher   **Jeannene Hansen**   **Bob Schlesinger**

ROSS BOOKS
p.o. box 4340
Berkeley, Calif.
94704

Copyright © 1982 Unterseher, Hansen and Schlesinger

Unterseher, Fred
    Holography handbook.

    Bibliography: p.
    Includes index.
       1. Holography - Handbooks, manuals, etc.
2. Photography, Abstract - Handbooks, manuals, etc.
I. Hansen, Jeannene, joint author.
II. Schlesinger, Bob, joint author.
III. Title.
TA1542.U57   621.36'75   80-17161
ISBN 0-89496-018-0
ISBN 0-89496-017-2 (pbk.)

Credits for cover photographs:

Front cover: "Cubes" image created by Fred Unterseher. Photo taken at holografix studio by Jeannene Hansen. Linda Charlop at the table.

Back Cover:

Back cover photos. Clockwise from upper left:

1.) "Embryo", white light transmission hologram by Dan Schweitzer. Photo © Nancy Safford, courtesy MOH.

2.) "Laser Disc" Chromagram Diffraction Grating by Vince DiBiase. 1979

3.) "Echo Chamber" white light transmission hologram by Rudie Berkhout, photo © Rudie Berkhout, collection MOH.

4.) "Cubes" Dichromate hologram produced at Dichromate Inc. by the authors. 1981. Photo by Franz Ross.

5.) "Still Life" by James Feroe, Open aperture hologram, 1979.

For those of you interested, this book was basically put together as follows:

Fred Unterseher: Original idea and concept for book. Overall editorial outline and organization.

Jeannene Hansen: Illustrations and layout.

Bob Schlesinger: Written material. Organization of material, photoreproduction, information research, some layout and illustration, overall proofreading.

Franz Ross: Lots more than a publisher usually does.

Lyrinda Snyderman: A special tip of the hat for her excellent drawings.

Mr. Unterseher, Holographic Pioneer

Ms. Hansen, Pioneer

Mr. Schlesinger, Holographic Pioneer

preface

"Why isn't there a book that describes how to make holograms that work - easily - and without a lot of mumbo jumbo? Does one need to be some kind of scientific whiz to do this? Does one need to be a *wealthy* whiz at that? Does one need to beware of huge lasers that are 'gonna getchya'?

We still continue to receive comments like these from folks. And yet, with the invention of holography more than 30 years behind us and the practical making of holograms with a laser almost 20 years old, these questions should not even have to be asked - but they are.

Holography is perhaps the most exciting revolutionary medium of the century, yet it is still one of the most easily misunderstood and mis-represented. Well, we're going to do something about this, and finally pull the plug on all the hocus-pocus. Enough of these holography texts which have so much math that you can't tell if you're reading the thing upside down or not, or those with incomplete information (or tech-niques that just don't work!).

Holography is easy. Anyone can make good quality holograms without any technical training. Holography is inexpensive, fun, and a dynamic outlet for creative expression. And this is the book that's going to prove it to you, or, we should say, show you how to prove it to yourself.

This book developed out of experiences with making thousands of holograms, and teaching hundreds of others to make countless more, over the past decade. We *know* they're easy to do. We think we've picked up some valuable pointers over the years - good shortcuts as well as quick ways around the common pitfalls. We describe not only how to make holograms with a minimum of effort, but how to recognize and correct problems that arise. That, we feel, is the job of a good workbook - the kind of book you'll actually *work with* in the studio, and use to help develop your own techniques and, we hope, to make new discoveries.

This is a do-it-first, ask questions later book. We take you through building or acquiring all necessary equipment*, setting it up, and then using it to make all kinds of holograms. *Then*, we run through a bit of theory, which we feel is a lot easier to understand if you've actually done the work yourself. Incidentally, the theory is light and easy to read, in keeping with the style of the whole book. Those readers interested in pursuing holographic theory in greater detail will find our annotated bibliography at the end of the book helpful (In this way, this book becomes an excellent practical companion to other textbooks for use in the classroom).

---

*Incidentally, prices for holographic equipment, like everything else these days, changes rapidly and almost always upward. So prices which appear in some sections may soon be out of date. Still, the cost of establishing a holographic lab is comparable with that of a photo darkroom and can be relatively inexpensive.

Later sections of the book discuss holography as art as well as the broader implications of how the process can be utilized conceptually to explain the workings of various natural phenomena.

Consider this book a tool, or simply a means towards developing a personal approach to the medium. The optical setups we describe are not necessarily meant to be the last word on the subject. There are many ways of achieving the same result. We did find them to be the easiest to begin with, but hope that they will stimulate creative new approaches, as familiarity with holographic techniques develops with time. Subsequent editions of this book may reflect new ideas that we, or others (perhaps you!) may contribute in the future.

The holographer's experience is truly evolutionary in nature and we hope this book will prove a valuable companion throughout the process, whether it be during the early stages, or with advanced work.

These are the early days of holography. We might compare ourselves to photographers working prior to 1860 - performing visual magic in a vast field of untapped potential, or to a present day astronaut about to embark on a journey to a completely new environment. We welcome you, as a fellow pioneer, to join in the excitement of being involved in it from the beginning.

Photo Credit: NASA

In this book, you will periodically find references to the "KISS" philosophy. This stands for Keep It Sweet and Simple and is meant as a gentle reminder that the simplest method is often the best.

table of contents

materials

basic holography

advanced holography

appendix

introduction

"What in the world is holography?" This phrase is heard less and less these days as the word "hologram" shows up in science fiction books and movies, and true holograms are used for commercial displays, or may be viewed in museums and galleries, and purchased as jewelry and gifts.

The magic of the hologram offers us a never ending sense of wonderment. The hologram itself is but a thin flat piece of film or glass, yet we treat it as we would a window. Viewing a hologram is rather like peering through a portal into a nonexistent three dimensional world of unbelievable realism. As you move your head up and down or side to side, you can actually see around different parts of the image. The 3-D scene is not a trick, either. Some other techniques, such as stereophotography, give the illusion of dimension from flat photos. Holography is the first *truly* 3-D image-making process, since it recreates what light waves actually do after reflecting from a real object. An explanation of why a hologram does this will be found later in this book.

A number of distinctly different types of holograms exist. Some, such as acoustical or sound holograms, do not even use light. For our purposes, we shall limit ourselves to optical holograms which we may use to create incredible visual displays.

Of these, perhaps the best known are of the laser viewable transmission variety. Light from a laser is spread out by a lens to light up the holographic plate. The viewer stands on the side of the plate opposite the laser, so that the light is "transmitted" through the hologram, and peers into the "window" to see the dimensional scene beyond.

The reflection hologram, although it must be made with a laser, can be viewed with ordinary white light. The hologram is seen with the light source and the viewer on the same side of the plate, the light reflecting from the hologram to the observer.

Some specially made transmission holograms can also be viewed in white light, with very bright images as a result. Rainbow and open aperture holograms operate in this way. These types can also be turned into "pseudo-reflection" holograms by placing a mirror on the back (This is the type supplied in the back of this book.)

Other kinds of holograms include the cylindrical type, which allows one to see all the way around an object; the holographic stereogram which allows the 3-D imaging of moving live subjects or outdoor scenes; the pulsed hologram for true holographic portraiture, and true color holograms made with red, green, and blue laser light. As one might expect, some types are more easily done than others. We shall discuss all and demonstrate how to make all but the few that require elaborate equipment.

what a hologram is not

Some devices are commonly called holograms but aren't. Princess Leia popping out of R2D2 in *Star Wars*, and the chess pieces were made to *look* like holograms but they were not. They were the result of matt-screening techniques often used in filmmaking. The images in the haunted house in Disneyland are also made with more conventional optical techniques. One technique often used is a double parabolic mirror, which makes a real object appear to be located elsewhere (This technique is used in a currently popular conversation piece, which consists of a box with a hole on top containing a coin apparently floating in space.) Also, holograms are *not* made by crisscrossing laser beams in space, with images appearing in midair where the beams intersect. We are going to show you how to make *real* holograms!

what it is

The excitement of viewing a hologram is only exceeded by the thrill of actually making one. The hologram must be made on a certain type of optics table to insure that everything remains motionless during exposure. We demonstrate how to build the table easily and cheaply. We then take a laser (of the inexpensive variety that, by the way, will not blast holes in you or in your loved ones), and split the beam into two parts.

One part will light up the object, the other will act as what is called a reference beam. Together, they will combine or interfere at the holographic plate and be recorded during the exposure. With some holograms, including the first ones we outline, the beam need not even necessarily be split to do this. The hologram is then processed using methods similar to those in photographic procedures. After processing, the hologram need only be illuminated with the proper light (depending on which type of hologram it is), and the object will appear magically displayed in three dimensions.

short history of holography

The first hologram was conceived of and produced by Dr. Dennis Gabor, a researcher at the Imperial College in London, long before the invention of the laser. In 1948, he published a paper entitled, "Image Formation by Reconstructed Wavefronts". He was the first to realize the awesome capabilities of the phenomenon of interference of light. For this he received the Nobel prize in Physics in 1971. Gabor was originally interested in storing an image illuminated with very short waves, such as x-rays, and playing it back with very long waves. In this way, he felt, great magnifications and improved imaging of the electron microscope might be achieved.

Gabor's major problem was the inability to find the proper source of light for his experiments. Not only was not a useful source of

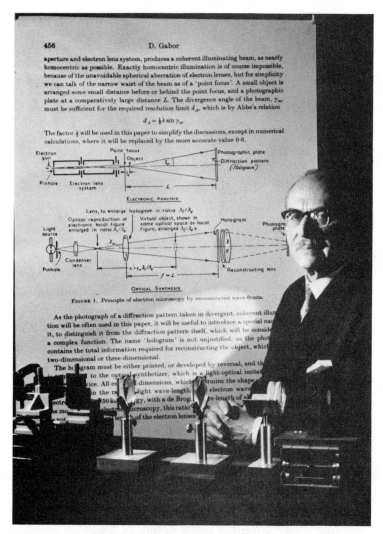

Portrait of Dennis Gabor. Photo by Fritz Goro, Time-Life Inc. Courtesy MOH (Museum of Holography)

x-rays available for his work, but neither was a device emitting the proper kind of visible light. Gabor's early holograms were crude, yet he was able to achieve results by filtering the light carefully, in an attempt to produce what the laser would supply 13 years later. Still, this light was not "pure" enough. His holograms, were very different from those usually demonstrated today, and were incapable of storing 3-dimensional scenes.

Dr. Dennis Gabor receives Nobel Prize from the King of Sweden, 1971. Photo from Nobel Prize Committee, Courtesy MOH.

Except for the work of a few researchers, very little was published in the field during the next decade. In 1960, the laser was invented. Experimenters familiar with Gabor's work, including Denisyuk in the Soviet Union and Leith and Upatnieks in the U.S. at the University of Michigan, applied this special new light, along with some rather novel optical techniques, to produce the first practical holograms. In the next several years, the scientific journals were filled with articles, as numerous laboratories jumped

Y.N. Denisyuk. Photo by Dr. S.A. Benton. Courtesy MOH

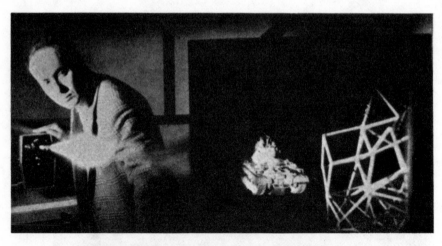

Emmett Leith, head of University of Michigan team that developed laser holograms, shows sample.

into what was seen as one of the most exciting inventions of this century. Hundreds of patents were issued as the big labs competed to develop everything from holographic television (not yet a reality) to holographic testing of airplane wings under stress (a practical application of the technique currently used). At the same time, everyone continued to marvel at the magic of nonexistent 3-D scenes appearing out of nowhere from a flat piece of film or glass.

One of the most common assumptions made about the historical development of holography, is that early holograms required a laser for viewing, whereas more recent breakthroughs allow the use of ordinary light. Actually, two different types of holograms, the laser viewable transmission hologram, and the white light viewable reflection hologram, were developed almost simultaneously, although independently. In the U.S., most of the work was with transmission holograms in the beginning, as these appeared most suitable for a variety of laboratory applications, and were most often seen in demonstrations of the technology. More recently, reflection holograms have gained in popularity, as techniques for producing better quality results have been perfected.

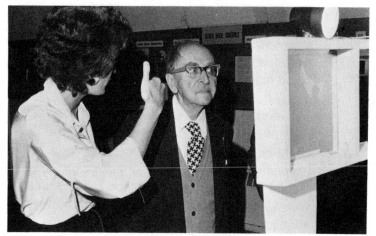

Rosemary Jackson, director of Museum of Holography and Dennis Gabor at MOH. Photo © by Brad Cantos.

In the late 60's, interest and money (defense spending), began to wane at the big labs, and although innovations continued, the original enthusiasum generated in the scientific community for practical uses appeared to be on the way out. Fortunately for us, the next chapter in the history of holography did not depend on the big labs.

Before the late 60's, it was assumed that the equipment needed to make holograms was priced far beyond the budgets of most home experimenters. The technology had been born of the professional optics lab; thus, it was taken for granted that thousands of dollars' worth of cumbersome equipment was needed to make even the simplest hologram. All this changed, however, after an ex-research scientist and laser pioneer by the name of Lloyd Cross, who

Lloyd Cross, photo © Sharon McCormack, courtesy MOH.

had been associated with Leith and Upatnieks at the University of Michigan, teamed up with a few artists. They had no access to the usual government or industrial mega-grants usually associated with holographic research, but found they could make holograms which were as good or better than those made by the large labs.

How did they do this while being so poor that they often ate their food stamps before they could be redeemed? The secret lay simply in understanding some basic principles of holography, and using a little common sense. It was possible to build a holographic lab, in many ways superior to those costing many thousands of dollars, *out of scrap materials!* (Technocrats watch out! Do you suppose there is a hidden lesson in all this?)

In front of San Francisco School of Holography, 1975. From left: Lloyd Cross, Michael Kan, T.H. Jeong, Emmet Leith. Photo by Sharon McCormack.

Along with the realization that the technology could be simplified, was the understanding that it could be taught, not only to the postdoctorate theoretician, but to Mr. & Ms. J.Q.Public. Most photographers were not able to derive, mathematically, the scientific principles at work in their art; there was no need to expect holographers to do the same.

With this in mind, Lloyd moved to the west coast and, along with several artists (including one of the authors of this book), established the San Francisco School of Holography in 1971.

Hundreds of persons with no science background took courses in practical techniques and were able, in turn, to teach hundreds more to make fine quality holograms. Small basement studios appeared with independent experimenters and artists rapidly filling the vacuum left by the exodus of the large labs. Significant improvements in technique were due to the efforts of many "little guys". A healthy diversification

Members of Multiplex Company, 1973. Top: Pam Brazier, Middle: Michael Fisher, Lloyd Cross, Michael Kan, Bottom: Peter Claudius, Dave Schmidt. Photo courtesy Multiplex Co. SF, CA. and MOH.

was occurring as well, with artists investigating the aesthetics of the new medium, entrepreneurs forming small businesses to manufacture holographic products, and the scientific community continuing to publish as well as to offer curricula in holography as part of physics or other studies.

It is in this spirit that we write this book. We shall demonstrate that, armed with only a practical sense of the fundamentals, anyone can become involved in the above activities.

In the last decade holography has, we believe, evolved well beyond its origin as a novel laboratory curiosity. It has fought its way to acceptance by the art "establishment" as a true medium in its own right. Businesses have begun professionally to develop its commercial potential. It's clear that the hologram will play a major role in a great many applications. We hope this book will be a useful guide to developing your own ideas in a field which should burst wide open in the next few years. We wish you the best in getting in on the action now.

"Hand in Jewells" by Robert Schinella. Displayed by Cartier, the international jeweler, at 5th avenue and 52nd street in New York City in 1972 and again in 1979. The holographic image appeared to project through Cartier's window out over the 5th Avenue sidewalk. One woman, visably upset, claimed it was the work of the devil and attempted to attack the image with her umbrella. Most others, however, simply stopped to gaze in wonderment.

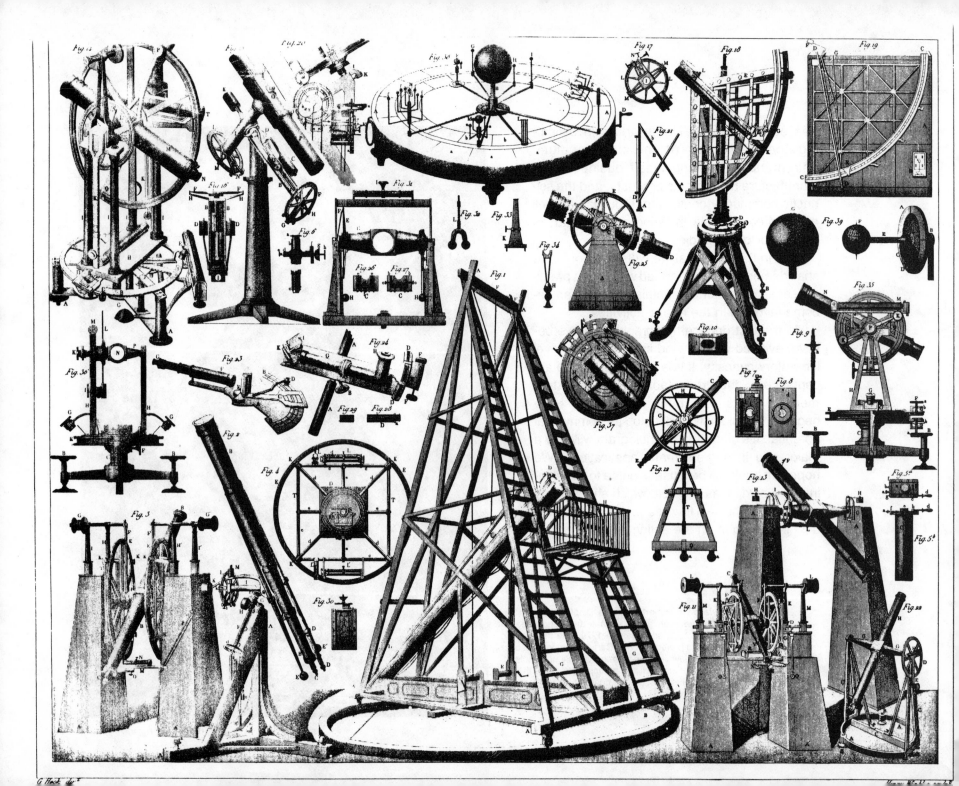

materials

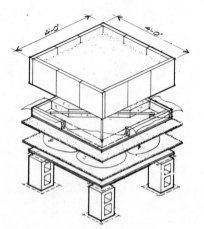

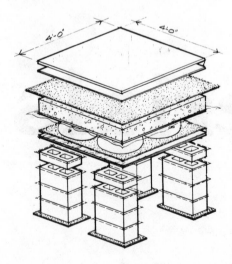

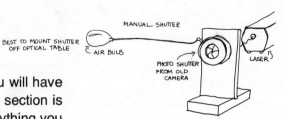

BEST TO MOUNT SHUTTER
OFF OPTICAL TABLE

MANUAL SHUTTER

AIR BULB

PHOTO SHUTTER
FROM OLD
CAMERA

LASER

introduction

To make holograms properly, you will have to get hold of some equipment. This section is devoted to building or acquiring everything you will need. We begin with those items we suggest you build to develop a holographic studio at very low cost. This includes the optics table, with three different design choices available; setting up the darkroom where the holograms will be "developed", and building optical mounts, which will hold the mirrors, lenses, etc., in place on the table. Next are those items which should be purchased, including the laser, mirrors, lenses, and beamsplitters to fit into the mounts you will build, and then the light meter. Following that, are items which can be purchased for better-quality or advanced types of holograms. For those who would rather buy than build a table, components, etc., a selection of conventional equipment is offered next. Finally, we cover the various recording materials (films and plates) used to make holograms (what we suggest you buy and where to get them.). After completing this section, you will be all set to make your first hologram.*

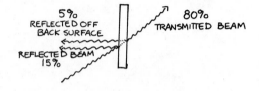

5%
REFLECTED OFF
BACK SURFACE

80%
TRANSMITTED BEAM

REFLECTED BEAM
15%

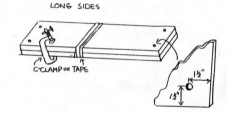

LONG SIDES

C-CLAMP OR TAPE

1½"

1½"

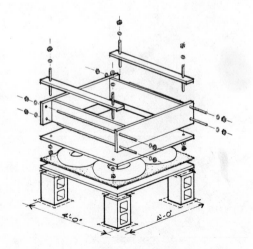

*Note: We strongly urge you to read the *entire* materials section before actually building or buying anything!

optics table construction

Stereograph of the San Francisco Earthquake
A bad time to make holograms.

stability

Holograms must be made in an environment entirely free of vibration or any type of movement. The goal is this: we wish to have a surface upon which we can mount optical components, and which is in some way isolated from the influence of the everyday rumble of life around it.

Heavy, low frequency sounds, such as those produced by passing traffic, or your neighbor practicing his drums, tend to travel easily along the surface of the earth, setting most other things (including your holographic components) oscillating almost inperceptably. This is often enough to ruin a hologram.

The problem is solved by using a vibration isolation system consisting of a very heavy table resting on an air support structure. Air is one of the best vibration eliminators available, and extremely heavy tables tend to be lazy and difficult to move.

Conventional optics systems usually utilize large granite or metal slabs supported by elaborate pneumatic devices to accomplish this (usually at a cost of thousands of dollars). We will discuss techniques which will result in an excellent holographic stability system, and will cost next to nothing.

An ingenious system was developed by Gerry Pethick, Lloyd Cross, and others, and used extensively at the San Francisco School ot Holography. The optics table consists simply of a heavy sandbox set upon ordinary automobile tire inner tubes. This "sand-based system" is still one of the most versatile around. Not only is it inexpensive, but it greatly simplifies the manipulation of optical components to moving lenses and mirrors mounted on pieces of plastic pipe around in the sand.

It's easy to see why the sand system works so well. A typical sand table will have almost a ton of sand, floating on inner tubes.

This is an extremely stable mass (for our purposes all the sand can be considered one massive object). The components, when set into the sand, become an integral part of this mass. The sand itself is such a poor vibration transmitter, that very little sound can make its way up to the components. And sand is cheap. For those holographers without five digit budgets, it is highly recommended.

where to put the table

Vibration isolation is a relative term. All matter is constantly vibrating to some degree. Also, anything will vibrate *noticeably* under the proper condition (as those of you who have lived through a major earthquake will attest).

It's therefore important to locate the sand table in a fairly quiet area. Nothing beats a cave.

In case this is impractical, try the ground floor, garage, or basement of your house. A concrete floor is best. Stay away from wood floors or upper floors of a building if at all possible. Buildings tend to amplify vibrations (the higher up one goes, the worse they become). This is not to say it's impossible to make holograms in an apartment building, and city dwellers should not be discouraged. But if you do live on the 26th floor, you're going to have to be quite a bit more careful. If you live on a high story of a building next to a train depot or an iron foundry, you had probably better forget it. Set up somewhere else*.

It is important to insure that the room used can be completely darkened. Remember, the holograms will be exposed on the table. The plates and film used are sensitive, not only to laser light, but to much of the ordinary light as well. A good way to keep this in mind is to imagine that the entire holographic setup is located inside a huge camera. If a camera with film is opened to the light, the film, of course, will fog and be ruined. So too with the sand-table room, which must be made completely dark while the holographic plate (or film) is being used prior to being developed.

---

*If you wish (and we recommend doing this), you may visually determine whether an area is suitable for a table by building a simple optical device called an interferometer. We will show you how to do this at the end of the chapter.

A room may be made dark in a variety of ways. A couple of layers of completely opaque material may be used to cover a window (black polyethylene is good for this). Light leaks can be covered with black electrical tape. A good way to find light leaks is to turn off all the lights, let your eyes get accustomed to the dark for a few minutes, and then look for spots where light gets through from the outside.

One need not be an absolute perfectionist with respect to making a room light-tight. Holographic emulsions are much less sensitive to light than ordinary photographic films. However, it is best to be on the safe side, and make the room as dark as possible.

design

Following are two different designs for sand tables. The first, or "tension sand table", developed by Lloyd Cross, is a wooden table held together with metal rods under stress, and filled with sand. The second table uses a concrete slab base, with mortared sides which hold in the sand. A third design, the magnetic slab table, is then discussed. The magnetic table is not recommended for beginning holographers, as the optical components required are more expensive, and generally more difficult to position. It is, however, an excellent table for serious workers in the field.

pros & cons

The advantages of the tension table are appealing in some respects. It is inexpensive to build and the materials are readily available; it is simple to put together (or take apart), and consequently easy to transport to another location.

Now the disadvantages: wood is more "alive" than cement and tends to vibrate (using particle board, as opposed to plywood, will alleviate some of the vibrations but not all); also, wood under the pressure of the sand has a tendency to warp over a period of time, so the life of this table may be shorter.

The sand/slab table is recommended for those "noiser" locations where stability might be a problem, or for those holographers certain to do advanced work. This table is *not* as easy to dissemble or transport as the tension sand table. On the other hand, it is so cheap you can afford to leave it when you move.

The table you choose will depend on whether you want to sacrifice some stability for portability or vice versa. We suggest the tension table as a starting point for holographers who may be on the move, or are unsure about their long term involvement.

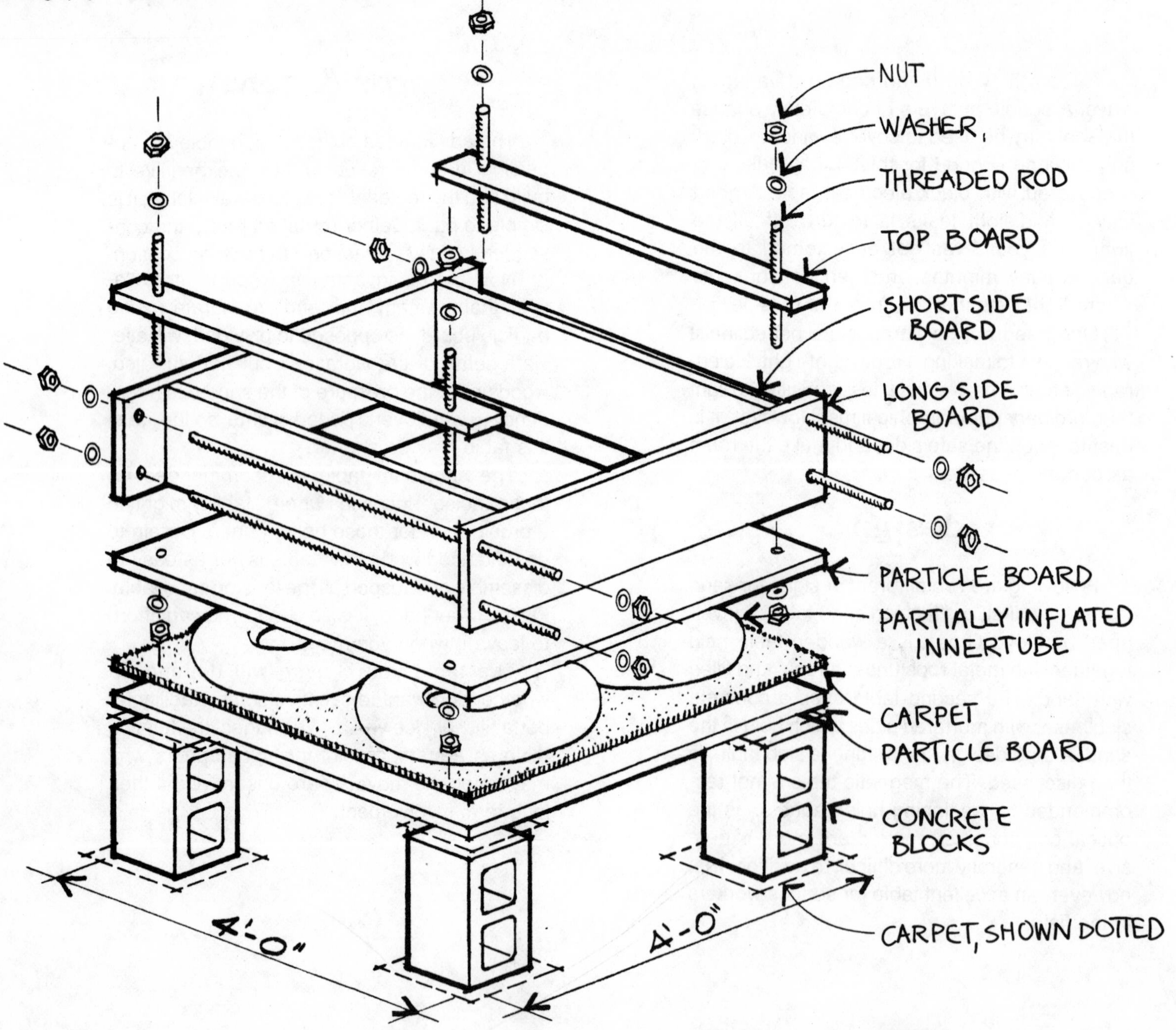

NUT

WASHER,

THREADED ROD

TOP BOARD

SHORT SIDE
BOARD

LONG SIDE
BOARD

PARTICLE BOARD

PARTIALLY INFLATED
INNERTUBE

CARPET

PARTICLE BOARD

CONCRETE
BLOCKS

CARPET, SHOWN DOTTED

4'-0"

4'-0"

tension sand table

A shopping list for a 4'x 4'x 1' table should include:

- Five construction blocks, each 8"x 8"x 16".
- Carpet: used (clean) pieces to be placed at the junctions of different materials (e.g., where inner tubes meet wood). You will need several 8"x 8" pieces, and two 4'x 4' pieces.
- Inner tubes: four smaller types (e.g., from a motorcycle, or Fiat, or Honda), with an 18"-20" diameter, would be ideal.
- Seventeen bags of #2 amber sand, or of a washed, clean variety (not the finest grain, but rather a #2 size grain of sand). Each bag of sand represents a cubic foot of area. A 4'x 4' table needs sixteen bags of sand (plus one for refilling later). Remember that 1600 lbs of sand is very heavy: two trips are much easier on your car's axle.

- Lumber:
Remember that particle board is preferred, to keep vibrations to a minimum. You will need:
- Two 4'x 8' sheets of plywood or particle board, ⅝" or ¾" thick.
- Two   1"x 4"x 4' Douglas fir, or pine, boards - good clear lumber in this case (to obtain these, cut one 1"x 4"x 4' sheet in half).
- Sixteen ⅜" nuts and sixteen ⅜" washers.

- Hardware:
- Four 15" lengths of ⅜" threaded rod; (this is also known as running stock: it is a solid ⅜" diameter rod, threaded its entire length. Five 6' rods will fill your requirements).

- Tools:
- One wood saw: hand, table, or circular saw variety.
- One drill, with a bit a little larger than the diameter of the threaded rod (7/16" will do).
- Two wrenches.
- To cut the threaded rods:
- One hacksaw.
- Soft aluminum, to protect the threads.
- One good file (nice but not necessary).

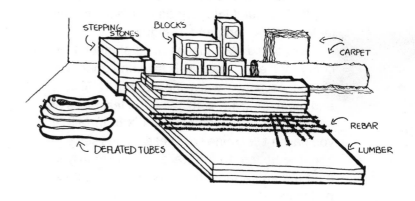

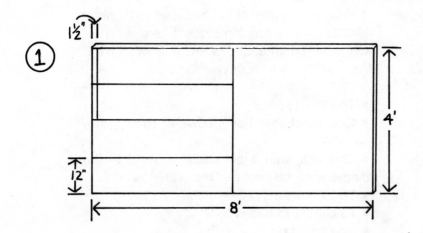

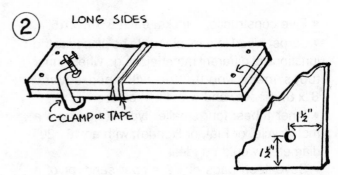

Following the drawings, cut both 4'x 8' sheets of plywood/particle board in half. This yields four 4'x 4' boards. Take one of these 4'x 4's and cut it into four equal pieces for the table walls (4' long and 1' high).

Two of the 4'x1' pieces need another 1 ½" cut off one end. When all is said and done, you should have two walls measuring 12''x 48'', and measuring 12''x 46 ½". One last piece of sawing: cut the piece of 1''x 4''x 8' in half, if you haven't yet done so. These pieces serve to hold the bottom to the walls and eventually to hold the entire table together. Read on, it explains itself.

Sort out the two long walls (48''x 12'') from the rest of your lumber and clamp or tape them together, with the edges flush. The secret here is to drill the holes all at the same time, so that they are perfectly aligned (this way, you drill 4 holes and end up with 8). The drawing shows a hole drilled at each corner 1 ½" in from each side.

Optional: While the boards are still clamped together, drill another hole in the exact center of the 4' length 1 ½" in from one side. A 5th rod will run through these holes for added stability.

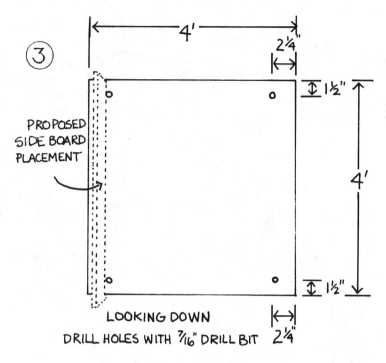

③

PRO POSED
SIDE BOARD
PLACEMENT

4'

2¼"

1½"

4'

1½"

2¼"

LOOKING DOWN

DRILL HOLES WITH 7/16" DRILL BIT   2¼

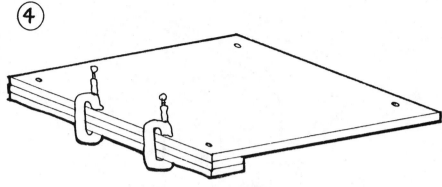

④

Drilling these next four holes in the bottom board (4'x 4') needs special attention. It is important that you follow the drawing exactly when marking the holes, and then recheck before you drill. The hole is at a point 1 ½" from one side and 2¼" from the other. Be sure that you can draw straight lines from one hole to the next and parallel to the sides. If you do it wrong, you'll understand why we tried to warn you.

If you have mastered the last step, this last bit of drilling will be a piece of cake.

Clamp the two 1'x 4' boards to the bottom board you've just drilled. This will serve as a guide for the two holes to be drilled (1½" from each end and 2¼" from each side). See drawing.

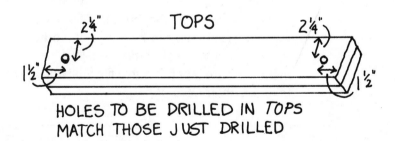

2¼"   TOPS   2¼"

1½"   1½"

HOLES TO BE DRILLED IN *TOPS*
MATCH THOSE JUST DRILLED

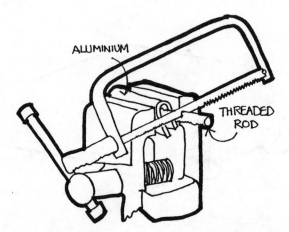

ALUMINIUM

THREADED ROD

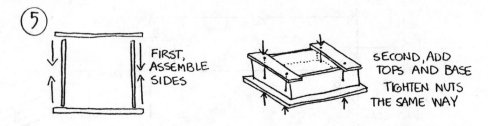

⑤ FIRST, ASSEMBLE SIDES

SECOND, ADD TOPS AND BASE TIGHTEN NUTS THE SAME WAY

If you must cut the threaded rods yourself, now is the time. 5 lengths of 50" rod are needed because these pieces will run the length of the 4 foot (48 inch) table with an inch for washers and nut at each end. The 5th rod is an additional brace to be used in the bottom of the table if needed. From the leftovers, cut four 15" long pieces. This affords plenty of leeway for nuts and washers at the ends of these rods.

Hacksaw instruction: when cutting the rods, sandwich them between soft aluminum in the vice so as to protect the threads. The aluminum will "squeeze" into the threads without hurting them. The nuts will start more easily if the ends of the rods are tapered with a file (just enough to take the sharpness off).

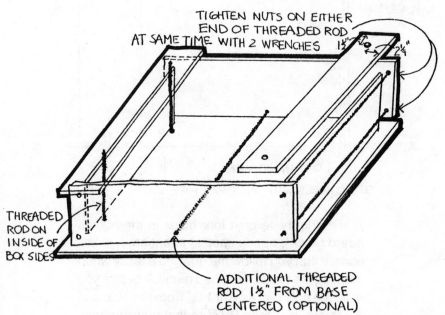

TIGHTEN NUTS ON EITHER END OF THREADED ROD AT SAME TIME WITH 2 WRENCHES

THREADED ROD ON INSIDE OF BOX SIDES

ADDITIONAL THREADED ROD 1½" FROM BASE CENTERED (OPTIONAL)

On top of the bottom board, arrange the two long walls opposite each other with a rod through each of the 4 holes, and a washer with a bolt slightly tightened on each end (8 washers and 8 bolts).

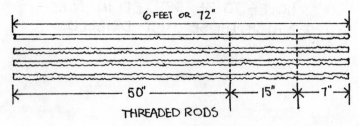

6 FEET OR 72"

50"   15"   7"

THREADED RODS

Slip the shorter walls between the long walls and inside the rods. Add a bolt and a washer to the end of each 15" length of threaded rod; poke through the 4 holes from the bottom so they are on the inside of the table.

Add the remaining two 1"x 4"x 4' boards to the top of the table; line up holes and rods; thread the verticle rods through the holes, and tighten the last four nuts and washers on top.

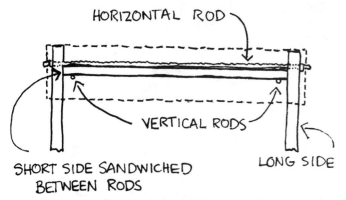

HORIZONTAL ROD

VERTICAL RODS

SHORT SIDE SANDWICHED
BETWEEN RODS

LONG SIDE

As you retighten the long rods and the short rods, you can see how the tension is squeezing, bracing, and working to hold each board perpendicular to another.

Adjust the 15" rods so that as little as possible (of rod, washer and nut) sticks out from the bottom of this box because it will be sitting on innertubes that can be punctured.

Optional: The fifth rod can also be added and tightened with nuts and washers. This will help prevent the structure from warping. If you've already drilled the holes for the rod, this is no longer an option for you (otherwise, the sand will slowly drain through these holes).

Arrange the five cement blocks on top of the carpet, as shown in fig. 1, page 38. Add carpet to the top of the block and position the two 4'x 4' pieces of particle board atop this. The second piece will add more weight to the base of the table. Add more carpet, then make a layer of innertubes. Add more carpet on top of all that. Now your newly constructed sand box (without sand) can be placed on top of these carpeted innertubes. Still, care must be taken so that the nuts and rods on the underneath side do not come into direct contact with the innertubes.

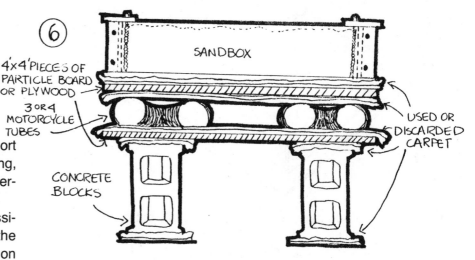

6

SANDBOX

4'x4' PIECES OF
PARTICLE BOARD
OR PLYWOOD

3 OR 4
MOTORCYCLE
TUBES

USED OR
DISCARDED
CARPET

CONCRETE
BLOCKS

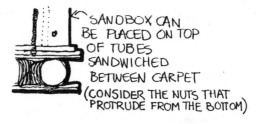

SANDBOX CAN
BE PLACED ON TOP
OF TUBES
SANDWICHED
BETWEEN CARPET
(CONSIDER THE NUTS THAT
PROTRUDE FROM THE BOTTOM)

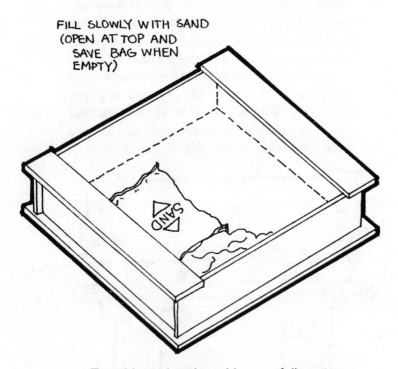

FILL SLOWLY WITH SAND
(OPEN AT TOP AND
SAVE BAG WHEN
EMPTY)

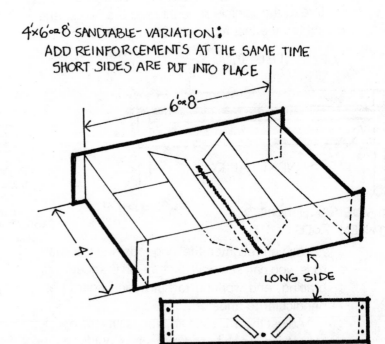

4'x 6' or 8' SANDTABLE- VARIATION:
ADD REINFORCEMENTS AT THE SAME TIME
SHORT SIDES ARE PUT INTO PLACE

6' or 8'

4'

LONG SIDE

To add sand to the table, carefully set one bag at a time into the middle of the table and cut the top of the bag. Pull it away slowly letting the and drain out of the top. The empty bags can be folded up and saved for moving the sand if it is necessary to relocate the table.

confessions of a 4'x8' tension table builder

Once you have acquired all the materials needed to build a Cross Tension Compression Table, it fits together like a kit. Two people can easily put it together in under an hour and you can have all your friends over for a party and dump the bags of sand in it in about 15 minutes.

The beauty of the table is that it can be moved should you ever have to relocate your studio, and you retain your investment. Equally, it is cheap enough to leave behind as an unusual conversation piece for any new tenant. I personally appreciate the portability of this table because I've moved mine three times. The following commentary is neither intended to be limiting nor all-inclusive. The design works, but feel free to experiment, enhance and improve.

1. *Here I am making a innertube sandwich using plenty of carpet pieces for bread. There is carpet between the concrete floor and the concrete blocks, between the blocks and the wood base, between the wood and the inner tubes, and on top of the tubes. I used 6 inch size inner tube, the type used in forklift tires because of the convenience of their size. The tubes should not be over inflated like a pool or river raft tube, but just to the point where the tube is fully expanded before the rubber begins to stretch. It has been suggested that valves be exposed which would enable you to re-inflate the tires as they do tend to go flat over a prolonged period of time. However, I just jack-up the table with a Datsun scissors jack and "change the tire" whenever necessary.*

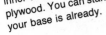

2. The pre-drilled plywood is laid down over the carpet covered inner tubes. Then the pre-drilled particle board is laid over the plywood. You can stand carefully in the center to see how sturdy your base is already.

3. Hold up the side panels and connect them with the pieces of threaded rod. Put on the nuts just far enough to keep the rod from slipping out. The sides will stand up by themselves.

4. Slip in the end panels. As you can see, the rods inserted in 3 are outside the box. You might also notice that holographers often have holes in their clothes because it is not yet a lucrativce field.

5. At this point, the 2'x4's have been connected to the horizontal rods, and the vertical rods are being inserted down through the top shelf. Be sure to use washers between the nuts and wood. Begin to tighten the nuts snugly but not too tight.

6. This is not a good way to fit the inner rods through the eyebolts, side pannels and 2'x4's. We had to do it this way with the inner rod because the walls prevented us from doing it as in 7.

7. The rod is slipped through the eye-bolt from one side to the other side. The nuts all around are tightened, including those under the table which pull the eye bolts down, giving additional tension to the inner rods. Any gaps betweeen the side or end panels and the bottom base boards should be filled with plastic wood or other filler, to prevent the sand from leaking out.

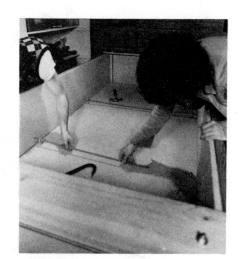

8. Inner panels were forced in between the side walls throughout the interior of the box. As the torque on the nuts provide an inner pull via the rods, these provided a resistance push force against the side walls. They were removed later when they got in the way of fully inserting the PVC pipe in the sand to position optical components. I recommend eliminating this step.

9. The floating box is now ready for sand, laser and holography. Notice that the vertical rods are inside the box in relation to the end panels. This box also works well in 4'x 6' size with one inner rod across the middle.

Happy Holography -C.B. Gaines

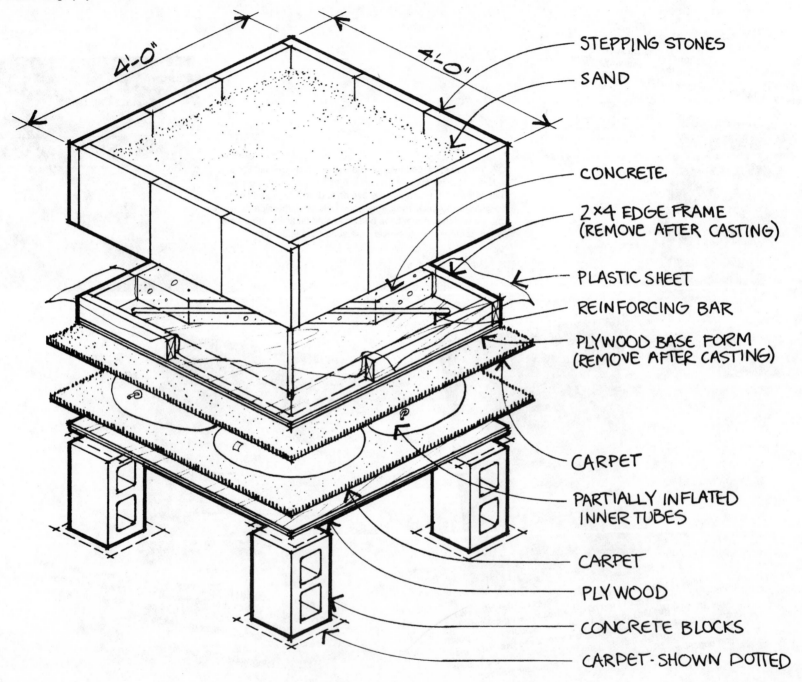

STEPPING STONES

SAND

4'-0"

4'-0"

CONCRETE

2×4 EDGE FRAME
(REMOVE AFTER CASTING)

PLASTIC SHEET

REINFORCING BAR

PLYWOOD BASE FORM
(REMOVE AFTER CASTING)

CARPET

PARTIALLY INFLATED
INNER TUBES

CARPET

PLYWOOD

CONCRETE BLOCKS

CARPET-SHOWN DOTTED

sand/slab table

Materials needed for a 4'x 4' sand-table:

- Five concrete blocks 8"x 8"x 16".
- One 4'x 8'x ¾" sheet of inexpensive plywood or particle board.
- Discarded carpet: lots of it, for covering two 4'x 4' surfaces with 4" fringe, plus tops and bottoms of all concrete blocks.
- Three or four small car inner tubes (e.g., from a motorcycle, or Fiat, etc.).
- Twenty-two linear feet of 2"x 4"s, to frame plywood for concrete mold. Have cut to the following dimensions:
    A) Two approx. 4' lengths.
    B) Two appprox. 3'x 10½ " lengths.
    C) One 5'-6' length.
- Polyethylene sheet, 4-6 mil, to fit inside mold for pouring concrete - extra large (10'x 10'), to provide an apron.
- Two approx. 4' lengths of reinforcing rod (known as rebar), or enough chicken wire (heavy gauge) for reinforcing the concrete; use both, if you like.
- Twelve approx. 2"x 12"x 16" stepping stones for framing sandbox.
- One bag of mortar mix and one qt. cement adhesive.
- Four bags of Redimix concrete.
- Sixteen bags of #2 amber sand, washed; (one bag is 1 cubic foot, is 100 lbs.)
- Miscellaneous: spool of wire, nails, trowel, bucket for mortar.

Materials needed for a 4'x 8' sand table*:

- Fifteen concrete blocks 8"x 8"x 16".
- Two approx. 4'x 8'x 3/4" sheets of inexpensive plywood or particle board.
- Six to eight small car inner tubes (e.g., from a motorcycle, or Fiat, etc.).
- Thirty-two to thirty-four linear feet of 2"x 4"s, to frame plywood for concrete mold. Have cut to the following dimensions:
    A) Two 4'x 4" pieces (ends).
    B) Two 8' pieces (sides).
    C) One 6' piece (leveling board).
- Polyethylene sheet, 4 - 6 mil, one 20'x 10' piece.
- One 20' length of reinforcing rod (rebar), or enough heavy-guage chicken wire to reinforce the concrete.
- Twenty-five stepping stones, 2"x 2"x 16", for framing sandbox.
- Eight bags of Redimix concrete.
- Thirty-two bags of #2 amber sand, washed; (one bag is 1 cubic foot, is 100 lbs).
- Miscellaneous: spool of wire, nails, trowel, bucket for mortar.

*For advanced setups, we recommend building a 4'x 8' table. The procedure is the same, only more materials are needed.

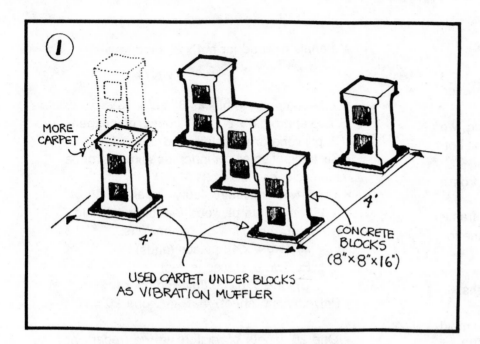

MORE CARPET

USED CARPET UNDER BLOCKS AS VIBRATION MUFFLER

CONCRETE BLOCKS (8"×8"×16")

4'

4'

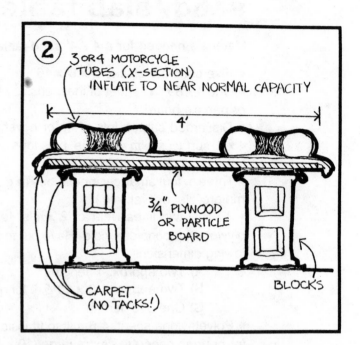

3 OR 4 MOTORCYCLE TUBES (X-SECTION) INFLATE TO NEAR NORMAL CAPACITY

4'

3/4" PLYWOOD OR PARTICLE BOARD

CARPET (NO TACKS!)

BLOCKS

(1) Before beginning, place a piece of polyethylene on the floor. A six foot square piece will save the floor and, for you, the clean-up later. Arrange the concrete blocks so one is at each of the four corners of a four foot square area and one is in the middle. The finished table will stand between 32 and 36 inches tall (you can use another layer of blocks with carpet between for increased elevation but usually the first 5 are sufficient). Place enough carpet underneath and on top of the blocks to cover them (usually an 8"x 8" square piece will do). From now on, carpet should be placed wherever layers come together.

(2) Position the 4'x 4' board atop the carpeted blocks, and add another layer of carpet over the entire top surface. Used carpet is inexpensive and works perfectly well, but be sure to check it for tacks, nails or staples that will inevitably puncture the next layer. The next layer consists of small inner tubes inflated to almost full capacity (not bulging).

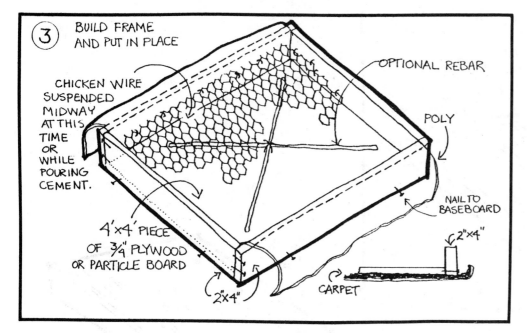

③ BUILD FRAME AND PUT IN PLACE

CHICKEN WIRE SUSPENDED MIDWAY AT THIS TIME OR WHILE POURING CEMENT.

OPTIONAL REBAR

POLY

NAIL TO BASEBOARD

4'x4' PIECE OF ¾" PLYWOOD OR PARTICLE BOARD

2"x4"

2"x4"

CARPET

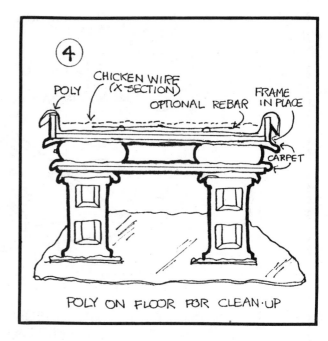

④

CHICKEN WIRE (X-SECTION)

POLY

OPTIONAL REBAR

FRAME IN PLACE

CARPET

POLY ON FLOOR FOR CLEAN·UP

(3) To the second 4' x 4' piece of plywood, attach the pre-cut lengths of 2''x 4'' to form a temporary box frame (as shown). These will later detach and serve only as part of the mold at this point. Turn the box over, and tack or staple carpet to the base at points where the staples will not come into contact with the inner tubes.

Place the completed frame, right side up, on top of the inner tubes. As a finishing touch, drape a piece of polyethylene in the box so that it hangs over the sides (this will prevent the moisture of the cement from soaking into the frame and warping it).

Hold onto the 5½' piece of 2''x 4''. It will be used as a leveling board for the concrete to come.

(4) At this point, it is wise to cut the chicken wire to size and wire two pieces of rebar (optional) into a cross, to prepare for placement in the frame during the pouring of cement. Set it aside.

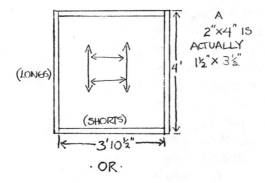

(LONGS)

(SHORTS)

4'

3'10½"

A 2"x4" IS ACTUALLY 1½"x 3½"

· OR ·

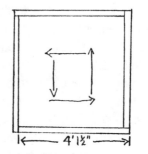

4'1½"

2"x4"s ALL THE SAME LENGTH

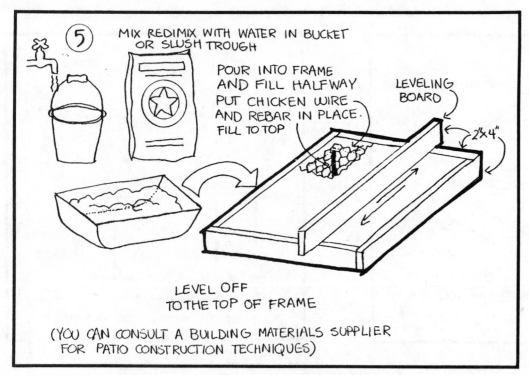

⑤ MIX REDIMIX WITH WATER IN BUCKET OR SLUSH TROUGH

POUR INTO FRAME AND FILL HALFWAY PUT CHICKEN WIRE AND REBAR IN PLACE. FILL TO TOP

LEVELING BOARD

2"x4"

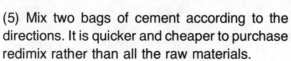

LEVEL OFF TO THE TOP OF FRAME

(YOU CAN CONSULT A BUILDING MATERIALS SUPPLIER FOR PATIO CONSTRUCTION TECHNIQUES)

⑤ₐ
OR

CONCRETE SLAB

(5) Mix two bags of cement according to the directions. It is quicker and cheaper to purchase redimix rather than all the raw materials.

Fill the frame half full and tamp it down. Carefully sink the chicken wire and the rebar slightly into the wet cement. Add the second bag of the mixed cement, and again tamp it down. The chicken wire and rebar "floating" in the middle of the slab add strength and stability. With the leveling board, level off as shown in drawing.

These same steps will render a patio slab (or foundation for an outhouse). If you are unsure about pouring cement, we suggest consulting your friendly building supply people. They usually have helpful hints and suggestions which prove invaluable.

Optional slab enhancement: while the slab is still wet, add small hooks or handles, taking into consideration where the stepping stones will be placed. This will make it easier to handle if you should have to move it at a later time.

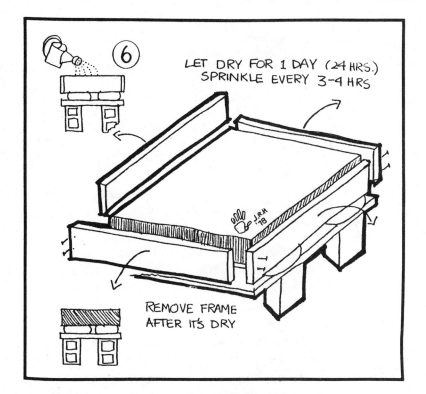

(6) Allow the slab to dry 24 hours. Sprinkle the top surface every 3 to 4 hours so it cures correctly (this will prevent cracking). Twenty-four hours later *carefully* knock off the 2x4s from the dried slab. Be careful - the cement is still damp and will chip easily. Also, cut away the polyethylene (again being careful not to puncture your inner tubes).

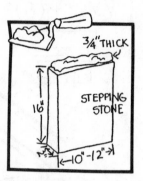
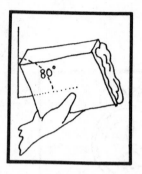
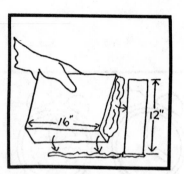

(7) Get acquainted with your stepping stones by arranging (spacing and squeezing) them around the top of the slab. They will be your retaining wall and should be about 12" high. Be sure in spacing the blocks to leave suitable room (about ½ inch between each) for mortar. Taking time to figure the spacing of the blocks will help you avoid the panic of wet mortar on the last block which is suddenly too big for the space that's left. You are now ready to mortar.

Mortaring these blocks is the most difficult task you'll come up against in building this table; it is also the messiest, so put on rubber gloves if you care about your hands (mortar is alkaline and tough on skin). Prepare ¾ of a bucket of mortar mix (always with clean water). This will afford you enough mix and time to affix several blocks before it sets up. Concrete adhesive can be a helpful addition to the mortar mix (see directions). To test rightness of mixture, hold a stone with mortar on it at almost a 80 degree angle to the floor. If the mortar doesn't fall off from being too wet or too dry, you've got it. Now you are ready to apply the mortar.

First, wet the edge of the slab with a paintbrush or sprinkling can. Now put a ¾" thick line of mortar along the edge (enough for several blocks to set in). Dip the first block in a bucket of water and put it in place. Dip a second block in the water and add a ¾" thick line of mortar to the side of the second block. Join it to the first block (each flush with the edges of the slab). Starting at a corner seems to work best. Follow this procedure, several stones at a time, progressing around the table (see drawing).

Helpful hints department: Nail a string to the wall on one side of the room and run it along the side of the table at the height of your first block. Then all succeeding blocks can easily be lined up with the first. Better yet: use your laser's beam instead of string!

Remember, help is as near as your local building supply person. Mistakes in judgement can be covered by filling in gaps with mortar, or breaking blocks in half and telling your friends you planned it that way. Help your project cure by sprinkling it with water periodically over a one day drying period.

⑦ MIX MORTAR MIX + 1 QT. MORTAR GLUE
(CONCRETE ADHESIVE)

FIGURE OUT
HOW MANY
BLOCKS ARE
NEEDED

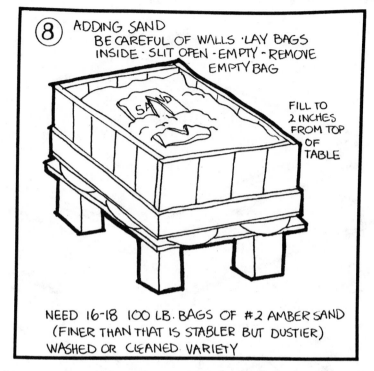

⑧ ADDING SAND
BE CAREFUL OF WALLS ·LAY BAGS
INSIDE · SLIT OPEN · EMPTY · REMOVE
EMPTY BAG

SAND

FILL TO
2 INCHES
FROM TOP
OF
TABLE

NEED 16-18 100 LB· BAGS OF #2 AMBER SAND
(FINER THAN THAT IS STABLER BUT DUSTIER)
WASHED OR CLEANED· VARIETY

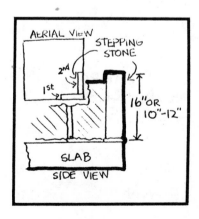

AERIAL VIEW
STEPPING
STONE

2nd
1st

16" OR
10"-12"

SLAB
SIDE VIEW

⑦a THIN CONCRETE
BLOCKS
8"x16"x 4"
OR
EVEN BRICKS
CAN BE USED
TO BUILD THE
WALLS OF
THE SANDBOX

SLAB

(8) Time to put the sand in. Carefully place a bag of sand in the table as shown. Slit open and release sand slowly (so as not to put too much quick pressure on the new walls). You'll find that one bag equals one square foot. If you are planning a move in the future, you may want to slit the sand bag open carefully at the top, so you can refill and reuse the bag later.

REMINDER: The goal, in case you've lost sight of it among all the concrete, is to achieve rest inertia. Simply stated, a heavy mass floating on air is still, and wants to stay that way.

Later, as your needs change, a second 4' x 4' table can be constructed alongside the first to gain all the advantages of a 4' x 8' table.

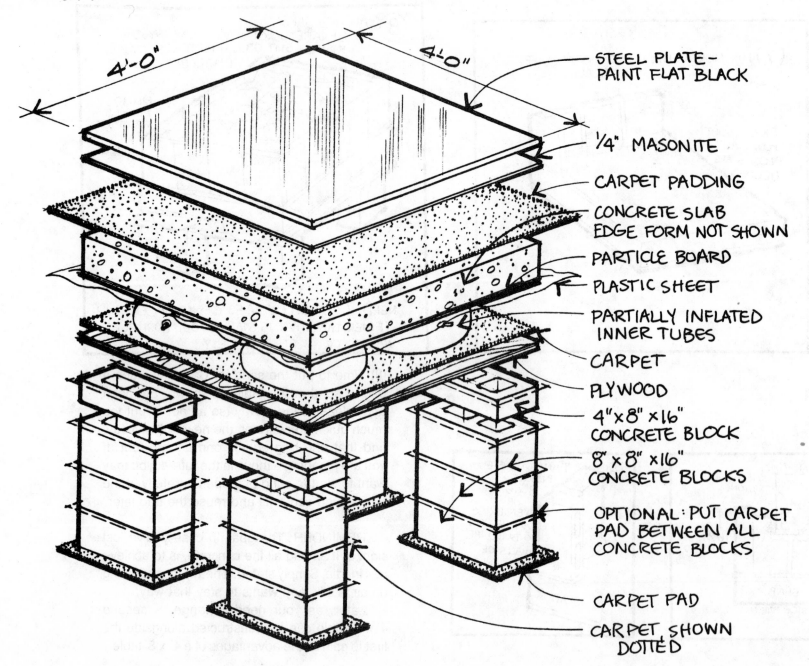

4'-0"

4'-0"

STEEL PLATE –
PAINT FLAT BLACK

¼" MASONITE

CARPET PADDING

CONCRETE SLAB
EDGE FORM NOT SHOWN

PARTICLE BOARD

PLASTIC SHEET

PARTIALLY INFLATED
INNER TUBES

CARPET

PLYWOOD

4" x 8" x 16"
CONCRETE BLOCK

8" x 8" x 16"
CONCRETE BLOCKS

OPTIONAL: PUT CARPET
PAD BETWEEN ALL
CONCRETE BLOCKS

CARPET PAD

CARPET SHOWN
DOTTED

metal/slab table

This table is simply a slab table without the sand, and is a cross between expensive professional systems and the sand system. The purpose of the metal top, which is steel, is to allow the use of magnetic mounts for the optical components. The concrete slab is positioned higher to allow placement at a comfortable level. Remember, you won't have 12" to 14" of sand on top of the slab for higher working surface. Thus, instead of using the 8"x 8"x 16" cinder blocks on end as with the sand/slab table, they are laid down with an 8"x 16" side against the floor for greatest stability and stacked three high. Carpet should be placed between the cinder blocks in the stack. We like to add a fourth block, 4"x 8"x 16", which will bring the table up to a very comfortable working height.

A wood frame is then placed on top of the blocks in the same manner as before, and a concrete slab poured. When the concrete is completely dried, and the frame removed, place a piece of felt or similar material to cover the top. A thin smooth steel sheet, which may be obtained, and usually cut to size, at a metal supply house, can then be placed on the felt (it's not necessary to glue it down), or one can make a felt, masonite, felt, steel sandwich. It's not important to get heavy gauge steel - we used the thinnest stuff we could find (.040 gauge), which cost us under $40 for a 4' x 8' sheet (your price may vary). Spray flat black paint on it after it's in place.

Magnetic setups are quite a bit more difficult to work with than sand systems, and are not recommended for beginners. The mounts are cumbersome, often difficult, and time-consuming to position. The advantages lie in the fact that the components are more rigidly fixed in place when they are set (which is preferred for commercial production or other situations where repeatable exposures are desired). Some advanced techniques, such as multiple exposure rainbow holograms or other composites are actually easier with this type of table. Expensive optical components also tend to last longer on a solid table, as they are not exposed to the unavoidable effects of sand blasting.

As an option, if desired, a top for a sand table may be built out of particle board, and the steel plate placed on top of it. This top can be placed or bolted onto the sand table when a metal top is needed, and then removed when the flexibility of sand is preferred.

Information about constructing component mounts for the metal-top table will be found at the end of the chapter on building optical components.

other table ideas

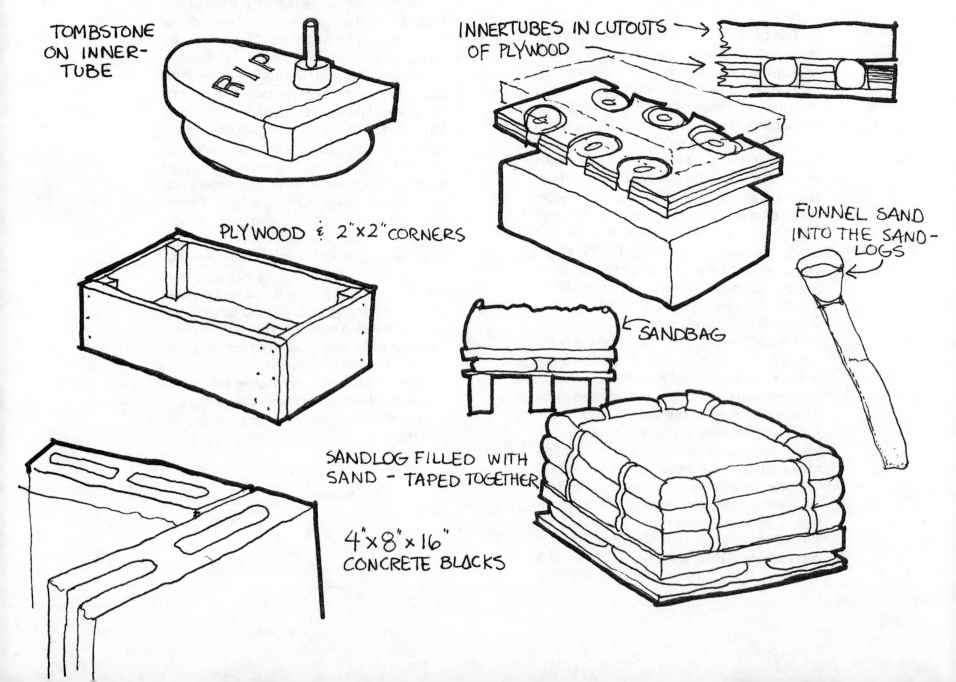

TOMBSTONE ON INNER-TUBE

INNERTUBES IN CUTOUTS OF PLYWOOD →

PLYWOOD & 2"x2" CORNERS

FUNNEL SAND INTO THE SAND-LOGS

SANDBAG

SANDLOG FILLED WITH SAND - TAPED TOGETHER

4"x8"x16" CONCRETE BLOCKS

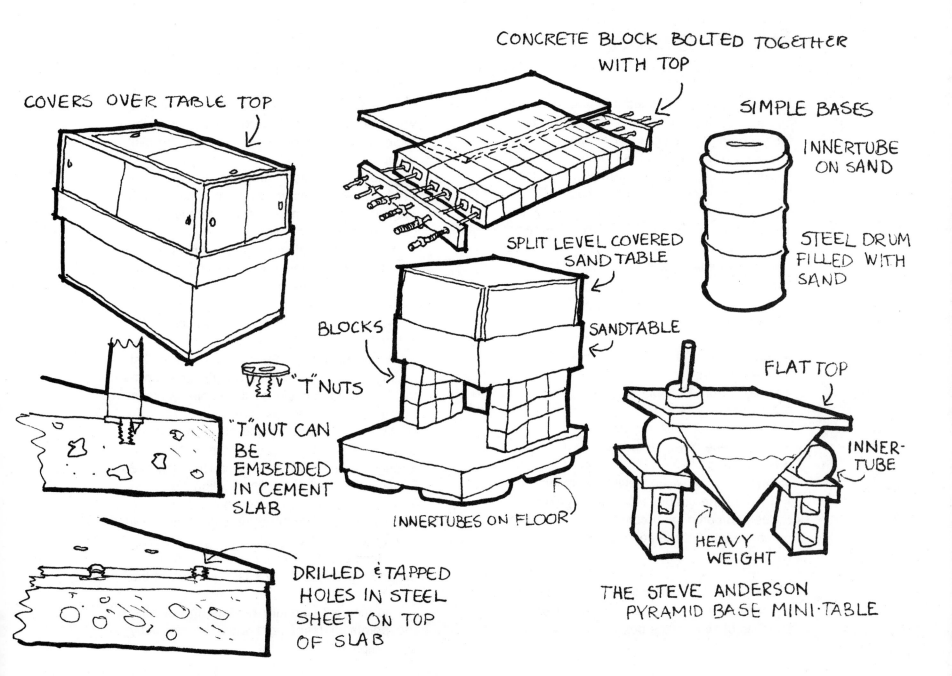

COVERS OVER TABLE TOP

CONCRETE BLOCK BOLTED TOGETHER WITH TOP

SIMPLE BASES

INNERTUBE ON SAND

STEEL DRUM FILLED WITH SAND

SPLIT LEVEL COVERED SAND TABLE

BLOCKS

"T" NUTS

"T" NUT CAN BE EMBEDDED IN CEMENT SLAB

SANDTABLE

FLAT TOP

INNER-TUBE

INNERTUBES ON FLOOR

HEAVY WEIGHT

DRILLED & TAPPED HOLES IN STEEL SHEET ON TOP OF SLAB

THE STEVE ANDERSON PYRAMID BASE MINI·TABLE

interferometer

This easily constructed interferometer can be used to determine whether an area is suitable for holographic imaging: in other words whether it is relatively free of vibrations and constant "noise". Testing an area before building a table is very wise, if you consider how heavy a ton of sand and a half dozen cinder blocks can be the *second* time you have to move them. This same setup can be used later to test the completed table (with the table taking the place of the small sand box).

You will need:
- Laser.
- Three front-surface mirrors, mounted.
- Beamsplitter.
- Lens, -11 or -13, mounted.
- An inner tube.
- A sturdy cardboard box.
- 2' x 2' piece of wood - same size as box.
- Some sand.
- White card, or wall on which to project light.
- To read the chapters on laser acquisition, optics and component construction and have purchased/built those needed above (obviously).

Place the wood, then the cardboard box, on top of the inflated inner tube, and add at least 4 - 5 inches of sand to the box.

Position the laser, either on, or off the table/box, so that the beam travels about 4 inches above, and parallel to, the surface of the sand.

The first transfer mirror, placed as shown, will direct the beam toward the center of the table/box. Adjust the mirror to maintain the level beam.

The beamsplitter, carefully positioned, will produce two beams, the first travelling through the beamsplitter, and the second deflected 90 degrees to mirror #1 as shown.

Intercept the deflected beam with mirror #1 and redirect the beam back through the beamsplitter. Its final destination can now be seen on the wall (or white card).

Measure the distance between mirror #1 and the beamsplitter.

A second mirror (2), placed into the first beam path, will bounce it back to the beamsplitter. Its position should be the *same distance* from the beamsplitter as mirror 1.

Two dots of light should be visible on the card, one reflected from mirror 1 (traveling through the beamsplitter), and one from mirror 2 (reflected off the beamsplitter). Align the two to form a single dot, by wiggling the mirrors in place. A single dot and possibly some fringes will now be visible on the wall.

A lens placed at either position B or C will expand the dot, to allow easier viewing of the fringes.

No fringes? Carefully readjust the mirrors; the points of laser light must be perfectly superimposed.

Allow the interferometer setup to settle for approximately 15 - 30 minutes. Depending on the stability of the environment, the fringes should eventually slow down, and appear to be motionless. Notice: movement or loud noise in the room, or exterior noise (trucks thundering by or buffalo stampeding), will upset the fringes, and set them in motion again.

Another visible phenomenom is drift, caused by settling of the components in the sand. Within the first 15 - 20 minutes of setting up the interferometer, the black and red lines will slowly march across the wall or card. Do not be dismayed, drift is a common occurence and the drift fringes will halt...in most cases.

If the fringes (regular or drift) never seem to rest, the area may be too noisy, and a different location may be the only answer.

If everything seems to test out, and the table is built, another interferometer can be set up in the sand of your new table. Several tests will give you an idea of how long settling times should be, of just how quickly the table recovers, and will generally prove the principles of rest inertia.

Tapping the table or stomping the floor will disturb the stability for several moments, in which case the fringes will appear to move rapidly or vanish completely. If the table is stable, however, the fringes should "regain their composure" and come to a standstill.

The black (or lack of) light you see (or don't see), is the result of one beam interfering with the other and canceling it out. The black is only possible when two beams are working together: block one of the beams with a card, and the fringes will vanish.

Body heat will also affect the beam's path. The heat from a single match will visibly set the fringes in motion if held in a beam path. The heat moves the air, which is also very disturbing to holography. Obviously, a drafty area is not the best for a sand table setup.

The fringes you see now are a magnification of the same type of fringes that make the hologram.

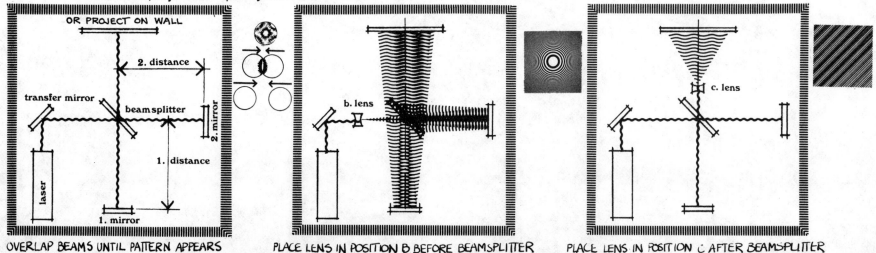

OVERLAP BEAMS UNTIL PATTERN APPEARS   PLACE LENS IN POSITION B BEFORE BEAMSPLITTER   PLACE LENS IN POSITION C AFTER BEAMSPLITTER

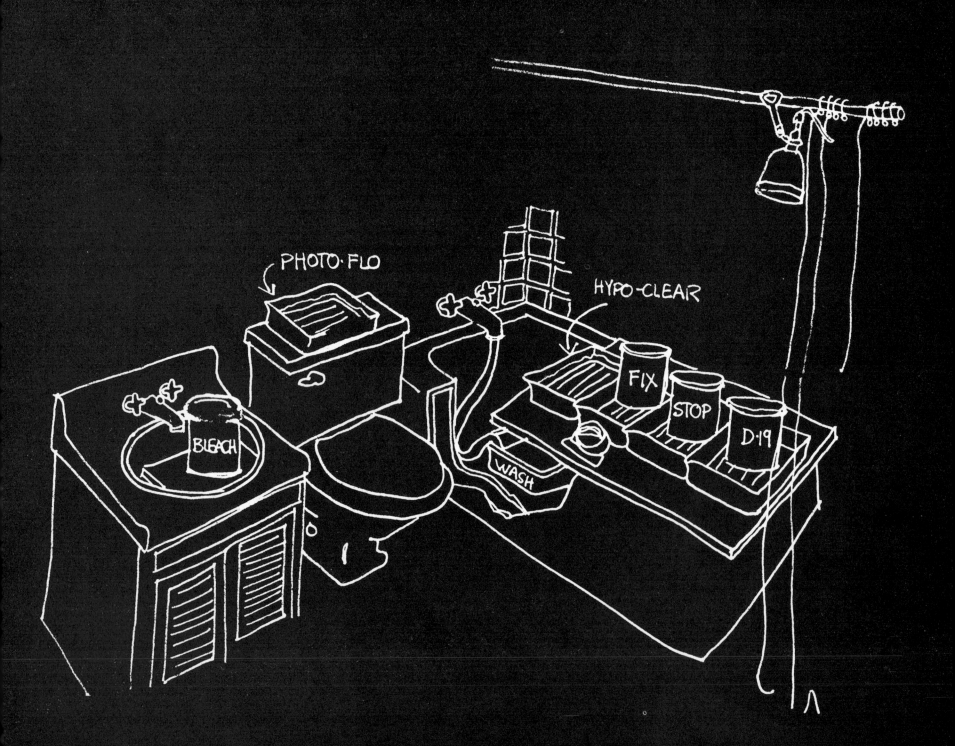

the darkroom

in this chapter:

**darkroom equipment
darkroom sink
darkroom chemicals**

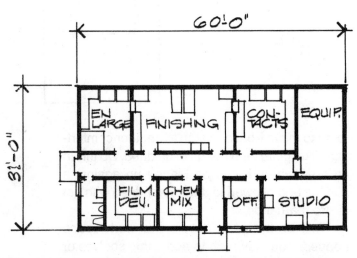

A Professional photography darkroom.
(can be used for holography)

The darkroom is nothing more that the word says - a room that is dark or one that can be made dark.

The same methods used to darken the optics table room can be used here. Although both rooms can technically be called darkrooms, we shall use this term for the area within which the hologram is processed or "developed". For those readers familiar with developing or printing photographs, you will see that there is nothing particularly unusual about a *holographic* darkroom. The function is the same: to provide a place for the processing of materials sensitive to light.

You may already have a darkroom, or the potential for one, without a lot of elaborate building. A bathroom or closet may be utilized. Dungeons are even better, if you happen to live in a medieval castle. Whichever you choose, locating the darkroom near the optics table room is preferred, since the hologram, after exposure, will be going directly from one room to the other and must be protected from light. Some people may wish to use the same room for both the table and the darkroom, yet most will probably wish to separate the two functions.

Some of the processing will take place in darkness and the rest will be done with the lights on. Another possibility, then, is to do the first part in the optics room, and the rest in a bathroom or other area with running water. Alternately, you can use a closet for the darkened area. In order to help you plan, the two parts can be described as follows:

1) Processing in the dark: Here you will need room for only three trays or buckets, a safelight, and yourself.

2) Processing in the light: You will want access to a sink or running water, a working surface for a few more chemicals, an electrical outlet, and a drying area.

Those with a little more ambition may wish to custom design a complete darkroom from scratch. There are any number of ways to do this, so we'll leave it to you to work out the most comfortable arrangement.

It may be helpful, at this stage, to run quickly through the processing in the darkroom, so that you can get a sense of how to lay things out. The exposed plate or film will go from developer to stop bath to fixer in the area that can be made fairly dark. You will have a safelight on, and will be pulling the plate out of the developer to check it's progress periodically by looking through it at the safelight (or at a white card on the wall nearby). You will thus want to mount your safelight in a spot which will facilitate this. After the fixer (third tray or bucket), the lights may be turned on or the plate brought to another area which can be lit.

Here, depending on the type of hologram, the plate will go through three or four more chemical steps and will be rinsed between each step in running water. You will want to allow enough

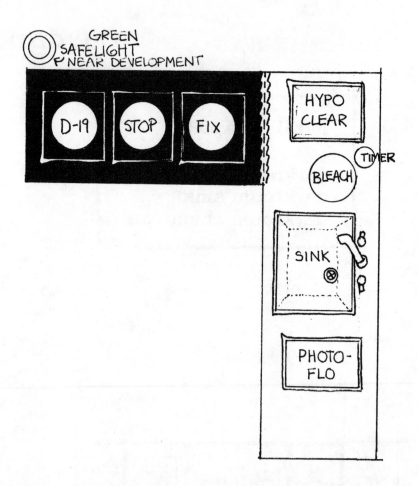

room to move from one step to another comfortably. The plate then needs to be dried. Do this in an area where it can be leaned up against a surface and allowed to drip-dry or dried with hot air from a hairdryer.

This pretty much covers the work space needed. You may wish to add some shelves or cabinets to store chemicals and supplies and you'll be in good shape.

darkroom sink

If you are building a darkroom from scratch, and have a good space with access to water and a drain, you might consider designing and building your own sink. Some holographers prefer a larger darkroom sink for greater latitude in processing designs. Large commercial sinks are often quite expensive but you can build a comparable one for under $20.

The trick is no more difficult than figuring out how to divide up a 4' x 8' piece of particle board. For example, say you'd like to build an 7 foot sink. Using only 1 piece of good particle board (at least ¾" thick), you might divide it up thus:

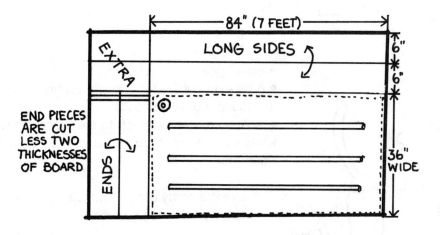

This will result in a sink 7 feet by 36 inches wide by 6 inches deep. Cut your pieces, making sure they are as straight as possible. Then lay them out as shown:

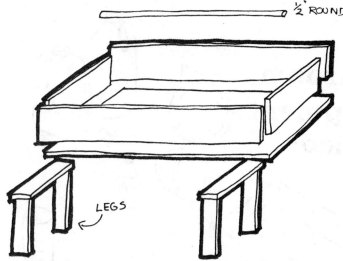

Go out and buy a standard drain, and then drill a hole of the proper diameter near one corner of the sink bottom. Those circle cutters that fit into an electric drill (see your hardware store) are nice to do this with.

Glue and nail the pieces into a box. If desired, "ribs" made from ½ round stock can be glued into the sink bottom, to prevent trays from sliding around, and to allow drainage. You can find the stock at any lumber yard - it is used for molding and is cheap.

Seal all joints with silicon, then seal everything with several coats of verathane. The sink can be mounted in place at its proper height by using 2" x 4" braces. This may be done either before or after sealing. Be sure to angle the sink slightly towards the drain hole so that water will run towards that corner. Cover everything with a few coats of enamel paint and then install (or get a plumber to install) the plumbing for the faucet and drain.

darkroom equipment needed:

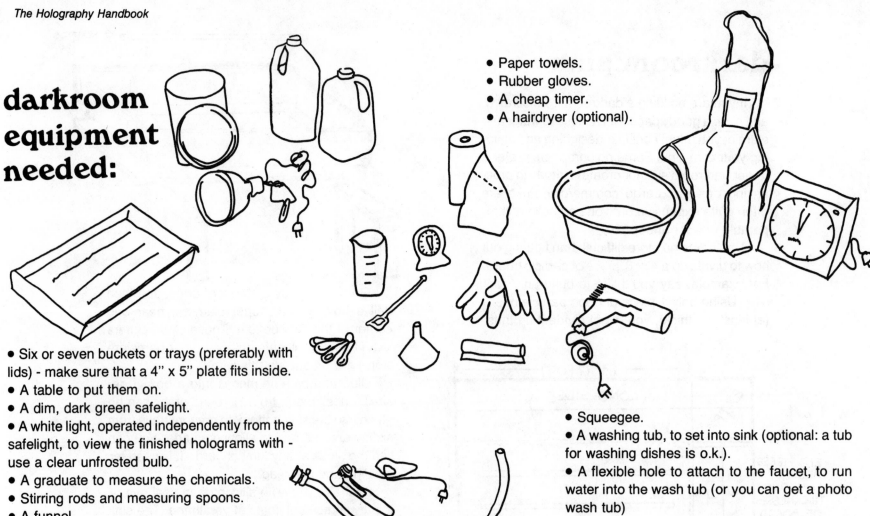

- Paper towels.
- Rubber gloves.
- A cheap timer.
- A hairdryer (optional).

- Six or seven buckets or trays (preferably with lids) - make sure that a 4" x 5" plate fits inside.
- A table to put them on.
- A dim, dark green safelight.
- A white light, operated independently from the safelight, to view the finished holograms with - use a clear unfrosted bulb.
- A graduate to measure the chemicals.
- Stirring rods and measuring spoons.
- A funnel.
- Storage containers for chemicals (water jugs are fine, but don't use them for water again).

- Squeegee.
- A washing tub, to set into sink (optional: a tub for washing dishes is o.k.).
- A flexible hole to attach to the faucet, to run water into the wash tub (or you can get a photo wash tub)

For an extra spiffy darkroom, you may want to consider the following options:

- A hot aquarium heater to heat developer (occasionally desirable).
- An apron to keep clothes clean.
- A good timer (Gra Lab is best).
- A film cutter (a paper cutter is fine).
- An automatic $10,000 film processor, if you're *really* into it.

A great squeegee can easily be made by cutting a slot in a piece of PVC pipe (at least 5 inches long) and sliding in a windshield wiper blade (see drawing). The cheaper, more flexible, wipers are best, as they are less likely to scratch the emulsion than the standard photographic squeegees are.

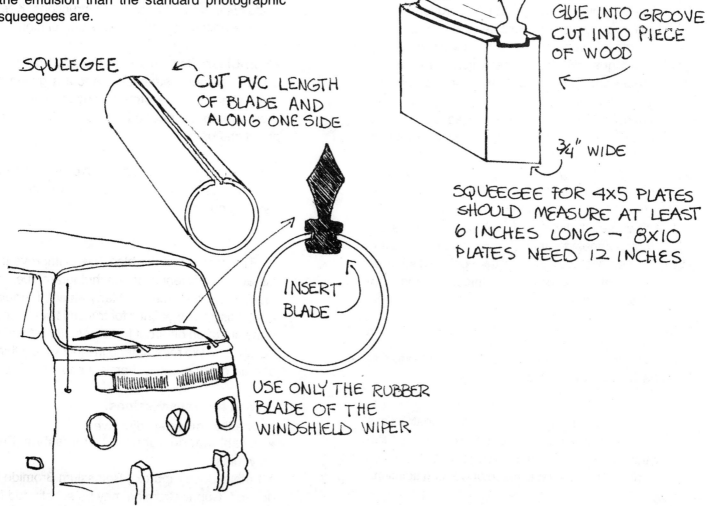

SQUEEGEE

CUT PVC LENGTH OF BLADE AND ALONG ONE SIDE

INSERT BLADE

GLUE INTO GROOVE CUT INTO PIECE OF WOOD

¾" WIDE

SQUEEGEE FOR 4X5 PLATES SHOULD MEASURE AT LEAST 6 INCHES LONG — 8X10 PLATES NEED 12 INCHES

USE ONLY THE RUBBER BLADE OF THE WINDSHIELD WIPER

darkroom chemicals

There are a number of different developers, bleach formulations, etc. currently used by holographers and a veritable revolution in formulations is in the midst of taking place. We expect some of the greatest improvements in holography over the next several years to be a result of better chemistry and processing techniques. We list those which are currently most effective and popular as of this printing, yet recommend that the current literature be consulted for new methods as they arise. Future editions of this book will be updated with new information as it arrives on the scene.

We begin with a listing of chemicals which have generally proved themselves to be reliable over many years. (An exception is the new bleach listed for reflection holograms, which appears due to the significance of its improvement over an earlier type). A listing of some of the alternate chemicals currently in use follows.

These basic chemicals can be purchased at a photo-supply store:

Developer: Kodak D19. Mix as per instructions and then dilute 4:1 (water:developer). The dilution will result in finer grain results. Start with enough for at least a gallon of working solution.

Stop Bath: Any commercial photo stop bath such as Kodak Glacial Acetic Acid will do. Mix as per instructions. Make enough to fill the container you are using.

Fixer: Use any commercial fixer. We prefer Kodak Rapid-Fix. Mix as per instructions. Keep about a gallon or more of working solution.

Hypo-Clear: Removes the fixer. Any brand will do. Mix as per instructions. About a gallon of Hypo-Clear can be stored concentrated, and then diluted as needed (as described in instructions on box).

Photo Flo: Any brand is good. We prefer Kodak 200. Mix as per instructions. Make enough to use as you need it.

The purpose of the bleach is to improve the diffraction efficiency of the hologram (i.e., to make a brighter image). Many bleach formulations exist, and bleaches for transmission holograms are often different from those for reflection holograms. We recommend the following, which can be obtained at a chemical supply house.

Bleach for transmissions:
In one liter of water, dissolve:
• 1 tablespoon (approx.) Potassium Ferrocyanide.
• 1 tablespoon (approx.) Potassium Bromide (If desired, Cupric Bromide may be substituted for the Potassium Ferrocyanide.).

Bleach for reflections:

PBQ Bleach: This is a new bleach, recently developed by Nick Phillips et.al.. The magic ingredient is p-benzoquinone which is abbreviated to P.B.Q. This bleach and its vapor are toxic, so wear gloves while mixing and using, and avoid breathing vapors. Use in a very well ventillated area (all bleaches should be treated with caution)*. It is much safer than the Mercuric Chloride bleach which used to be the standard for reflection holograms (see alternate bleaches).

In one liter of water, dissolve:

Potassium Bromide . . . . 30 grams
Borax . . . . 15 grams
Potassiun Dichromate . . . . 2 grams

Just prior to use, add:

P.B.Q. . . . . 2 grams

After the P.B.Q. is added to the above mixture, the bleach must be used within 15 minutes (see reflection processing).

All chemicals should be stored tightly capped when not in use (funnelling into a plastic or glass jug is fine for some). We prefer to mix the bleaches in the containers they are to be stored in, to avoid splattering the stuff around while pouring from one container to another.

These chemicals, like any used in photoprocessing, have a limited lifetime (they will not last forever) and so will have to be replaced periodically. The frequency of this depends upon the chemical, and/or on how often it's used. The D-19 developer and the transmission bleach mentioned above will go a long way, either on the shelf, or with much use. The reflection bleach, as noted, has a very short lifetime. The fixer is usually replaced a little more often, depending on use, and this can be determined by using an indicator solution to check on it (available at a photo supply store). Generally, most will last weeks or even months.

We do recommend using new batches of stop bath, hypo-clear, and photo-flo each day you work.

As a general rule, follow good standard darkroom procedure in mixing, handling, and storing of all chemicals. Be sure to label all containers and keep everything well out of reach of children.

*Some chemicals are more toxic than others. A good comparison of chemicals for holography and their effects can be found in Holosphere, Vol. 8, Dec. 1979.

alternate darkroom chemicals

developers

Neofin Blau (blue) developer: Manufactured by Tetenal Photowerk - Hamburg and Berlin, Germany - and available from better photo suppliers.

This developer makes holograms of excellent, low-noise quality. We personally prefer it to D-19, although it becomes quite a bit more expensive to use (as the solution is used once only and then discarded). We especially recommend it for use with rainbow holograms - both masters and copies, where low noise becomes extremely important. The developer is supplied as a concentrate in several small vials. Cost can be kept low by using only part of a vial at a time, diluted to make enough solution just to cover the plate used.

GP-8 Developer: This fine grain developer was invented in the Soviet Union and is evidently in wide use there today. A discussion of the action of the developer by Nick Phillips of the Loughborough University of Technology, England, may be found in the May 1979 issue of *Holosphere*. The materials for this mixture are available at good chemical supply houses:

Phenidone . . . . 0.026 grams
Hydroquinone . . . . 0.650 grams
Anhydrous Sodium Sulfite . . . . 13 grams
Potassium Hydroxide . . . . 1.38 grams
Ammonium Thiocyanate . . . . 3.12 grams
Distilled Water . . . . 1 liter

The developer should be used once only. Best results occur if the plate is overexposed to some extent.

GP-61 Developer: This formulation is currently recommended by Agfa-Gavert for transmission holograms made with their new HD series emulsion. We have not, as yet, been terribly impressed with results from this or from the GP-62 mixture which follows, yet include it so you may experiment:

Distilled Water . . . . 700 cc
Metol . . . . 6 grams
Hydroquinone . . . . 7 grams
Phenidone . . . . 0.8 grams
Anhydrous Sodium Sulfite . . . . 30 grams
Anhydrous Sodium Carbonate . . . . . 60 grams
Potassium Bromide . . . . 2 grams
Sequestrene Agent . . . . 1 gram
Additional water to make one liter

GP-62: This developer is used with reflection holograms (along with some special bleaches), to achieve playback in approximately the same wavelength colors as used to make the hologram (e.g., red images with HeNe lasers), without using triethanolomine (see reflection hologram processing). We have found that the process does work, although the images are not especially bright. The developer formula appears below, in two parts which are mixed just prior to use:

Part A:

Distilled Water . . . . 700 cc
Metol . . . . 15 grams
Pyrogallol . . . . 7 grams
Anhydrous Sodium Sulfite . . . . 20 grams
Potassium Bromide . . . . 4 grams
Sequestrene Agent . . . . 2 grams
Additional water to make one liter

Part B:

Distilled Water . . . . 700 cc
Anhydrous Sodium Carbonate . . . . . 60 grams
Additional water to make one liter

To make working solution, mix one part of solution A to two parts water and one part of solution B.

Parts A and B are stable as separate solutions; but the mixture is usable for only one to two hours.

Caution: Always use gloves while mixing and using this developer as Pyrogallol is poisonous and may cause skin irritation.

Kodak D-8 Developer: Some holographers have had success with this rather than D-19. We have not tried it, yet understand that T.H.Jeong of Lake Forest College (also Intergraf) has worked out a successful technique, which, we hope, will be published soon.

For low noise results, Steve Benton utilizes a developer he calls PAAP:

Ascorbic Acid (Vitamin C) . . . . 18 grams
Sodium Hydroxide . . . . 12 grams
Sodium Phosphate Dibasic . . . . . . 28.4 grams.
Distilled Water . . . . 1 liter
To the above add, just before use:
Phenidone . . . . 0.5 grams.

Notes:

1) If you have trouble locating some of the chemicals at a local supply house (especially hydroquinone, metol, phenidone and pyrogallol), they may be obtained from Eastman Organics, Rochester, N.Y.

2) Processing techniques using GP-61 and GP-62 are different from the usual methods. This will be discussed in the basic transmission or reflection section respectively.

other bleaches

Although the transmission and reflection hologram bleaches mentioned previously are excellent in our opinion (they are readily available and easy to use), there are alternatives. Different bleaches will effect the emulsion in different ways. Some will yield brighter images at the expense of more "grain" or "noise". Others produce cleaner images which are not as bright, or cause the emulsion to become unstable, resulting in a hologram with a short lifetime. Nick Phillips has been extensively investigating the effects of a number of bleach formulations on the emulsions, and some findings were presented at the Society of Photo-optical Instrumentation Engineers (SPIE) conference in May 1979.

Depending upon the formulation and emulsion used, the bleach step can either occur after fixing (standard bleach) or immediately following the developer or stop bath (reversal bleach). The sequences are described in the transmission and reflection processing procedures. Our preference for transmission holograms is to use a standard bleach procedure, since often the decision as to whether to bleach is made after inspecting the finished hologram. Reflection holograms must be bleached, otherwise the image will be quite dim. The P.B.Q. bleach is used as a reversal bleach. Another reflection bleach, using Mercuric Chloride (described below) is used as a standard bleach.

Other Bleaches:

Mercuric Chloride Bleach: This has been the standard reflection hologram bleach and used reliably for years. It is extremely toxic, however, so great care must be taken in handling - both in mixing and use. Some people still prefer this to P.B.Q. bleaches - it has unlimited shelf life, is easy to use, and gives bright results.

In one liter of water, dissolve:

- 1 tablespoon (approx.) of Mercuric Chloride.
- 1 tablespoon (approx.) of Potassium Bromide.

Variations on P.B.Q.: Nick Phillips also has used the following formula:

In one liter of distilled water, dissolve:

```
Potassium Bromide. . . . . . . . . . . . . .30 grams
Boric acid. . . . . . . . . . . . . . . . . . . . .1.5 grams
P.B.Q. . . . . . . . . . . . . . . . . . . . . . . .2 grams
```
and use within 15 minutes of mixing.

Jeff Milton claims to have solved the instability problems associated with the use of P.B.Q. Although not fully tested as of this printing, we include it for experimental purposes:

In one liter of distilled water, dissolve:

Sulphric acid. . . . . .1 gram
(note: acid is *always* added to water, *never the other way around*)
Potassium Bromide. . . . . .5 grams
Methyl Paraben. . . . . .2 grams
  (Methyl Paraben is a preservative)
Hydrogen Peroxide. . . . . .4 grams
(Hydrogen Peroxide is also used to increase shelf life. This stuff is often supplied in solution so you will need to figure how many grams of pure H_2O_2 is there)
Potassium Alum. . . . . .5 grams
(Potassium Alum is an emulsion hardener)
P.B.Q. . . . . .1 gram

Optional: add 1 gram phenosafranine as a desensitizer. Milton claims the above mixture has a long shelf life (several months minimum).

Bromine Bleach: The use of this bleach can result in high diffraction efficiencies. However, the image can often be fairly noisy. The hologram is either bleached in dilute bromine solution (bromine water), or allowed to react with the fumes (bromine vapor) of concentrated bromine in a closed container. Bromine is highly toxic and caustic, and *extreme care* should be exercised in handling or breathing (avoid breathing). Some serious workers do prefer the results of this bleach to many others. The bleach can be used with both transmission and reflection holograms. It is usually used as a reversal bleach.

GP-431 Bleach: This bleach is ordinarily used along with the GP-61 developer for transmission holograms:
Water . . . . 600 cc
Ferric Nitrate 9-Hydrate . . . . 150 grams
Potassium Bromide . . . . 30 grams
Dissolve 0.3 grams of phenosafranine in 200 cc of methanol and then add to the above mixture. Then add more water to make one liter. When you are ready to use this bleach, dilute one part bleach to 4 parts water.

Our favorite recommendations:

After trying many of the published formulations, we prefer the following for high quality work:

For transmission holograms:
Neofin or D-19 developer.
Potassium Ferrocyanide bleach.

For reflection holograms:
Neofin or D-19 developer.
P.B.Q. Reversal Bleach or Mercuric Chloride standard bleach.

alternate fixer

Benton* claims that rapid fixers can often attack the developer silver while removing the undeveloped silver halide. He prefers to use the following formulations (first a non-hardening fixer (used in the normal fixer step in the processing sequence) and later a hardening bath, just prior to the final water rinse.

Non-Hardening Fixer F-24

Water at 125°F. (50°C.) .... 500 ml.
Sodium Thiosulfate:5H$_2$O .... 240 grams
Sodium Sulfite, desicatted .... 10 grams
Sodium Bisulfite .... 25 grams
Cold water to make .... 1 liter

Use at 68° F. or less
Hardening Bath
Water .... 500 ml.
Formaldehyde (37%) .... 10ml.
Sodium Carbonate (monohydrate) ... 5 grams
water to make .... 1 liter.

*Benton, Steve "Photographic Materials and Their Handling" Handbook of Optical Holography, p. 360 (see bibliography)

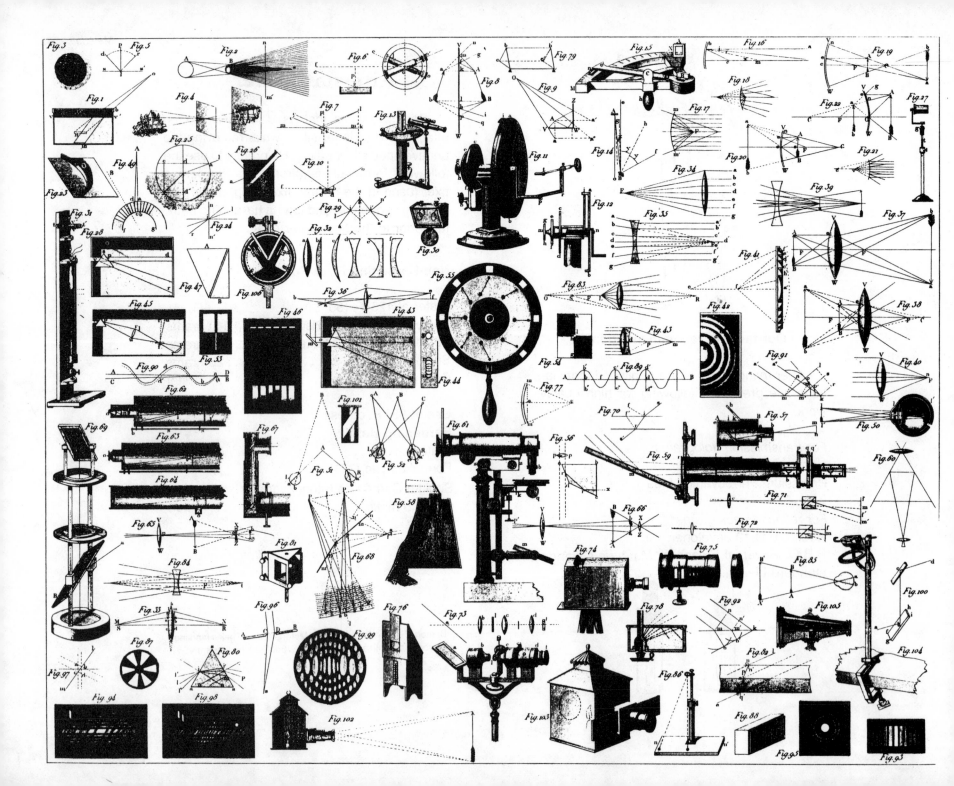

optical component construction

The optical mounts, like the table, can be constructed easily with a minimum amount of effort and money. The K.I.S.S. philosophy, is used to its fullest extent, so there is no reason to hesitate in improvising.

The satisfaction realized from creating the components should be extremely rewarding. In addition, obtaining a first hand knowledge of the construction allows one to better know their limitations. A redesigned or customized component for a particular application also becomes easier to achieve.

Similar components can be purchased, but in most instances they are over-designed, over-automated, and overpriced. This type of component, though specifically manufactured for magnetic tables, can be adapted for use with the sand table. The K.I.S.S. components presented herein work as well, if not better, than the types initally created for holography by commercial manufacturers.

These components can be fitted with a wide variety of optics. A selection of suitable mirrors, lenses, and beamsplitters will be found in the following chapter on purchasing materials. In addition, at the end of that chapter, complete kits, including recommended optics for beginners, advanced, and school or professional studios, will be listed.

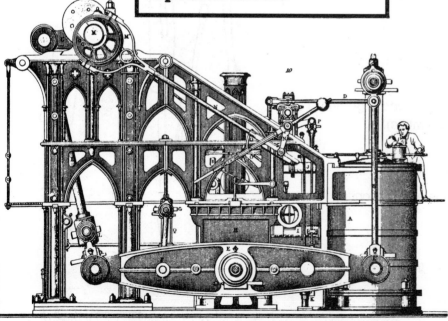

K.I.S.S.   Keep It Sweet and Simple!

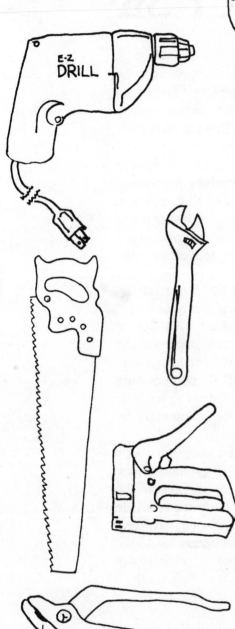

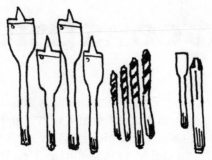

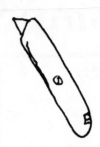

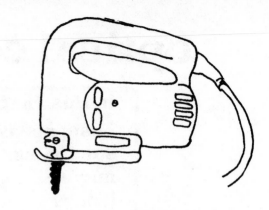

tools

- Hand saw or electric saw

- Drill with ¼" drill bit (assorted sizes)

- Pliers

- Screwdriver

- Vice is nice but a heavy foot will do

- Ruler

optional

- Variable beamsplitter ¼"x 6" wedge gradually silvered (from Edmund Scientific).

- Glass cutter

- Black paint to spray pipe and masonite

- Sand paper

- Rat-tail file for enlarging drilled holes

- Band saw for fancy pipe cuts

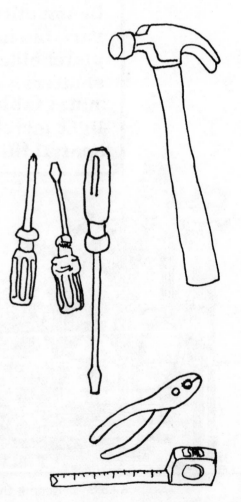

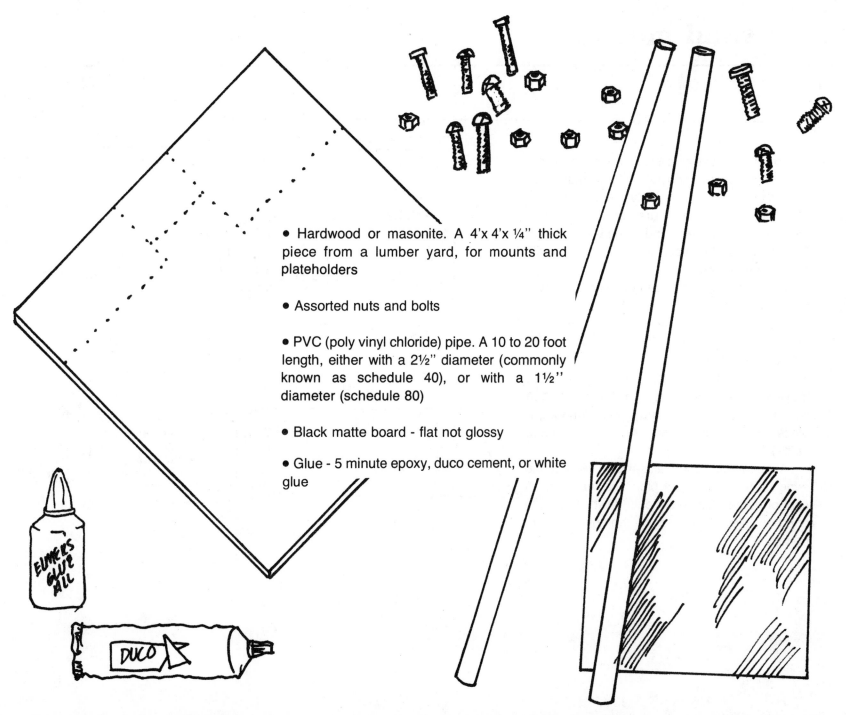

- Hardwood or masonite. A 4'x 4'x ¼'' thick piece from a lumber yard, for mounts and plateholders

- Assorted nuts and bolts

- PVC (poly vinyl chloride) pipe. A 10 to 20 foot length, either with a 2½'' diameter (commonly known as schedule 40), or with a 1½'' diameter (schedule 80)

- Black matte board - flat not glossy

- Glue - 5 minute epoxy, duco cement, or white glue

sand mounts

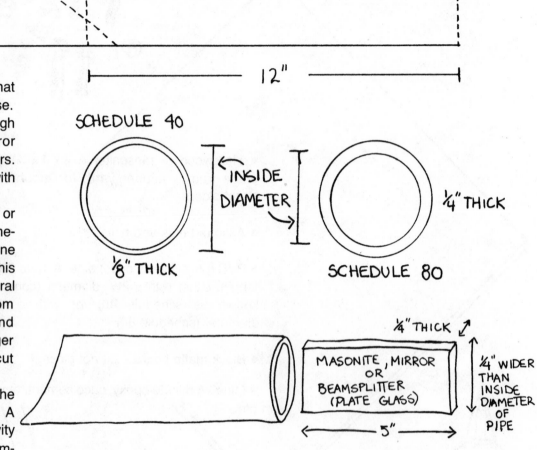

There are several preliminary projects that can be done at once and put aside for future use.

A bit of sawing will supply you with enough PVC pipe to manufacture lens and mirror mounts, beamsplitters, and some plateholders. Cut as a spike, the pipe will enter the sand with much greater ease.

Whether you've decided on schedule 40 or 80, an exact measurement of the inside diameter of the pipe must be taken to help determine the width of the masonite. Simply add 1¼" to this measuement (1¼" plus I.D.) and saw several strips this width. Five inch lengths cut from these strips will be appropriate for mirror and lens boards (the need may arise for longer pieces of masonite, so leave extra uncut strips).

Although not essential, sanding will take the edge off the freshly cut pipe and masonite. A coat of matte black paint cuts down reflectivity and possible light scatter on the finished components.

CUT A DOZEN 12 INCH LENGTHS

CUT A 2 FOOT LENGTH STRAIGHT
AND HALVE IT ON A DIAGONAL

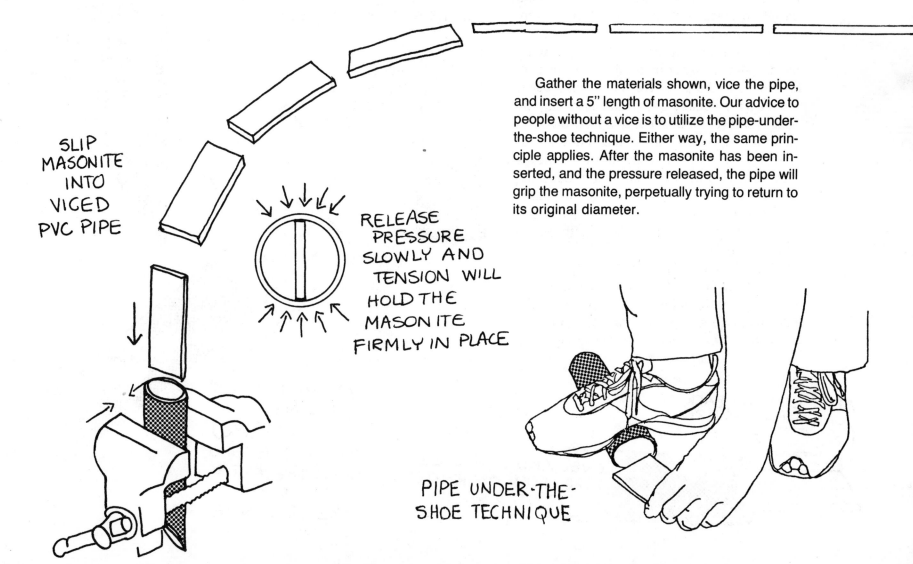

SLIP
MASONITE
INTO
VICED
PVC PIPE

RELEASE
PRESSURE
SLOWLY AND
TENSION WILL
HOLD THE
MASONITE
FIRMLY IN PLACE

Gather the materials shown, vice the pipe, and insert a 5" length of masonite. Our advice to people without a vice is to utilize the pipe-under-the-shoe technique. Either way, the same principle applies. After the masonite has been inserted, and the pressure released, the pipe will grip the masonite, perpetually trying to return to its original diameter.

PIPE UNDER-THE-
SHOE TECHNIQUE

mirrors

Front surface (mirrors) become useless to a holographer if they acquire fingerprints or scratches. Handle them carefully, especially while attaching them to the mounts.

The mirrors are often shipped with a protective coating over the surface which should be left in place until the component is completely assembled.

One of a number of types of glue may be used to attach the mirror to either the masonite or directly to the pipe. We recommend using silicon as it applies easily and also may be easily removed, if desired, at a later date. An alternate non-glue procedure consists of wedging the mirror directly into the pipe as was done with the masonite (see preceeding page) or cutting a slot in the top of the pipe to cradle a larger mirror. Part of the mirror surface area will be wasted if these methods are used, however.

After the mirror is mounted, the protective coating may be removed by gently sticking a piece of adhesive tape to one corner and carefully lifting up and pulling the coating across the mirror. Remember not to touch the mirror surface with your fingers.

FINGERPRINTS RUIN MIRRORS

MIRROR CAREFULLY VICED INTO PIPE

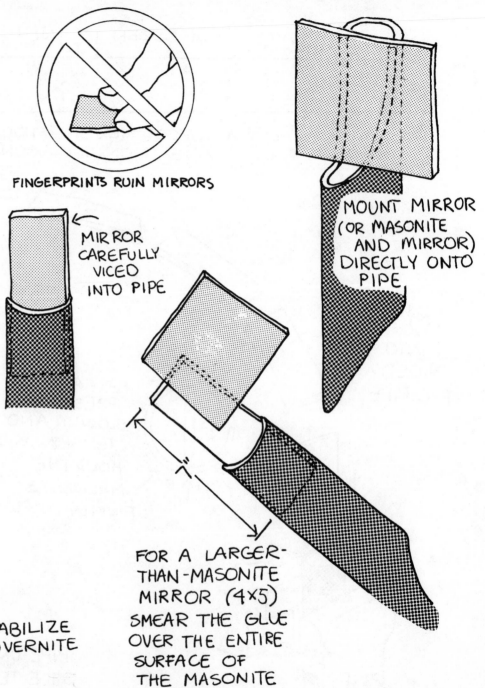

MOUNT MIRROR (OR MASONITE AND MIRROR) DIRECTLY ONTO PIPE

FOR A LARGER-THAN-MASONITE MIRROR (4×5) SMEAR THE GLUE OVER THE ENTIRE SURFACE OF THE MASONITE

APPLY GLUE TO BACK OF SMALLER MIRRORS

LEVEL AND STABILIZE PIPE FOR OVERNITE DRYING

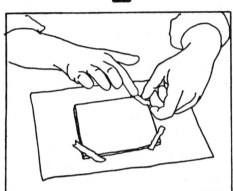

1

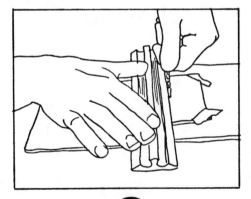

2

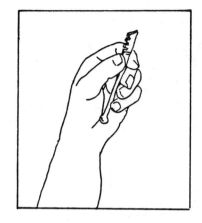

It may be less expensive to cut a larger mirror into several smaller ones with a glass cutter.* Tape a coated mirror down on a flat surface, scribe a straight line, and tap the glass from the other side as shown (suggestion: practice with an equal thickness of scrap glass). In this way, smaller mirrors of desired sizes can be obtained from a larger mirror.

*easiest with thin mirrors.

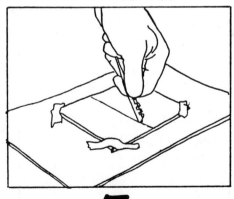

3

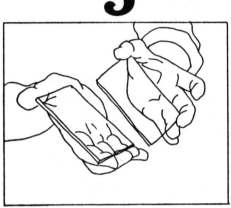

5

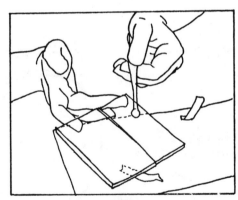

4

larger mirror mounts

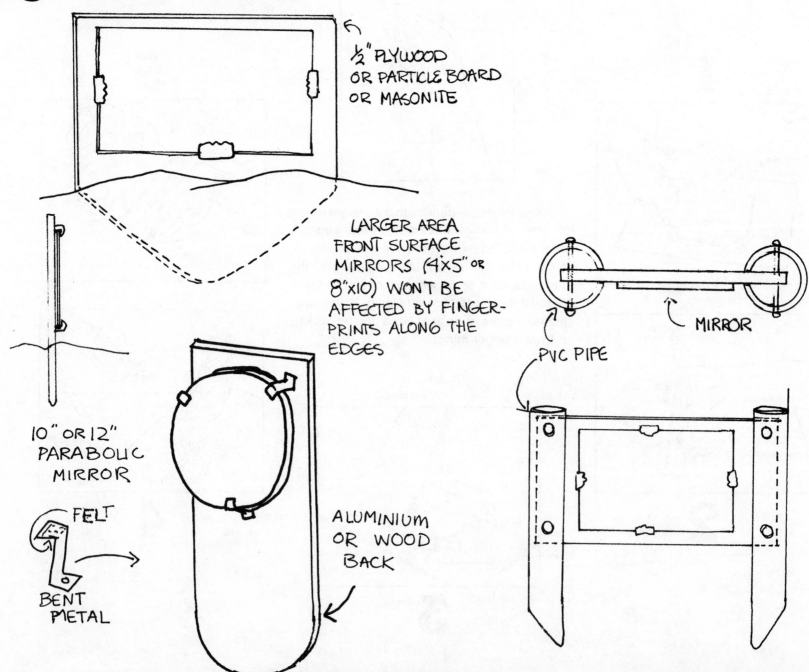

½" PLYWOOD
OR PARTICLE BOARD
OR MASONITE

LARGER AREA
FRONT SURFACE
MIRRORS (4"x5" OR
8"x10) WON'T BE
AFFECTED BY FINGER-
PRINTS ALONG THE
EDGES

10" OR 12"
PARABOLIC
MIRROR

FELT

BENT
METAL

ALUMINIUM
OR WOOD
BACK

MIRROR

PVC PIPE

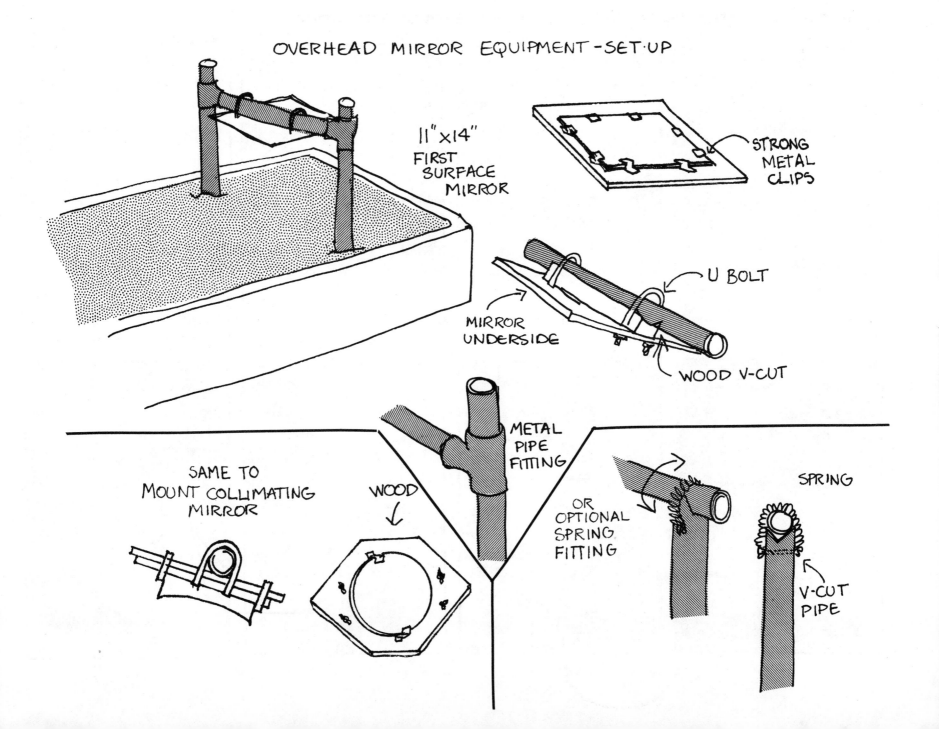

OVERHEAD MIRROR EQUIPMENT - SET·UP

11" x 14" FIRST SURFACE MIRROR

STRONG METAL CLIPS

U BOLT

MIRROR UNDERSIDE

WOOD V-CUT

SAME TO MOUNT COLLIMATING MIRROR

WOOD

METAL PIPE FITTING

OR OPTIONAL SPRING FITTING

SPRING

V-CUT PIPE

Making a mirror mount for those of you with power tools:

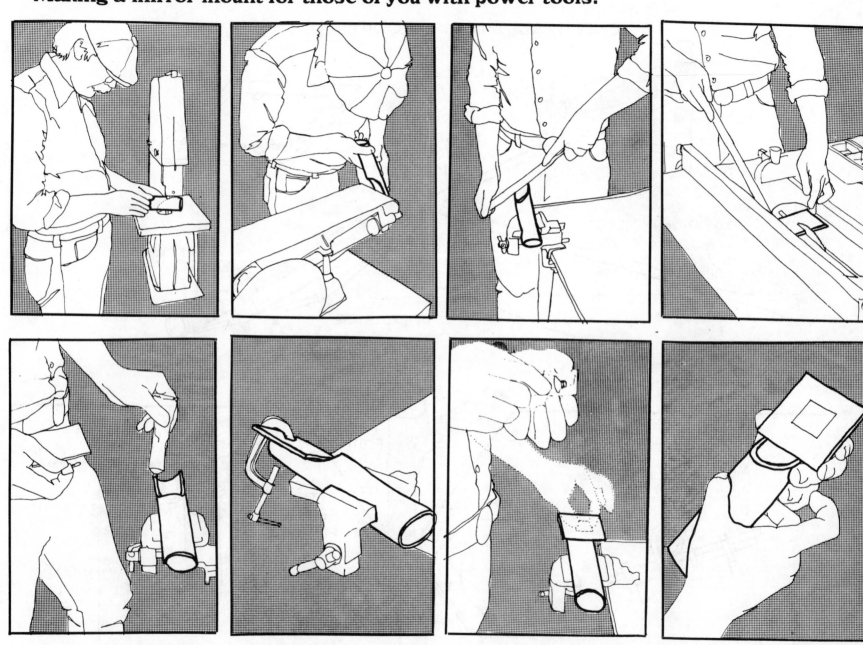

lenses

A hole with a diameter 1/8 inch smaller than that of the lens is drilled into the masonite. The hole is made smaller so that a surface remains for gluing. (It is also possible to mount the lens so that it fits flush into the hole, however it is often difficult to prevent light leaks around the lens using this method.) The masonite is then viced into the PVC pipe.

Care should be taken to keep the lens clean during the mounting process. Lenses are generally easier to clean than mirrors, yet it is best to always handle lenses by their edges. The lens should be clean before it is glued in place, as it is often difficult to reach some parts of the surface after mounting. If you do find it necessary to clean a lens, use the special lens-cleaning fluid and paper available at photo stores (they might also show you how to do it properly).*

To glue the lens, first center it with respect to the hole while making certain there is contact with the masonite all the way around. Spots of glue may then be applied around the edge and the assembly left to dry in a horizontal position.

Variations shown include a masonite cut out for a larger lens, an extender lens for odd-to-reach angles, a direct pipe-mounted lens and several types of interchangeable mounts. Fancy and pretty are hard to resist, but they all perform the same function. Remember K.I.S.S.

*Mirrors can also be cleaned by substituting acetone for the lens fluid. However, the quality of a cleaned mirror is never as good as one carefully kept clean in the first place.

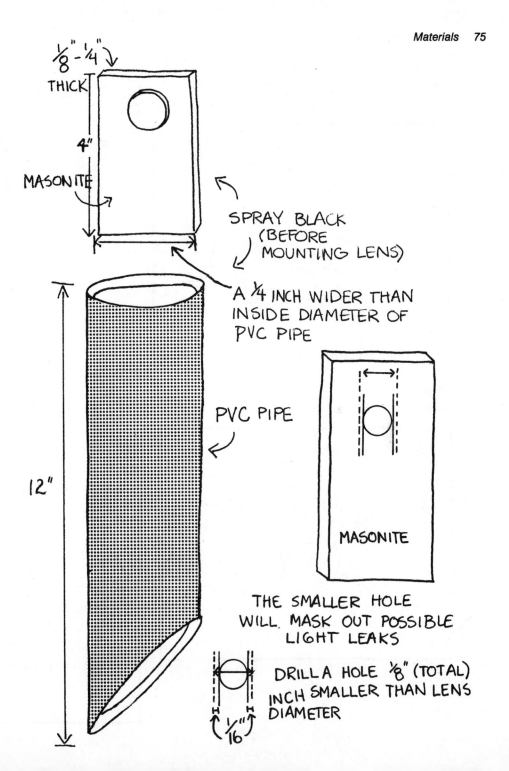

$\frac{1}{8}" - \frac{1}{4}"$ THICK

4"

MASONITE

SPRAY BLACK (BEFORE MOUNTING LENS)

A ¼ INCH WIDER THAN INSIDE DIAMETER OF PVC PIPE

PVC PIPE

12"

MASONITE

THE SMALLER HOLE WILL MASK OUT POSSIBLE LIGHT LEAKS

DRILL A HOLE ⅛" (TOTAL) INCH SMALLER THAN LENS DIAMETER

$\frac{1}{16}"$

1

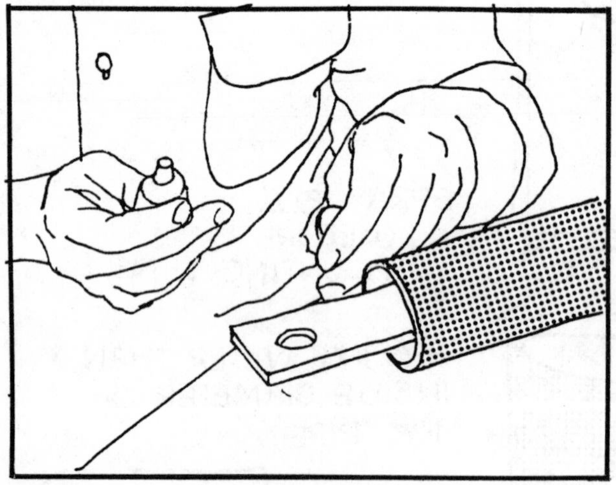

2

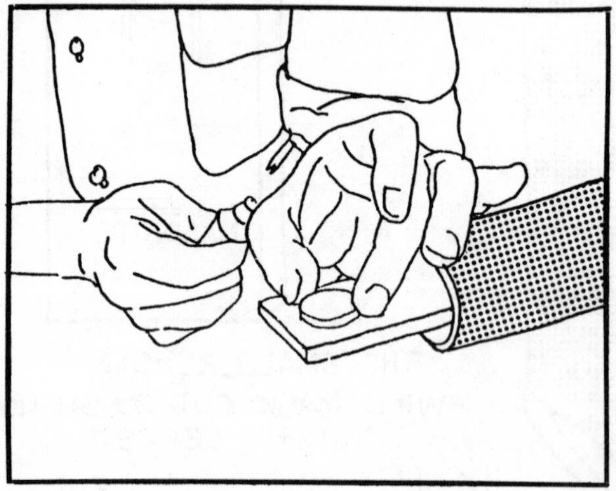

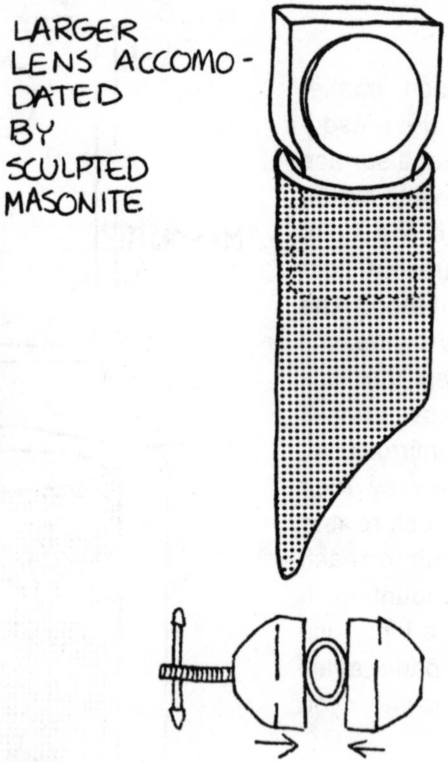

3

LARGER LENS ACCOMODATED BY SCULPTED MASONITE

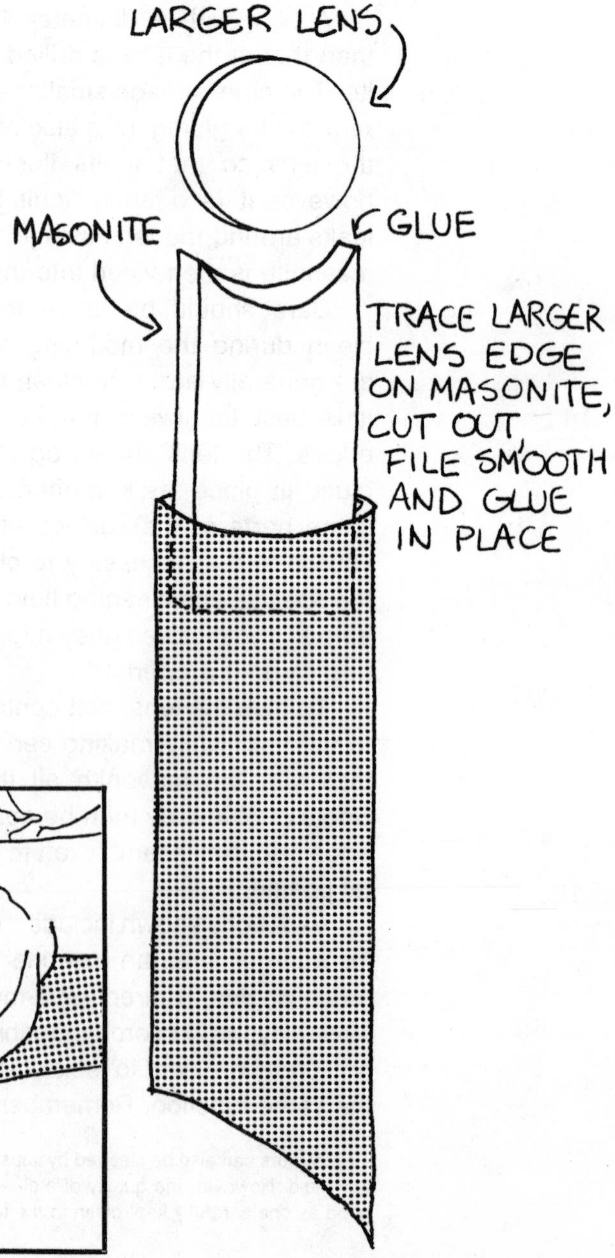

LARGER LENS

MASONITE

GLUE

TRACE LARGER LENS EDGE ON MASONITE, CUT OUT, FILE SMOOTH AND GLUE IN PLACE

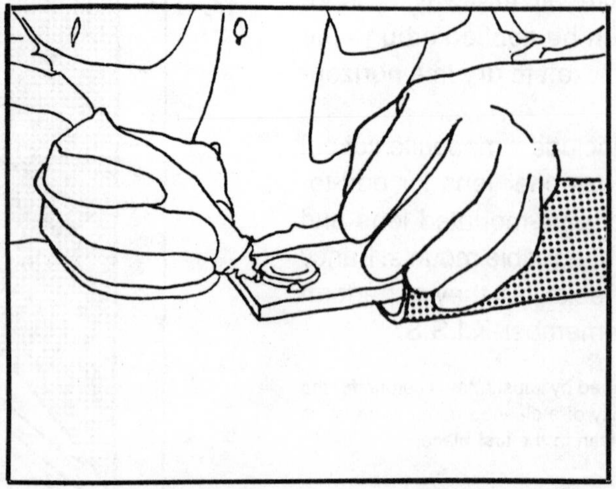

OPTIONS

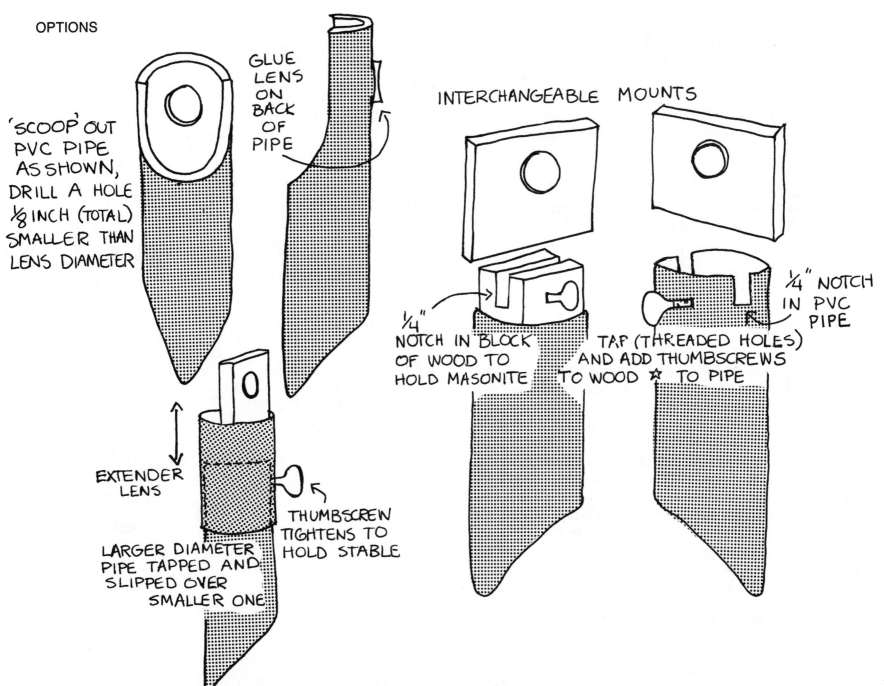

'SCOOP' OUT PVC PIPE AS SHOWN, DRILL A HOLE ⅛ INCH (TOTAL) SMALLER THAN LENS DIAMETER

GLUE LENS ON BACK OF PIPE

EXTENDER LENS

LARGER DIAMETER PIPE TAPPED AND SLIPPED OVER SMALLER ONE

THUMBSCREW TIGHTENS TO HOLD STABLE

INTERCHANGEABLE MOUNTS

¼" NOTCH IN BLOCK OF WOOD TO HOLD MASONITE

TAP (THREADED HOLES) AND ADD THUMBSCREWS TO WOOD ☆ TO PIPE

¼" NOTCH IN PVC PIPE

beamsplitters

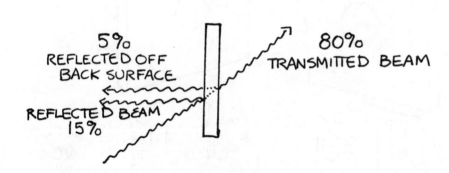

5%
REFLECTED OFF
BACK SURFACE

REFLECTED BEAM
15%

80%
TRANSMITTED BEAM

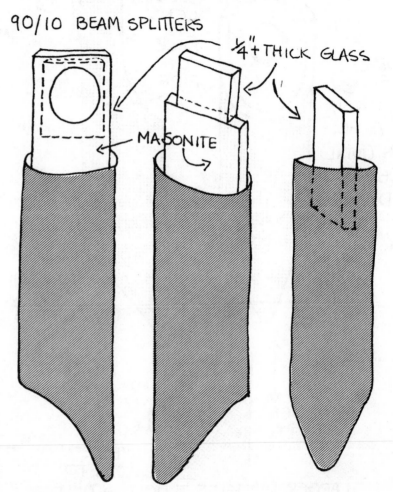

90/10 BEAM SPLITTERS

¼"+ THICK GLASS

MASONITE

For simple holography two specific beamsplitters are needed, a 90:10 and a 50:50, but for more advanced holography a variable beam-splitter is invaluable.

The most simple to construct is the 90:10 which is merely a ¼" thick piece of plate glass wedged into a length of pipe. Although called a 90:10 beam-splitter, the light distribution (after some absorption), is actually 80 percent transmitted light, 15 percent reflected back off the front surface of the glass, and another small percentage reflected off the second surface. It is important that the glass be at least ¼" thick or the reflected light will get in the way of itself.

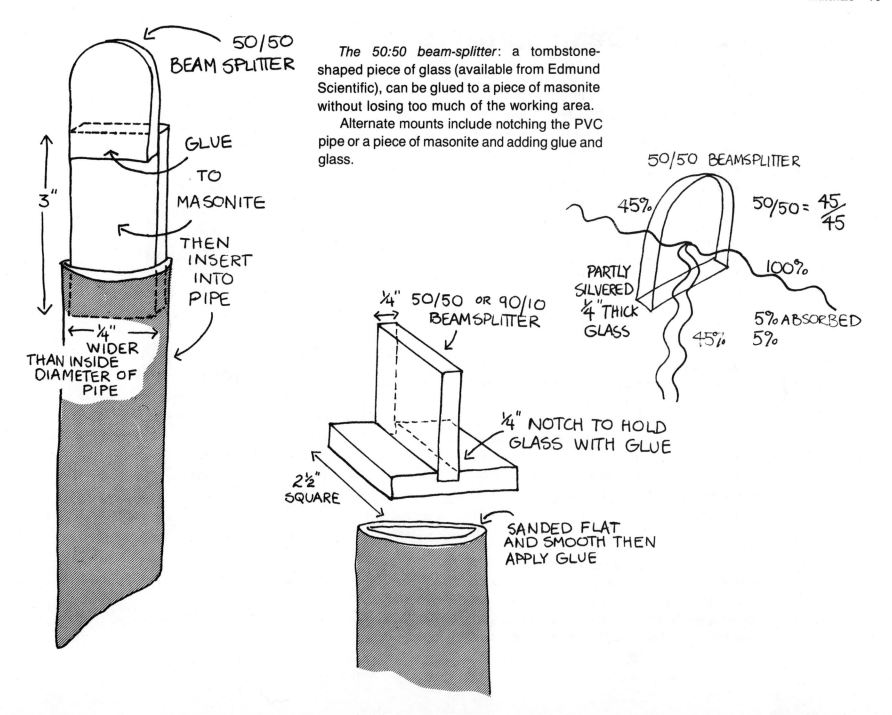

50/50 BEAM SPLITTER

GLUE TO MASONITE THEN INSERT INTO PIPE

3"

¼" WIDER THAN INSIDE DIAMETER OF PIPE

The *50:50 beam-splitter*: a tombstone-shaped piece of glass (available from Edmund Scientific), can be glued to a piece of masonite without losing too much of the working area.

Alternate mounts include notching the PVC pipe or a piece of masonite and adding glue and glass.

50/50 BEAMSPLITTER

45%

$50/50 = \dfrac{45}{45}$

100%

PARTLY SILVERED ¼" THICK GLASS

45%

5% ABSORBED 5%

¼" 50/50 OR 90/10 BEAMSPLITTER

¼" NOTCH TO HOLD GLASS WITH GLUE

2½" SQUARE

SANDED FLAT AND SMOOTH THEN APPLY GLUE

variable beamsplitters

The variable beamsplitter is simply a beamsplitter which may be used to adjust the ratio of intensities of reflected to transmitted beam. It consists of a mirror silvered gradually from about 5% to approximately 98% reflective. These beamsplitters are offered in either a disc format (much preferred, although expensive) or a rectangular "wedge" with a linear variation. Edmund Scientific Co. offers a low cost wedge which may be mounted in a variety of ways and simply moved along the long axis to change ratios. The wedge can be converted into a circular beamsplitter, as shown, so that turning the masonite disc will cause the beam to travel through different spots on the beamsplitter.

Unfortunately, the wedge substrate is rather thin, and light interfering between the front and back surfaces at times may form undesirable fringe patterns. This is not critical for beginning setups, yet is often unsuitable for higher quality work. The variable attenuated disc, as described in the section on purchasing materials, is, in our opinion, essential for advanced or serious holography.

Innovative and competent optical fabricators may wish to try solving the wedge substrate problem by affixing the thin piece to a ¼" thick piece of glass with a good optical cement*. Remember to leave the coated side of the beamsplitter on the outside if you do this. During use, the coated surface should always intercept the beam before the glass surface does.

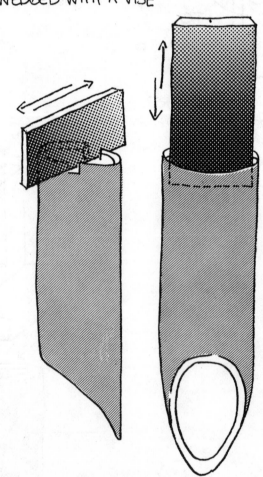

GRADUALLY SILVERED MIRRORS EPOXIED TO PVC PIPE OR CAREFULLY WEDGED WITH A VISE

*such as Norland UV adhesive (see suppliers listing).

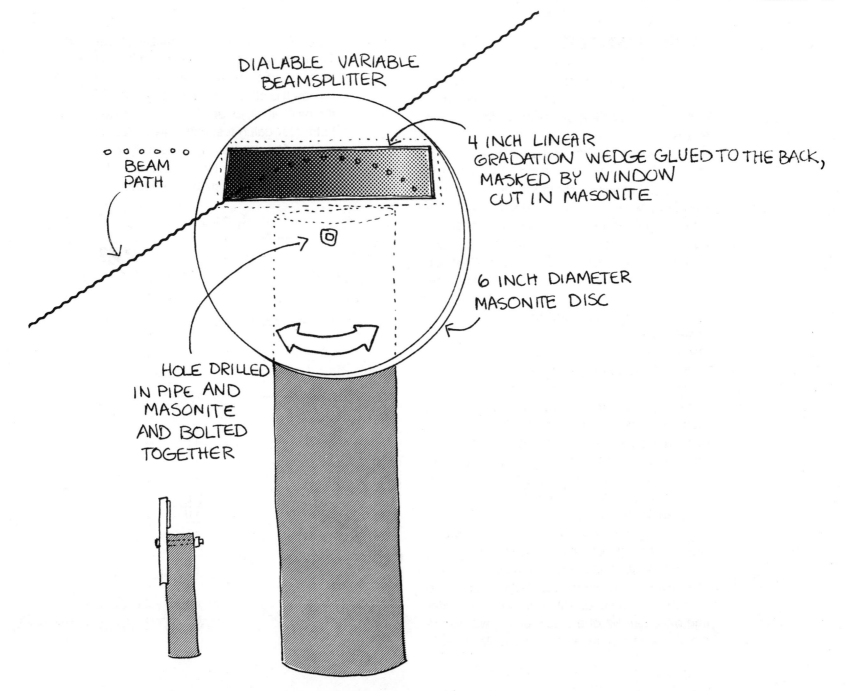

DIALABLE VARIABLE BEAMSPLITTER

BEAM PATH

4 INCH LINEAR GRADATION WEDGE GLUED TO THE BACK, MASKED BY WINDOW CUT IN MASONITE

6 INCH DIAMETER MASONITE DISC

HOLE DRILLED IN PIPE AND MASONITE AND BOLTED TOGETHER

plateholders

In this chapter we shall discuss techniques for construction devices to hold the holographic plate or film in position on the optics table for the exposure.

Plateholder design probably ranks among the most innovative for components. The holders are fairly simple for the basic holograms. However, when one starts playing with advanced composites, new techniques often require new plateholders. Remember, holography is so new, and developing so rapidly, that holders for some types of holograms could not be purchased even if you wanted to. Also, we believe that those described here are more versatile, yet less expensive, than any you will find elsewhere.

The most commonly used holographic recording materials are supplied on 4"x 5" glass plates. The construction of one 4"x 5" holder, then, is all that is required for all of the basic holographic setups (except for the image plane hologram, where two are required). The same basic design can then be applied to other plateholders of other sizes and modified for other applications.

The holder is basically a three-piece U-shaped sandwich, with the inner piece having a slightly larger cutout than the two outer pieces. The holder has a tapered base which allows it to be set into the sand. Although the plateholder and base can be of a single piece, we recommend making a separate removable base in order to increase the versatility of the holder. The holder can then be used in either a horizontal or a vertical position on the table, depending on how it is attached to the base, or used without the base at all, as an insert into a larger 8"x 10" holder for doing test exposures. You should be aware of the overall thickness of the holder if you plan to use it for this purpose, and consider designing the two sizes together.

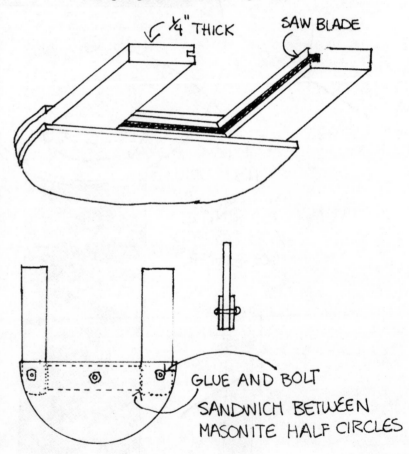

Constructing the holder:

You will need:

- Two pieces 8"x 10" masonite ⅛" thick or less (or tempered aluminum).
- 8"x 10" cardboard, plastic, or what have you
- 4"x 5" glass plate
- glue or bolts to put it together

The plate should fit snugly in the holder without flopping around, yet should be loose enough to allow the plate to be inserted and removed easily; remember, you will load the plate in the dark. To make the holder, you will need to use an actual 4"x 5" plate as a template. Since one must first build a plateholder before making a hologram, this means using an unexposed plate (you will ruin it by exposing it to light, but it's worth it to insure a proper fit). Make sure you open the box of plates in total darkness, remove one, and then close the box before turning on the lights.

The plate will first be used as a guide for determining the thickness of the spacer, or inside part of the three-piece sandwich. The spacer must be slightly thicker than the plate. To make one, place the plate on a flat surface. Alongside it, build up a stack of material such as cardboard, plastic sheet, paper, or some combination of all, so that the thickness of the stack is just a bit more than that of the plate. The components of the stack should now be glued together and trimmed to 8"x 10".

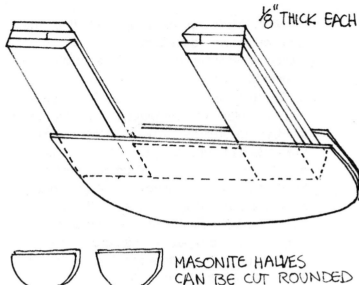

⅛" THICK EACH

MASONITE HALVES CAN BE CUT ROUNDED OR POINTED

The plate is now used as a guide to cutting out a portion of the stack to make the U-shaped spacer. Place the plate on top of the stack, center it lengthwise along one edge and simply trace around it. Cutting out this portion will leave you with the desired spacer.

Now, using the two outer pieces, cut out U shapes in each which are approximately ⅛" less (inside) all the way around than the spacer (i.e., the dimensions of these cutouts should be about 3⅞"x 4¾").

You may, at this point, slightly bevel the U-shaped edges on both of the outer pieces, on one side. This is to reduce the shadow the edges of the holder may cast onto the plate.

The three pieces can then be clamped, drilled and bolted together or glued; (try to center the spacer opening between the two outer shells as well as possible.).

Next, drill two holes along the bottom and sides as shown; (both sets of holes should be equidistant from each other). These are for attaching the holder to a base.

The plateholder base should be made so that the holder can be placed easily into the sand, yet securely held in place after positioning. We have had success with a U-shaped or tapered piece of wood or aluminum which is bolted to the holder. This can simply be cut to the shape shown, and holes drilled to match the holes in the plateholder. If the plateholder holes are properly spaced apart, the same base can be bolted to either the bottom or side of the holder. To start with, bolt the base to the bottom. Using a bolt with wing nuts will facilitate quick removal or changing of the base. As a final step, spray the entire holder with flat black paint.

Holders for 8"x 10", 30 x 40 cm, or any other size plates can be made in the same way, simply by using the desired plate as a template. If you do not want to ruin a large expensive plate to do this, have a glass cutter make you one of the size and thickness required; (see the box of plates for the thickness measurement). For people who are sure that they will be doing advanced work, we recommend having at least two 4"x 5" holders and two 8"x 10" holders, with at least one of the 4"x 5" holders capable of being inserted into the 8"x 10" holder.

Many of the setups utilize the plateholder tilted in the sand at an angle. In these cases, to insure stability, a longer base, sometimes 10" to 14", may be required. It is a simple matter to construct one of these bases and to use it interchangeably with the regular base when needed.

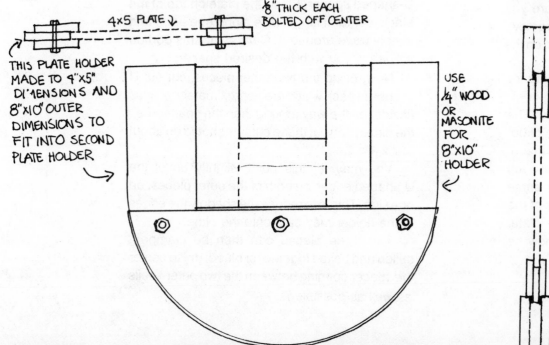

4x5 PLATE

⅛"THICK EACH BOLTED OFF CENTER

THIS PLATE HOLDER MADE TO 4"x5" DIMENSIONS AND 8"x10" OUTER DIMENSIONS TO FIT INTO SECOND PLATE HOLDER

USE ¼" WOOD OR MASONITE FOR 8"x10" HOLDER

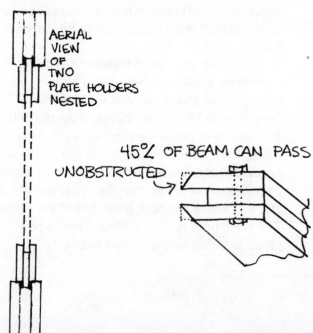

AERIAL VIEW OF TWO PLATE HOLDERS NESTED

45% OF BEAM CAN PASS UNOBSTRUCTED

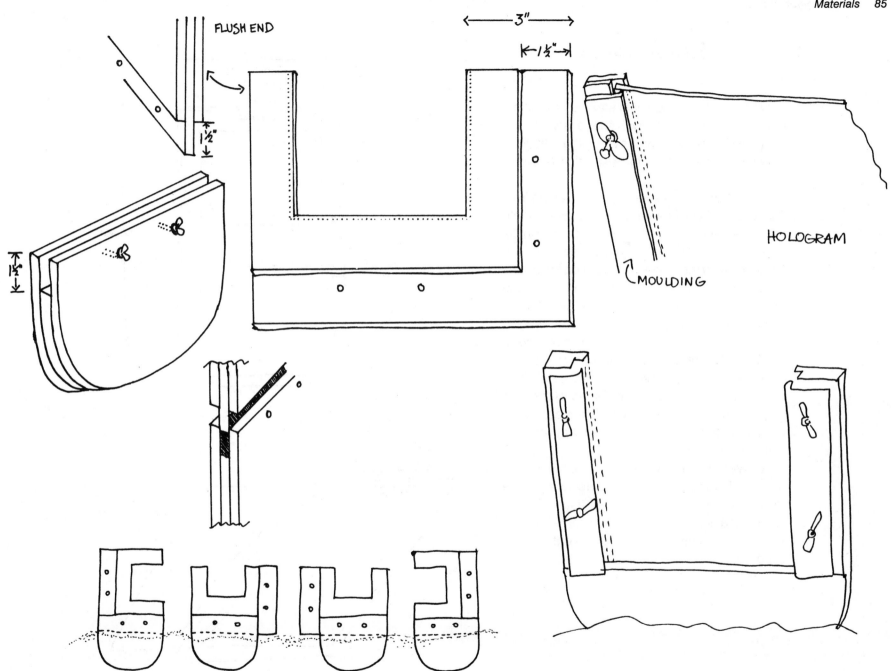

FLUSH END

3"

1½"

½"

½"

HOLOGRAM

MOULDING

Various methods of mounting plateholder.

filmholders

Those holographers just starting out will not need to be concerned with these holders immediately, as we recommend working with glass rather than film until some expertise is developed.

The easiest filmholder to use is the type from an old 4"x 5" view camera. The film slides into the holder, and a cover plate keeps the film in the dark until it is used (Your photo salesman can show you how they work). To position the holder on the optics table, a sandwich similar to the plateholder, yet wide enough to hold the film can then be removed and replaced as if it were a plate. Precut 4"x 5" film is available (see recording materials) or it may be cut to size, using a paper cutter in the darkroom with a safelight.

Another method of using film is to clamp it between two pieces of glass. The glass sandwich can then be slid into an appropriate plateholder.

Curved film must usually be taped to a piece of curved aluminum or plastic. Because of the relative instability of film, it must usually be affixed to the holder and allowed to settle for quite some time (sometimes even overnight), before exposure.

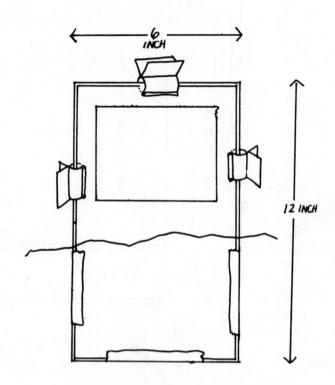

6 INCH

12 INCH

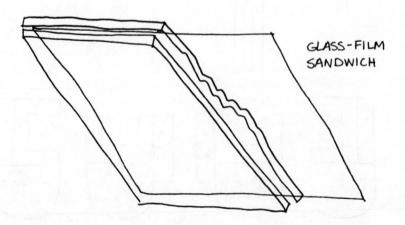

GLASS-FILM SANDWICH

index matching holders

Patterns known as Newton Rings are often considered undesirable for advanced high quality transmission holography. These are caused by light interacting between the front and back surfaces of the glass plate or film. Some plates are supplied with an antihalation (AH) backing to cut down on this effect; most, however, are not. A solution is to use a plate or film holder with an index-matching fluid between the holographic plate or film, and a dark black plate.

The back plate can be fashioned from a piece of black glass, plastic, or metal. The plateholder can be simply an adaption of one of the designs mentioned previously. The plateholder opening is made to accommodate the thickness of the holographic plate and back plate together. Clamps are attached to hold the plates together by bolting one side of 90 degree metal molding to the holder, as shown. The other edge of the molding will press against the plates holding them together.

The indexing fluid can be applied either in the holder, or before the plate is inserted. If the first technique is used, the back plate can be made as a fixed part of the plateholder. The holographic plate is inserted by hinging the bottom edge against the back plate and then using a dropper, depositing a bead of xylene (the indexing chemical) along the common edge between them. The holographic plate is carefully hinged up to bring the two together while carefully working the liquid up between them. It is best to apply xylene to the glass side of the plate, not the emulsion side.

Once the sandwich is made, the clamps can be tightened to hold the two in place. The whole action can be done under a dark green safelight but it is best to practice first with a piece of glass in normal light to get the hang of it.

Another way to make the sandwich is to put the two together outside the holder and then to insert the completed sandwich. This is done by placing the back plate on a flat surface and placing a couple of drops of xylene on the surface. The holographic plate is hinged down carefully so as not to cause bubbles,* and the fluid will spread itself evenly between the two. They are then placed into the holder.

After the exposure, the xylene can either be wiped off the back of the plate with a paper towel, or evaporate, or simply dissolved out in the developer.

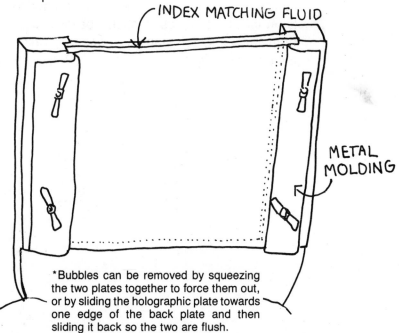

INDEX MATCHING FLUID

METAL MOLDING

*Bubbles can be removed by squeezing the two plates together to force them out, or by sliding the holographic plate towards one edge of the back plate and then sliding it back so the two are flush.

vacuum platen

For professional systems, a film vacuum platen can be constructed by using a ¼" aluminum sheet with many holes drilled throughout with a very small drill. The back of the platen is made by gluing a thin spacer around the edges and afficing another plate to it, to allow a cavity between the two flat plates. An opening for a hose can be drilled into the back plate, and a good seal made for the hose and sandwich by covering all joints with silicon material. The hose can be attached to the intake of an air compressor or a modified vacuum sweeper. If a piece of film covers all of the holes in the platen, it will be held securely when a vacuum is drawn. The platen should be sprayed black to reduce the effects of back scatter (If you do use air equipment, you should be sufficiently experienced as a holographer to know that the compressor should be located far from the table for stability reasons - which means long hoses).

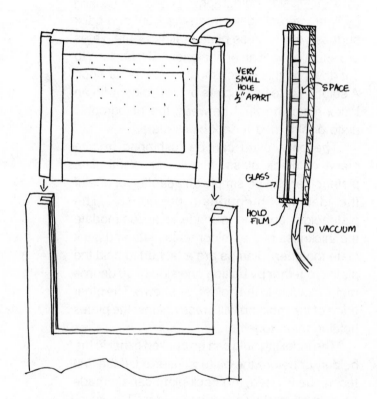

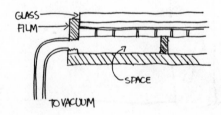

shutters

The shutter used in holographic setups serves the purpose of blocking the beam, and then opening for the required exposure time. The function is essentially the same as that of the camera shutter in photography: to control when and for how long light reaches the film/plate.

Home made shutters can easily be constructed in a variety of ways:

Construction of "card in sand" variety: this shutter involves essentially no forethought, expense, manual dexterity or advanced engineering degrees. A cardboard or masonite card simply sits in the sand blocking the beam, is pulled out for the exposure, and then replaced after the desired time period. The major disadvantage of this type of shutter is its tendency to disturb the stability of the system during the exposure ...One of the following remote shutters is preferred:

slides up, allowing the beam to pass until the string is released. If the laser is not placed on the optics table, it is a simple matter to mount the shutter just in front of the beam opening. If the laser is used on the table, the shutter must be mounted in such a way as not to add vibrations to the system during operation, i.e., it cannot be attached to the table. The shutter can be placed on a separate base and allowed to stick out over the optics table in front of the beam. It can also be mounted on an arm attached to the wall, or even suspended from the ceiling.

A slightly more sophisticated version of the above is to use a bulb and shutter from an old view camera. This type of device can sometimes be located after a little photo-detective work, and then mounted in a manner similar to that of the guillotine shutter. The bulb is squeezed, causing the shutter to open, and remains so until the pressure is released.

BEST TO MOUNT SHUTTER OFF OPTICAL TABLE

MANUAL SHUTTER

AIR BULB

PHOTO SHUTTER FROM OLD CAMERA

LASER

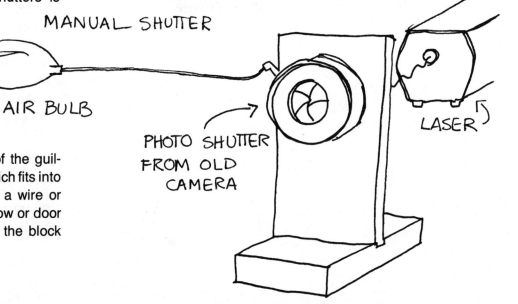

The simplest remote shutter is of the guillotine type. A wood or metal block, which fits into two, U-channel slots, is attached to a wire or string running up and over to a window or door opening. When the string is pulled, the block

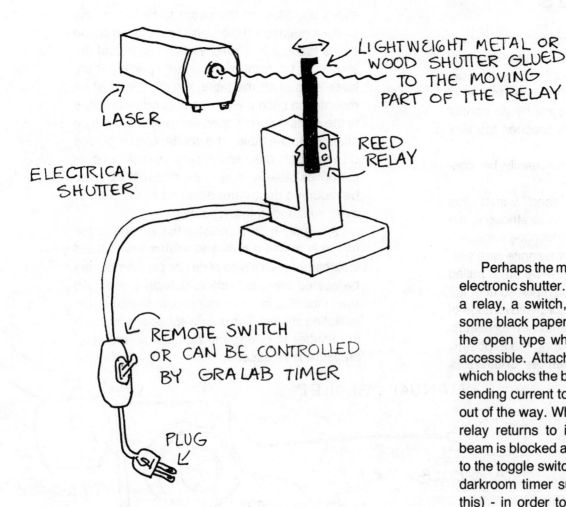

LIGHTWEIGHT METAL OR
WOOD SHUTTER GLUED
TO THE MOVING
PART OF THE RELAY

LASER

REED
RELAY

ELECTRICAL
SHUTTER

REMOTE SWITCH
OR CAN BE CONTROLLED
BY GRA LAB TIMER

PLUG

Perhaps the most sophisticated design is the electronic shutter. This can be constructed using a relay, a switch, a small piece of wood, and some black paper. The relay used should be of the open type which has the movable section accessible. Attached to this is the wood/paper which blocks the beam. The shutter operates by sending current to the relay, causing it to move out of the way. When the current is shut off, the relay returns to its original position, and the beam is blocked again. As an option, in addition to the toggle switch, a timer can be wired in - (a darkroom timer such as Gra Lab will work for this) - in order to determine automatically the exposure (much the same way as the timer used on a photo enlarger works). In fact, one may even cannibilize an old enlarger-timer for this purpose.

metal table mounts

If you are using a metal rather than a sand system, the components will need to have magnetic bases rather than those made of plastic pipe. Professional bases are available, usually at great cost, from a number of manufacturers. Good quality bases can be obtained, at one-third to one-tenth the cost by going, not to the optical manufacturers, but to the machinist supply house. Magnetic bases are used as pointers or placers by machine shops and are very similar to optical bases. They are every bit as good. We often work with bases manufactured by Mitutoyo. The optics can simply be bolted or glued to one of the rods attached to the base.

Alternatives to this are to build your own bases using magnets, (try Radio Shack), or simply to use very heavy flat bases which stay put on the table surface.

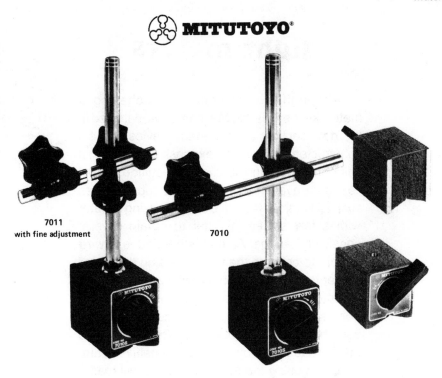

MITUTOYO

7011
with fine adjustment

7010

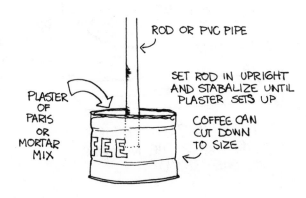

ROD OR PVC PIPE

SET ROD IN UPRIGHT AND STABALIZE UNTIL PLASTER SETS UP

PLASTER OF PARIS OR MORTAR MIX

COFFEE CAN CUT DOWN TO SIZE

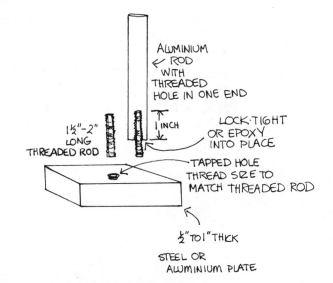

ALUMINIUM ROD WITH THREADED HOLE IN ONE END

1 INCH

1½"-2" LONG THREADED ROD

LOCK-TIGHT OR EPOXY INTO PLACE

TAPPED HOLE THREAD SIZE TO MATCH THREADED ROD

½" TO 1" THICK

STEEL OR ALUMINIUM PLATE

light meters

Although we recommend purchasing a meter like the Science Mechanics model listed in the next section, a homemade device may be constructed easily and inexpensively by using a volt-ohmmeter and a cadmium-sulfide photo-cell, both available at electronic supply stores such as Radio Shack. We suggest using a meter with a few scales in order to obtain several sensitivity ranges. All that is needed is to wire the photo-cell into the two outlets on the meter (either lead from the photocell may go to either receptor). To calibrate, set the meter to read "ohms" and place the photocell close to a very bright light. Rotate the knob until the meter reads zero on the most sensitive scale.

When using the meter, the scale will read a *lower* number for a *brighter* reading. This is potentially confusing, yet is one of the peculiarities of this type of meter. Remember, a reading of 100 will be twice as bright as a reading of 200.

Using the various scales, you may read the meter easily in a variety of conditions. For example, the needle may either not register at all, or go right off the scale; under these conditions, you can easily move to a more appropriate scale, to cause the meter to register correctly.

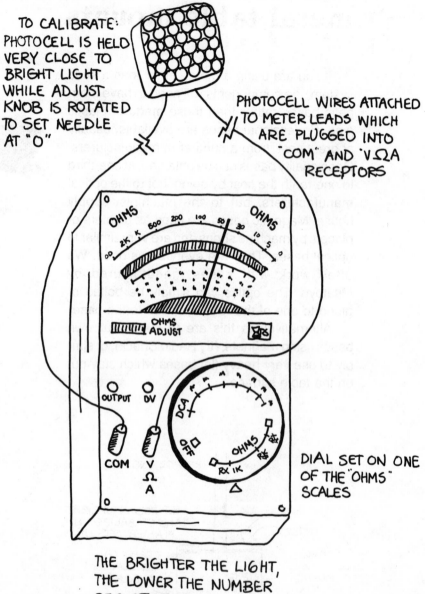

TO CALIBRATE:
PHOTOCELL IS HELD VERY CLOSE TO BRIGHT LIGHT WHILE ADJUST KNOB IS ROTATED TO SET NEEDLE AT "O"

PHOTOCELL WIRES ATTACHED TO METER LEADS WHICH ARE PLUGGED INTO "COM" AND "VΩA" RECEPTORS

DIAL SET ON ONE OF THE "OHMS" SCALES

THE BRIGHTER THE LIGHT, THE LOWER THE NUMBER REGISTERED

spatial filter

A Design for an Inexpensive Spatial Filter

by
Lawrence D. Brooks
Optical Sciences Center
University of Arizona
Tucson, AZ

Table. *Descriptions of Individual Parts of Spatial Filter Assembly (letters keyed to Figure 2).*

A. Frontspiece: 1/4 x 1 3/4 x 2 inch aluminum. A 15/16 inch hole is located 3/4 inch from the top. The side and top are tapped to receive X and Y. A hole is drilled and partially tapped at 45º to receive H. A hole is drilled to receive Z and is counterbored on the inside to receive G.

B. Support Base: 1/2 x 1 3/4 x 2 1/2 inch aluminum. A 1 inch wide x 1/4 inch deep channel serves as ways for microscope objective holder base C.

C. Microscope Objective Holder Base: 1/4 x 1 1/2 x 0.998 inch aluminum (should be about 0.002 inch narrower than channel of B). Tapped 8-32 halfway (drilled rest of way to receive Z. Two holes drilled in the side to receive I.

D. Microscope Objective Holder: 3/8 x 1 x 1 1/4 inch aluminum. Tapped 0.800-inch 36 AMO to receive standard microscope objective, tap centered 1/2 inch below top.

E. Pinhole Assembly Holder: 3/4 inch diam. x 1/4 inch steel disk; 3/8 inch through hole; 1/2 inch diam. (or just large enough to receive F) flat shoulder bore, 0.23 inch deep. A V-groove is cut, centered around the edge.

F. Pinhole Assembly: Pinhole substrate is attached to Edmund's P-41,990 ring magnet (Melles Griot precision pinholes

We recommend purchasing, rather than building, the spatial filter (unless you have access to a good machine shop). For those of you who do, or who may find it cheaper to have someone do it for you than to buy a finished one, the following design is included:

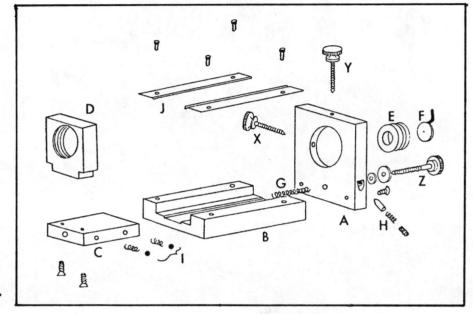

Figure 2. Modified version, Model 2.

---

are recommended because they are available on a 9.5-millimeter-diam. substrate). Handle attached to front of ring magnet.

G. Longitudinal Tension Spring: Slips around Z and provides for smooth focusing.

H. Spring Pin Assembly: Spring loaded, pointed pin held in by set screw; mounted at 45º to sides of A.

I. Transverse Tension Springs and Bearings: Two springs and two 1/8 inch ball bearings fit in side holes of C.

J. Slide Hold-Down: For Model 1, a "drawer clip" (available from the Kennedy Manufacturing Co.) was used. For Model 2, 0.010 inch beryllium copper

spring stock was cut and stamped into the shape shown in Figure 2.

X., Pinhole Positioners: For Model 1, 8-32
Y. brass knurled thumbscrews were used. For Model 2, 10-32 brass knurled thumbscrews were turned, and 8-40 threads were chased on them. The ends of all positioners were tapered to fit into the groove around E. Tip of the taper must lie on axis of the screw.

Z. Focus Adjust: For Model 1, a knurled 1/2 inch rod of aluminum was press-fit onto an 8-32 hex-head machine screw. For Model 2, a press-on "Shearloc" instant thumbscrew cap was used in place of the rod. Two washers (a small nylon one and a larger steel one) were then used to make the cap stand away from the frontspiece for easier turning.

(The above article is reprinted from Holosphere)

purchasing materials

There are some items which are better bought than built. This applies particularly to the laser, the heart of the holographic system, assembly of which is beyond the expertise of most readers. This is usually true for the spatial filter as well, (unless you have it built for you by a machinist). The actual unmounted optics, (including mirrors, lenses, and beamsplitters) must, of course, be purchased. We recommend purchasing rather than building a light meter, although plans for one were included in the last chapter.

For some, it may be preferable to buy some or all of the items we recommended building. If this applies to you, you'll be happy to note that examples of professional optical mounts, tables, plateholders and shutters will also be found in this chapter.

In addition, we recommend consulting copies of either the "Laser Focus Buyer's Guide" or "Optical Spectra Purchasing Directory" (See chapter on "Suppliers Addresses").

if you don't want to build it, buy it!

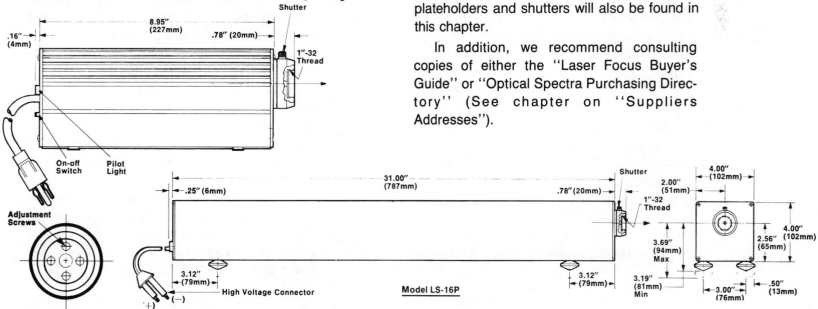

Model LS-16P

acquiring a laser

Since the invention of the laser in 1960, hundreds of different models and types for thousands of applications have been developed. There are solid state lasers, gas lasers, continuous wave lasers, pulsed lasers, ion lasers, heavy industrial cutting lasers, dye lasers, etc. They come polarized or randomly oriented, air or water cooled, in a large range of lasing materials, powers, colors, sizes, designs and prices. With all this, how does one begin to select a suitable instrument for holography?

Lasers vary in price from about $100 on up to hundreds of thousands or more. Fortunately, lasers suitable for work with optical holography, as described in this book, fall into the bottom end of this scale.

By far the most common laser produced today emits a continuous red beam from a gas tube filled with a mixture of helium and neon. The laser consists of this specially designed tube (which actually looks a lot like a neon light from an "Eat at Joe's" sign) and a proper power supply.

Whereas other CW (continuous wave) gas lasers such as Argon or Helium-Cadmium start at about $5,000., a good helium neon (HeNe) laser suitable for holography can be had for a few hundred dollars.

There are a lot of manufacturers of helium-neon lasers these days. Some people are even buying or reconditioning old tubes and building their own power supplies. From whatever the source, there are a few qualities the laser must have to be suitable for making holograms. In addition to being a CW gas laser (as are all helium-neons available), the laser must be rated in what is called the tem_{oo} mode (this refers to coherence which is discussed later). *This is extremely important - you will probably not be able to make holograms if your laser does not operate in this mode.* Most lasers you will come into contact with will have this feature.

Perhaps the questions asked most are what power laser to get and by what manufacturer. Helium neon lasers are rated in milliwatts and are available from about ½ milliwatt up to 50 milliwatts. The greater the power, the less exposure time you will need for making holograms and the more versatile it will be for complex setups and larger holograms. You have to pay for power, though. A laser in the 2-5 mw range is fine for starters; ½ to 1 mw is a little weak but will work. If you can at all afford it, go higher. Serious holographers go for powers of 10 to 50 mw. These lasers cost in excess of $1,000, however.

The laser's rated power is usually the minimum the tube will perform during the guarantee period. The actual power may be a bit higher but will drop as the laser ages. Good tubes can have lifetimes of 10,000 hours or more - enough for years of use.

Selecting a manufacturer is perhaps the trickier decision. For one thing, within a certain power range there may be a wide variety of prices (i.e., 5 mw lasers range from a few hundred dollars to well over $1,000). The better ones are going to cost more, but they are worth it if you can afford it. Some features to look for on better lasers include adjustable or removable mirrors (which allow the tube to be realigned periodically), polarized output and separate power supply.

You will find that lasers are available in both a rectangular and a cylindrical format. The major difference is simply that a rectangular housing will sit still, while a cylinder will roll around. Since you will probably not wish your laser to roll off the table, nor to get sand in the housing, you will have to build a mount for a cylindrical model.

If you find it difficult to make a choice among all those lasers out there, we can make some recommendations based on our own experience (addresses are at the end of materials section):

Current Best Buys:

(0.5 to 16 mw range)

1) Hughes
2) Aerotech
3) Metrologic
4) Melles Griot
5) Ealing
6) Jodon (a little more expensive)
7) Coherent Inc. (good quality yet expensive)
8) Spectra Physics (the best quality although most expensive)

*Great savings on used lasers are very often available yet information about these is usually obtained by plugging into the grapevine. Try joining one or more of the organizations listed on page 399 - someone else may know of a good deal.

current best buys

(20 - 50 mw range)

These are expensive lasers usually only within reach of a school, a cooperative, someone with money, or someone with connections.

1) Jodon offers good deals on systems which are essentially of Spectra Physics design, yet at a savings.
2) Spectra Physics although expensive, makes probably the best product with the best service available.

lasers

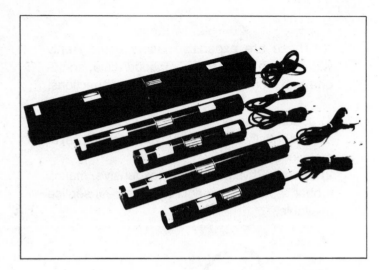

hughes

aerotech

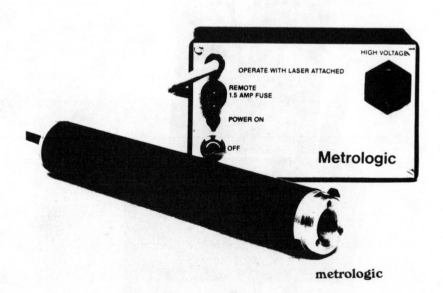

metrologic

metrologic

HN-2

HN-1, HN-2
HN-1S, HN-2S   Low Power HeNe LASERS
The HN-1 (1mw) and HN-2 (2mw) are integral laser and power supply packages designed for educational, industrial, and laboratory applications. The HN-1S and HN-2S have separate power supplies with the laser heads contained in a 2.0″ diameter, 14″ long sealed aluminum cylinder. All laser models feature long life cold cathode plasma tubes, regulated solid state power supplies, automatic starting, key lock switch and pilot lamp, plus all other BRH class IIIb laser requirements. Request Brochure 111A. Stock.

jodon

7, 10, 15 AND 20 MILLIWATT HeNe LASERS
Jodon's new line of HeNe Lasers was designed with simplicity, optical stability, reliability, and low cost as paramount design considerations. Key features include a thermally stable optical frame, interchangeable optics, differential screw mirror controls, a regulated solid state power supply, and an unconditional one year warranty. Request brochure 111B. Stock.

jodon

Spectra·Physics
Model 120
Stabilite™ Helium-Neon Laser

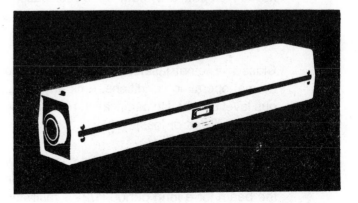

spectra physics

laser safety

Lasers used for most holography are not dangerous, yet they produce intense light, and some care should be exercised while using them. Laser light can, like the sun, cause eye-damage. Since you will more than likely only use a low powered laser, the light will not be hazardous unless you stare directly into the unspread beam. Never look down the bore into the laser or down the barrel of the unspread light, and take care (e.g., when redirecting the beam with a mirror) not to shine it directly into someone's eyes. One need not be frightened of this light, however. Once the light is spread out by a lens, it can be viewed directly. The unspread beam can be viewed from the side, just as long as you don't look at it "in line".

Few eye injuries from lasers have been reported in 20 or so years of use, since probability of the eye being positioned within the narrow beam is small.

The U.S Bureau of Radiological Health* has developed a system for classification of laser hazards, and all new lasers sold must be certified for its class. These are as follows:

Class I: Exempt lasers. These cannot, under normal operating conditions, emit any hazardous level of light. No certification required (usually ½ milliwatt or less).

Class II: Lasers which don't have enough power to cause retinal damage unless one stares into the beam for a long period (1/2 - 1 milliwatt)

*See ''Suppliers Addresses'' for additional information.

Cass IIIa: Lasers that cannot injure the eye with the normal aversion one would have to bright light, but may cause injury if the light is collected and put into the eye, as with binoculars (usually 1 - 2 milliwatts).

Class IIIb: Lasers that can produce accidental injury if viewed directly (usually 2 - 50 milliwatts or more. Most HeNe lasers used for holography are class 111b lasers).

Class IV: Lasers which not only produce a hazadrous direct or reflected beam, but can also be a fire hazard or produce a hazardous diffuse reflection. It is unlikely that most of you will ever use a class IV laser.

optics

The following optics may be used in conjunction with the sand mounts described previously, or any other mounts, to fabricate completed optical components. There are many sources for optics, and a wide variety of pricing and specifications to choose from. A sampling of the best sources is presented here. Those specific pieces most recommended will be referred to in the listing of suggested kits on page 124.

By far the most useful source for lenses and mirrors for holography is Edmund Scientific Co., and no one should be without one of their catalogues. Good optics are also available from Rolyn Optics, and from a number of other companies.

mirrors

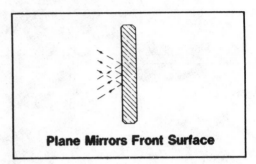

Plane Mirrors Front Surface

rolyn optics

PRECISION FIRST SURFACE MIRRORS- 1/10 WAVE

These mirrors are made of low expansion substrates and accuracies are certified to 1/10 wave. Edges are ground with slight safety chamfer and underside is fine ground. The first surface is aluminized with an overlay of quartz, to protect the surfaces. There are many uses in all lines of instrumentation research and production for an exceptionally high grade of Mirror.

STOCK #	Dia.	Min. T.	PRICE
60.0010	1.0″	1/4″	$ 36.75
60.0020	2.0″	1/2″	70.90
60.0030	3.0″	5/8″	109.22
60.0040	4.0″	3/4″	167.51

Other sizes to 12″ dia. on special order.

1/4 WAVE FIRST SURFACE, PYREX MIRRORS

STOCK #	Dia.	Min. T.	PRICE
60.0050	1.0″	1/4″	$ 19.00
60.0060	2.0″	1/2″	65.00
60.0070	3.0″	5/8″	67.50
60.0080	4.0″	3/4″	96.25

LIGHTWEIGHT FIRST SURFACE MIRRORS

Edmund Scientific Co.

Front surface mirrors are aluminized on the surface nearest the incident light. Light strikes this surface and is reflected without passing through any glass. Light losses are minimized & there is no secondary refraction caused by a glass-to-air surface. Front surface mirrors are more fragile, however, since the aluminized surface is vulnerable to scratching. Second surface mirrors have the aluminized surface at the back of the glass, protected from scratching. Household mirrors are commonly second surface mirrors because the light loss & ghost image are not critical to their use.

Number	Price	Size mm	Thick mm	Comments	Number	Price	Size mm	Thick mm	Comments
31,000	$ 2.10	1.5 x 1.5	0.1	Alum. both sides	31,017	4.65	46 Dia.	6.5	
31,001	2.10	3.0 x 3.0	1.0		30,840	32.95	48 x 66	9.5	¼ Wave Ellipse Pyrex
31,002	2.25	6.0 x 6.0	1.0						
31,003	2.35	8.2 Dia.	1.5		42,214	$ 4.75	49 x 70	3.0	
31,004	2.45	9.0 x 9.0	1.0		41,519	4.10	49 x 49	0.8	Very Thin
40,238	2.10	9 x 19	1.8		31,424	3.50	51 x 51	3.0	
31,418	2.50	9.5 x 11.2	1.2		31,018	2.25	51 x 73	3.0	22 mm Cut Corner
31,005	2.45	12 x 12	1.0		40,040	4.75	51 x 76	6.0	
30,546	2.10	14.5 x 19	1.8		42,384	5.35	52 Dia.	5.0	
31,006	2.55	15 Dia.	1.5		40,773	4.75	57 x 70	3.0	
31,417	2.75	16.8 x 22.7	1.0		41,131	44.50	57 x 81	9.5	¼ Wave Ellipse Pyrex
30,875	4.40	16 x 23	6.0	Flat to ¼ wave	41,742	4.75	60 x 131	4.0	Bev. Edge Cut Corners
31,007	2.55	18 x 18	1.0						
30,625	2.55	18 Dia.	0.2	Ultra thin	41,166	5.90	60 Dia.	1.0	
31,008	2.55	21 x 21	1.0		40,324	5.00	63 x 76	3.0	
40,286	2.55	21 x 26	6.0		31,019	3.10	63 x 235	9.7	Seconds
31,419	2.75	21.7 x 28.8	1.6		42,385	6.30	64 Dia.	3.0	
30,621	3.45	21 Dia.	1.2		42,213	5.00	65 x 70	3.0	
30,836	10.95	22 x 30	6.3	¼ Wave Ellipse	41,620	4.40	67 x 67	1.0	
30,428	2.75	22 x 35	3.0		30,209	8.70	67 x 83	5.0	
31,420	2.75	22.9 x 25.3	1.7		42,563	52.95	67 x 94	9.5	¼ Wave Ellipse Pyrex
31,426	3.50	23.2 x 37.5	1.7						
40,288	10.50	25 x 35	6.0	Flat to ¼ wave	42,212	5.35	70 x 97	3.0	
30,626	2.60	25 Dia.	0.2	Ultra Thin	42,211	6.30	70 x 194	3.0	
31,009	3.10	26 Dia.	2.8		31,020	7.30	75 x 351	6.0	
31,010	4.75	28 x 305	17.4	Very Thick	40,041	5.35	76 x 102	6.0	3″ x 4″
31,011	3.60	29 x 57	2.3		42,564	69.95	76 x 108	9.5	¼ Wave Ellipse
31,012	3.40	30 x 30	1.5		41,621	4.75	81 x 100	1.0	Very Thin
30,622	2.60	30 Dia.	1.2		31,021	12.50	100 Dia.	10.0	
30,429	3.60	32 x 40	3.2		40,042	9.10	102 x 127	6.0	4″ x 5″ Bev. Trapezoid
30,205	10.95	32 x 45	9.5	¼ Wave Ellipse	31,022	8.20	119 x 203/248	6.4	Plate Glass Matrix
31,421	4.00	34.9 x 38	3.2		41,320	13.80	127 x 178	1.6	
31,013	3.60	36 x 36	1.5		40,043	10.95	127 x 178	6.0	
31,422	4.00	38 x 38.5	3.2		31,023	12.00	127 x 203	3.0	Plate Glass Matrix
31,425	4.00	38 x 50.7	3.2		31,024	9.45	133 x 158	3.0	Plate Glass Matrix
30,258	17.95	38 x 54	9.5	¼ Wave Ellipse	31,025	8.20	149 x 171/260	6.4	Trapezoid
31,423	4.00	38 x 57.1	2.2		41,321	24.00	204 x 254	1.6	Plate Glass
31,014	5.00	38 x 251	17.2	Very Thick	40,067	21.50	204 x 254	6.0	Plate Glass
31,015	3.60	39 x 48	3.0		85,036	34.20	254 x 356	6.0	Plate Glass
31,016	4.75	39 x 96	3.0		31,026	17.40	254 x 406	6.0	Seconds
30,623	3.60	40 Dia.	1.0		85,206	44.00	406 x 406	6.0	Plate Glass
581	4.60	43 x 46	3.0		85,207	88.00	508 x 609	6.0	Plate Glass
547	4.60	43 x 140	3.0						

lenses

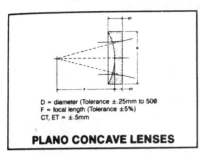

D = diameter (Tolerance ±.25mm to 50Ø)
F = focal length (Tolerance ±5%)
CT, ET = ±.5mm

PLANO CONCAVE LENSES

A Plano-Concave Lens is a divergent or negative lens, thicker at the edge than in the center and flat on one side. It forms a virtual image and is frequently used to extend the effective focal length of a converging lens when they are used as a pair — see formulae in front of the catalog for computation method. (These lenses are of the highest commercial quality, not surplus.). We suggest that negative focal lengths be stated on order.

STOCK #	D-mm	F-mm	CT	ET	PRICE
12.0005	25.0	—25.4	1.2	10.0	$ 16.11
12.0010	18.0	29.0	3.6	6.4	11.28
12.0015	25.0	50.0	1.7	4.9	7.12
12.0020	30.0	50.8	1.9	6.5	7.15
12.0025	40.0	63.5	1.5	8.2	7.87

STANDARD PLANO-CONVEX

STOCK #	D-mm	F-mm	CT	ET	PRICE
10.0004	2.5	3.0	2.0	1.4	$ 17.15
10.0005	4.3	8.1	1.6	1.0	4.37
10.0007	7.6	9.7	2.3	0.5	6.92
10.0010	5.0	10.0	1.4	0.8	4.14
10.0015	7.5	13.0	2.0	0.9	6.14

Plano-Concave Lenses (PCV)
PCV lenses have one flat & one inward curving face. PCV lenses have a negative F.L. & act as reducing or dispersing lenses.

Number	Price	Dia.	F.L.	Number	Price	Dia.	F.L.
94,926	$3.00	3	−7	94,816*	$2.60	35	−53
94,835	2.60	15	−13	94,763	5.30	54	−56
94,857*†	3.25	15	−15	95,805	2.60	5	−61
95,389	2.60	5	−20	95,408	3.40	33	−62
31,453	2.50	13	−22	94,437*	2.60	33	−73
95,391	2.60	13	−23	95,411	2.60	16	−74
94,843*†	2.60	17	−23	94,815	2.60	42	−84
94,718	2.60	20	−24	95,416	2.80	22	−85
95,454	2.60	16	−28	95,420	2.60	17	−100
95,396	2.60	18	−30	95,439*	3.00	17	−100
94,429*	3.25	18	−33	95,418	2.60	13	−116
94,856	2.60	31	−33	94,636*	2.60	45	−130
95,430*	3.70	16	−34	95,425	2.60	12	−148
94,719	4.35	32	−34	94,837*†	16.38	20	−193
95,400	2.60	17	−36	30,986	6.24	21	−222
94,834	3.00	40	−40	31,452	3.50	42	−238
95,402	2.60	14	−46	94,927	2.00	51	−246

rolyn optics

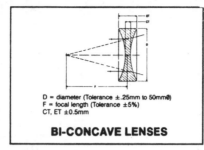

D = diameter (Tolerance ±.25mm to 50mmØ)
F = focal length (Tolerance ±5%)
CT, ET ±0.5mm

BI-CONCAVE LENSES

A Bi-Concave Lens is a divergent or negative lens, thicker at the edge than in the center. All ROLYN Bi-Concave Lenses are symmetrical having equal radii on both sides. These lenses form a virtual image and are frequently used to extend the effective focal length of a converging lens when they are used as a pair — see formulae in front of catalog for computational method. NO SURPLUS. Please specify negative focal length on orders.

STOCK #	D-mm	F-mm	CT	ET	PRICE
13.0050	20.0	−20.0	1.0	6.5	$ 10.22
13.0100	10.0	22.5	1.1	2.2	10.22
13.0150	20.0	30.0	1.1	5.9	11.02
13.0200	20.0	40.0	1.5	3.6	11.02
13.0225	25.0	50.0	1.5	4.5	11.38

edmund scientific

Double-Concave Lenses (DCV)
Have 2 inward curving faces, sometimes symmetric & sometimes with one curve sharper than the other. DCV lenses have a negative F.L. & act as reducing or dispersing lenses.

Number	Price	Dia.	F.L.	Number	Price	Dia	F.L.
30,956*†	$2.60	8	−8	30,958*†	$2.60	37	−41
30,957†	2.60	14	−9	95,500*	2.60	11	−43
94,844*†	2.60	10	−11	95,466	2.90	20	−43
94,838*†	2.60	8	−12	95,472	2.70	25	−48
95,442	3.65	12	−13	94,501*	2.60	25	−48
94,845*†	2.60	17	−18	95,474	2.60	13	−50
95,447	2.80	13	−21	95,476	2.60	19	−50
95,453	2.90	23	−27	95,503*	2.90	18	−52
95,458*	3.00	19	−29	95,800	3.00	33	−67
95,796	2.90	26	−31	95,483	2.60	12	−68
95,797*	2.90	13	−32	95,485	2.60	19	−70
95,497*	3.65	17	−34	94,429	3.25	53	−78
95,461	3.00	20	−35	94,507	2.60	19	−79
94,498*	2.60	22	−37	94,846*	8.15	77	−87
95,799	2.80	26	−37	94,847*	8.15	72	−95
94,463	2.60	33	−38	94,839*†	3.00	29	−144
94,498*	3.00	33	−38	94,928	3.00	37	−153

edmund scientific

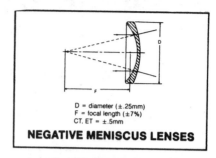

D = diameter (±.25mm)
F = focal length (±7%)
CT, ET = ±.5mm

NEGATIVE MENISCUS LENSES

A negative or divergent meniscus lens is also convex-concave, but is thinner at the center than at the edges. Otherwise description is similar to positive meniscus. Functionally similar to plano-concave lenses.

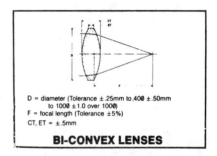

D = diameter (Tolerance ±.25mm to 400 ±.50mm
to 1000 ±1.0 over 1000)
F = focal length (Tolerance ±5%)
CT, ET = ±.5mm

BI-CONVEX LENSES

A Bi-Convex lens is a convergent or positive lens thicker in the center than at the edges. ROLYN Bi Convex Lenses are all symmetrical, having equal radii on both sides. Aberrations in such lenses are at a minimum when the object/image ratio is 1:1. Typical appli cations are: magnifiers, objectives, some condensing systems.

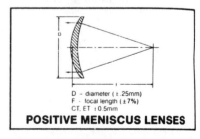

D = diameter (±.25mm)
F = focal length (±7%)
CT, ET ±0.5mm

POSITIVE MENISCUS LENSES

A positive or convergent meniscus lens is a convex-concave lens thicker at the center than the edges. They are felt polished and are used universally in the Ophthalmic Industry where convention dictates that lens power to be specified in Diopters. For your convenience we have also specified the focal length in millimeters for all ROLYN Meniscus Lenses.

beamsplitters

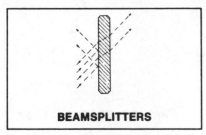

BEAMSPLITTERS

Beamsplitters are semi-mirrors which reflect part of the incident energy (absorb a relatively small part) and transmit the rest.

Mirror Type
Very neutral color characteristics. Interference coating extra curable. Glass parallel to 20 seconds of arc. Surface flat to approximately 10 wavelengths.

Number	Size (mm)	Reflectivity	Transmission	Price
31,416	12x19x1	10	90	$ 5.00
31,415	15 Dia. x 2	10	90	4.50
31,414	18x30x2	50	50	5.50
31,413	20x37x1	50	50	6.00
31,725	22x30x2	30	70	5.50
31,412	25x28x1	30	70	6.50
31,437	25x38x3	25	75	5.50
31,436	25x38x3	50	50	5.50
31,435	25x38x3	75	25	5.50
31,411	25x76x2	50	50	3.50
31,434	51x76x3	25	75	9.75
31,433	51x76x3	50	50	9.75
31,432	51x76x3	75	25	9.75
578	67x83x5	30	70	12.60
31,410	77 Dia. x 2	20	80	7.50
61,097	127x178x3	30	70	12.50

Variable Density Beam Splitter
3" inconel coated. Density varies linearly w/length from .02 (96% trans.) to 2.0 (1% trans.) Used to adjust spectrophotometers, monochromators & lasers.
No. 41,960

variable beamsplitters

CIRCULAR WEDGE

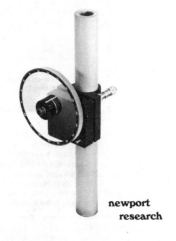

newport research

jodon
VBA-200
VARIABLE BEAM SPLITTER
A variable reflectivity neutral density aluminum mirror designed for use in optical beam splitting or attenuator applications. This instrument eliminates the need for numerous step filters often required in optical systems. Transmitted beam is continuously variable from 0.7% to 90% and reflected beam is variable from 8.5% to 85% of incident beam. An interferometrically stable, low loss device complete with adjustable height stand. May also be mounted on standard optical rails. Request brochure 103. Shipping weight 17 lbs. Stock.

newport research

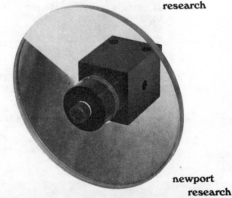

newport research

spatial filters

Almost all lenses and mirrors have small defects, scratches, or bits of dirt which will show up in the hologram. The spatial filter "cleans up" the beam, greatly improving the quality of the image.

This device consists of a microscope objective, a precision-made pinhole, and a fine focusing mechanism. Spatial filters can be built by those familiar with their design and having access to good machining equipment. However, most of us will wish to buy one. We recommend those made by Jodon, Noll or Newport Research (do *not* purchase the kind of spatial filter that attaches to the laser - that's not where it is going to be used!). Additional objectives or pinholes may be ordered from other sources if desired (Pinholes, by the way, can be made by poking a tiny hole in aluminum foil, but it's next to impossible to control the size, and one has to be very good or very lucky to pull this off. The best pinholes are made professionally.)

Spatial Filter Model NSF-2

noll

newport research

jodon

objectives & pinholes

ACHROMATIC OBJECTIVES

Excellent for both visual and photomicrographic use. Give remarkably crisp image, free of residual color. All are designed for tube lengths of 160 mm and cover glass thicknesses of 0.18 mm. Standard thread.

Number	30,045	30,046	30,047	30,048	30,049
Power	5X	10X	20X	40X	60X
N.A.	0.10	0.25	0.40	0.65	0.85
Price	$25.95	$33.95	$46.95	$57.95	$72.95

Different sized microscope objectives are used with different pinholes for various applications. There is a relationship between the diameter of the beam, the power of the objective, and the pinhole diameter, which must be considered for the system to work properly. Below is a chart of these relationships. For most of our work with most HeNe lasers we used a 10x objective with a $25\mu m$ pinhole.

PINHOLE/OBJECTIVE SELECTION GUIDE

Objective	Select Pinhole (D-μm)	E.F.L. (F-mm)	Max. Input Beam Size (mm)	Calc. Pinhole Dia. (μm)
M- 5x	50	30.0	9.5	40.0
M-10x	25	14.8	7.5	18.7
M-20x	10	7.5	4.8	9.5
M-40x	5	4.3	3.7	5.4
M-60x	5	2.9	3.0	3.7

jodon

collimating mirrors

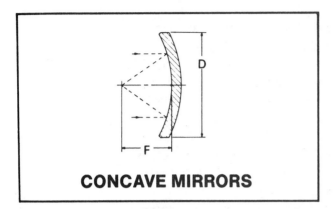

CONCAVE MIRRORS

These mirrors are important for making certain types of advanced holograms by controlling the convergence/divergence of the light. They are the same kind of parabolic mirrors used in reflector telescopes. By far the best buy we've found is from Coulter Optical, with prices far below the going rates. We suggest a 10"- 14" diameter with a relatively low f/number. (If in doubt and wishing to save money, try a 10" f 5.6; we have successfully used one of these for years.)

These mirrors are quite massive, and must be securely mounted to use on the sand table (see mirror mounting in last chapter); alternatively a good mount should be purchased.

The whole purpose of Coulter Optical Company is to bring to the amateur astronomer "optimum optics at minimum costs." Although our prices are considerably lower than you are used to, we do not cut corners in quality to accomplish this. Whether for custom or production astronomical mirrors, raw materials are the finest, workmanship of the highest precision. If you are not completely satisfied, your money will be promptly returned.

PRODUCTION ASTRONOMICAL MIRRORS

Special equipment and techniques make Coulter production mirrors by far the least expensive of any other comparably-coated mirror. Resolving power and reflective ability are the highest.

BLANKS: Made from a premium quality fine-annealed blank of low expansion type Pyrex #7740 with a 1:6 ratio of thickness to diameter. Average Linear Coefficient of Expansion 32×10^{-7} per 1°C between 0° and 300°C.

OPTICAL SURFACE: Extremely smooth — no dog-bisquiting. Exceptionally clean. Focal lengths ± 3%.

OPTICAL FIGURE: Extremely smooth parabolic figure without any zones or breaks. Each disk exhibits a good diffraction ring — no turned edges or edge zones. Each mirror exhibits straight Ronchi bands edge to edge under the sensitive double pass auto-collimation test. Null figured. Free of astigmatism.

COATING: Aluminized with silicon monoxide.

Photo by Alfred University

PRODUCTION ASTRONOMICAL MIRRORS

coulter

PRICE LIST

6"f/5	$39.95 pp.
6"f/8	39.95 pp.
8"f/6	69.95 pp.
8"f/7	69.95 pp.
10"f/5.6	$105.50 pp.
12½"f/6	199.50 pp.
14¼"f/6	399.50 pp.

LARGE ADJUSTABLE MOUNT
MODEL NAM

noll

MODEL	DIAMETER
NAM-6	1.5" to 6"
NAM-9	1.5" to 9"
NAM-11	1.5" to 11"
NAM-13	1.5" to 13"

— Rigid Construction
— Adjustable for Mirror Diameter and Thickness
— No direct Metal-to-Glass contact
— Extremely Fine Orthogonal Tilt Adjustments (0.1 Arc Second)
— Vertical, Axial, and Transverse Adjustments Available

light meters

You get more for your money when you own the S & M Model A-3 Supersensitive Photo Meter. Darkroom techniques are mastered with accuracy and ease with this Darkroom Meter.

science & mechanics

A good light meter should be capable of reading over a wide sensitivity range for the type of light used. We can recommend three low-cost quality meters for use with HeNe lasers: the Science and Mechanics, the Metrologic, and the Spectra Lumicon (which is not shown) models.

Our personal preference is for the Science & Mechanics Model A-3 darkroom photometer. We use it extensively for our work. A handy exposure chart for calibrating this meter to the Agfa recording materials will be found at the beginning of each basic holographic setup.

The Model A-3 Photo Meter is named the "Darkroom Meter" because it is used extensively with our Special Easel probe for reading enlarger easel exposures.

This Darkroom Meter has a sensitivity control which is used to find the correct exposure for printing. The 4½ inch dial of the meter is self-illuminated with built-in battery lamps for easy, accurate reading in the darkroom.

You read on four sensitivity ranges; therefore, the meter dial and range selector switch give you an equivalent of 18 inches of total dial space for reading accuracy. The sensitivity of this instrument is sufficient to detect the light of a match 10 feet away in a darkened room. As a light meter, its

specifications are identical to those of our Model 102 Photo Meter.

This supersensitive and most versatile, battery-powered Photo Meter—6¾" x 5¼" x 2¾" in size and 2 lb. 6 oz. in weight—is supplied with its Standard probe, 4¼ inch diameter exposure calculator and attractive carrying case. The meter is designed with a battery-test circuit to indicate instantly the condition of the mercury batteries utilized. At present, there are four types of plug-in probes, each with three-foot long flexible cable, available for this unit. However, probes of different cable lengths can be supplied as desired.

These probes utilize cadmium sulfide photocells—the best available today—balanced for color, and the complete range of exposures from dim available light to full bright sunlight. The field of view of the Standard probe is approximately 43 degrees, equal to normal camera lens angles.

The A-3 can be used as a densitometer, to find the highlights and shadow densities of your negatives and the CORRECT paper for printing.

This Model A-3 Photo Meter has earned a reputation as a precision instrument for color photography, portraiture, ground-glass exposures, and available light photography.

MODEL A-3

45-230 Laser Power Meter (Photometer)

Measure beam intensity. Study polarization effects in laser light. Make BRH compliance measurements. Examine light pipe attenuation. Measure transmission through filters. Demonstrate the inverse square law. Determine beam diameter from the $1/e^2$ power limits. Demodulate audio modulated laser signals. Use the 45-230 Laser Power Meter for these and many other classroom and laboratory uses.

Accuracy is within 10% for each of eight sensitivity ranges. Full scale readings range from 0.003 milliwatts to 10 milliwatts.

The meter is calibrated for direct reading of the 633 nm wavelength of helium-neon laser light.

shutters

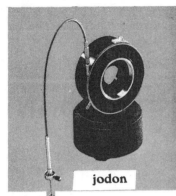

S-10B LASER SHUTTER
This self-cocking shutter assembly with heavy steel base and cable release is a versatile accessory for holography. The iris provides aperture control from 0.04" to 1.10" diameter with shutter speeds of T, B, 1 second and fractional seconds: 2, 4, 8, 15, 30, 60, and 125. An external electrical sync output is also provided. Shutter axis height is variable from 2.85" to 3.45" above the table using the adjustable stand. In addition, a standard 1.00" by 32 pitch external thread is provided for direct mounting to laser bezels. Shipping weight 5 lbs. Stock.

jodon

830
$180

newport research

Alternatively, look for old used camera equipment. You might be able to find an old view camera shutter, if you're lucky, and operate it with a cable release.

plateholders

jodon

PH-45A

MPH-45

MPH-45W

FH-500

noll

UPH-100

FT-35/70

DRT-45

APP-100/RT-45W

tables

If indeed you are a bold, ambitious, and wealthy holographer, a really good optical system is not complete without a commercial vibration isolation table. There are granite slabs, steel slabs and honeycomb slabs. These tables demand the highest quality magnetic mounts, they are optical benches that utilize sophisticated pillars with pneumatic interlocking compensation and .or balance systems ad infinitum. All of these "toys" can be bought separately or as a complete system for approximately $18,000 to $26,000.

This piece of advanced equipment is one of the luxuries that has been deemed overly indulgent by sand table holographers, but for those of you with extra cash, these tables are unsurpassed in quality and construction.

newport research

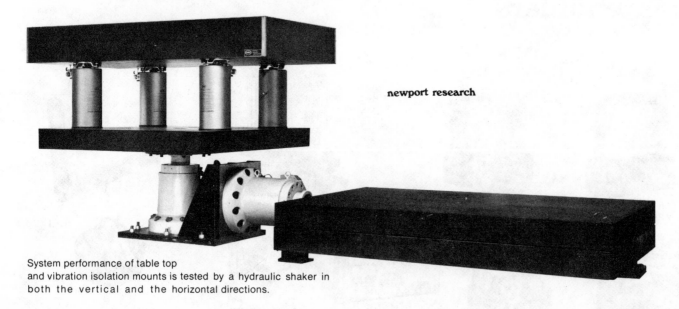

System performance of table top
and vibration isolation mounts is tested by a hydraulic shaker in
both the vertical and the horizontal directions.

ealing corporation

FILM SELECTION GUIDE

KODAK PRODUCT	EXPOSURE TO ACHIEVE D = 1.0 (ergs/cm²)								RESOLVING POWER @ TOC		GRANULARITY @ D = 1.0		CONTRAST	EMULSION THICKNESS	STANDARD BASE	DEVELOPMENT
	HeCd 3250	HeCd 4416	Ar 4880	Ar 5145	Nd:YAG 5320	HeNe 6328	Kr 6471	RUBY 6943	1000:1	1.6:1	48μm	6μm	γ	μm		
Spectroscopic Type 649-F Plate/Film	—	500	800	800	1000	900	800	5000+	2000+	—	<5	<10	5/4	17/6	0.040 Unb or Bkd/A5B	6-8′ D-19
Holographic Plate, Type 120 (−02 or −01)/Film, SO-173	—	500	—	—	—	400	400	400	2000+	—	<5	<10	5/4	6	0.040 Unb or Bkd/E4B	6-8′ D-19
Special Plate, Type 125 (−02 or −01)/Film, SO-424	20	80	50	50	100	—	—	—	1250	630	<5	13	4	7/3	0.040 Unb or Bkd/A5CB	6-8′ D-19
High Speed Holographic Plate, Type 131 (−02 or −01)/Film SO-253	—	20-35	40-65	25-35	20-30	5-8	3.5-6	1000+	1250	800	<5	14	7	9	0.040 Unb or Bkd/E4B	6-8′ D-19
Technical Pan Film, 2415	—	0.4	0.8	0.8	0.7	0.4	0.3	—	320	125	8	—	1-3	7.5	E4AH	6-8′ HC-110(D) 4′ D-19
AGFA-GEVAERT																
8E56HD-AHI Plate/Film	—	350	600	350	300	—	—	—	2500+	—	—	—	4	7	0.060 Bkd/E7B	6-8′ D-19
8E75HD-NAH Plate/Film	—	400	—	—	450	150	150	100	2500+	—	—	—	3	7	0.060 Unb/E7C	6-8′ D-19
10E56-NAH Plate/Film	—	60	30	20	20	—	—	—	1500+	—	—	—	7	7	0.060 Unb/E7B	6-8′ D-19
10E75-NAH Plate/Film	—	60	—	120	60	20	20	20	1500+	—	—	—	4	7	0.060 Unb/E7C	6-8′ D-19

recording materials

Both Kodak and Agfa-Gevaert manufacture materials suitable for making holograms. These belong to the silver halide group of light sensitive emulsions, as do most photographic films, and are as simple to handle and process. The important difference between holographic and photographic materials is resolving power, usually expressed in lines per millimeter. Whereas photographic films usually cannot resolve more than 50-100 lines/millimeter, between 1500-3000 lines/mm or more are required for holographic applications. Better forget Plus-X for making holograms - no way!

Some of the products are more versatile than others. Overall, the best seem to be Agfa Holotest materials (known as 8E75 and 10E75 for use with red lasers, and 8E56 and 10E56 for use with blue or green lasers). Most readers will be using 8E75 and 10E75 which is best for HeNe lasers. These emulsions offer the greatest sensitivity (important for short exposures with low power lasers), the greatest resolving power, reliability, durability, and availability. The 8E emulsion is really the only suitable material in the group for reflection holography.

The emulsions are available on both glass plates and film. Glass is preferred for most applications, especially reflection holography, and is a *must* for beginners because of its rigidity.

Kodak also makes materials suitable for holography. Most early holograms were made on Kodak 649F plates, which are still available today. They are much slower (less sensitive) than the Agfa plates, yet respond to all colors, and can be used to make full color reflection holograms (assuming you've got red, green, and blue laser light to play with).

Kodak produces SO 173 film which is also slow (not as slow as 649F), but is of excellent quality. This same emulsion is available on glass plates as well and labeled Type 120-01, or 120-02 plates. A faster emulsion is SO 253, although it has less resolving capability.. You may wish to try the Kodak products and make your own comparisons to those from Agfa.

The most popular size is a 4"x 5" plate, although 8"x 10" plates are available without much difficulty.

Some of the plates are supplied with an antihalation coating on the back (e.g. box is marked 10E56AH). This coating is valuable at times, when making transmission holograms, as it helps to cut down on reflections from the back surface of the glass or film. *These plates cannot, however, be used to make reflection holograms.*

Agfa periodically makes adjustments in the products offered. To the chagrin of many holographers, the company (for some unknown reason) discontinued its AH backing on red-laser sensitive materials. We sincerely hope they are reintroduced; keep asking for them.

The non-backed materials will be marked NAH, (as in 10E75NAH), or will have no special notation (10E75). These certainly can be used for transmission holograms.

Recently, Agfa introduced a new high-resolution emulsion, marked 8E75HD. The HD or high definition series has increased resolving capabilities, and makes generally brighter, crisper holograms. The 8E series is less sensitive (needs more light or longer exposure) than the 10E series, yet is best for serious work. Once again, the 8E must be used for reflection holograms. On the other hand, 10E75 is ideal for transmission holograms made with low-power lasers, and is a good material to start with. A safe bet is to begin with 1 box of 10E75 and 1 box of 8E75HD in the 4"x 5" size.

The sensitivity of photographic films is usually expressed as ASA numbers. For example, Plus-X is rated at ASA 120. Holographic materials are so much less sensitive than standard films that their ASA numbers would be only tiny fractions. Their sensitivities are usually expressed in micro-joules/square centimeters (or ergs/cm^2 for you old timers, although the "ergs" unit is being phased out). This is the actual energy per unit area needed to make the proper exposure.

Since most of us will not be using energy meters, we need not be overly concerned with this. We will be able to determine the proper exposure experimentally or, for those who purchased the recommended light meter, by using the precalibrated chart in this book.

The use of film to make a hologram is generally a bit trickier than using glass, because of the difficulty in keeping the film absolutely rigid during exposure. In general, film is only practical for transmission types of holograms.

A major advantage of film is its low cost, as compared with glass, which means that it is preferred for experimentation (test exposures, development technique). Film may be held flat, like a glass plate, or curved. It can also be easily cut, making it desirable for montage work or constructing unusually shaped displays. The settling time of film is usually longer than that of glass. Film can be sandwiched or clamped between two pieces of glass, but the best optical results are obtained when the emulsion is open to air. The edges or corners of the film should be securely taped to the platen, and allowed to stabilize. AH backed film will minimize the effects of reflections from the back surface of the platen.

HOLOTEST FILMS AND PLATES FOR HOLOGRAPHY

TYPE	SIZE	PACKING
8E56 HD AHI	4 X 5"	20 Plates
8E56 HD AHI	8 X 10"	10 "
8E75 HD NAH	4 X 5"	20 "
8E75 HD NAH	8 X 10"	10 "
10E56 NAH	4 X 5"	20 Plates
10E56 NAH	8 X 10"	10 "
10E56 AH Non Perf	70mm X 100'	Roll
10E75 NAH	4 X 5"	20 Plates
10E75 NAH	8 X 10"	10 "
10E75 NAH Non Perf	70mm X 100'	Roll

Emulsion Thickness = 7 microns

Glass Thickness - 4 X 5" = 1.5mm
                 8 X 10" = 3.0mm

AHI - Anti Halo is within the emulsion,
AH - Anti Halo backing.
NAH - Without Anti Halo backing.

P3 - Polyester Film Base Thickness = 7 Mil

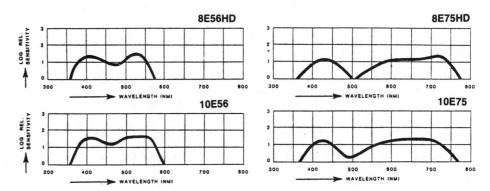

Darkroom Illumination

Holotest 8E56HD and 10E75 should be handled in complete darkness.
Holotest 8E56HD and 10E56 should be handled under dark-red safe-lights, such as Agfa-Gevaert R-4 safelight filter or equivalent.

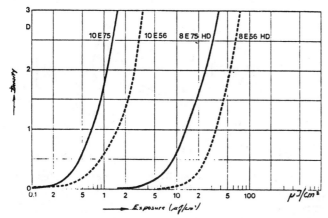

Amplitude Transmission Curves:

Since amplitude transmission plays such an important role in holography, the amplitude transmission curves for the holotest material are reproduced in Figure 2.

~ 0.5 μJ/cm² for 10 E 75
~ 10 μJ/cm² for 8 E 75 HD
~ 1 μJ/cm² for 10 E 56
~ 25 μJ/cm² for 8 E 56 HD

Technical Data:

1) **General Data:**

Emulsion	Grain Size	Resolving Power Lines/nm	Exposure Energy Micro-j/cm 2 *
8E56HD	35nm	5000	18.25
10E56	90nm	3000	1 - 1.5
8E75HD	35nm	5000	10
10E75	90nm	3000	0.5 - 0.75

* - Exposure energy determined for 56 material at a wavelength of 521nm and for 75 material at 627nm.

Density Curves:

The density curves for holotest materials are given in Figure 1. The exposures of holotest 8E75HD and 10E75 were produced at the principle wavelength of 627nm and of the holotest 8E56HD and 10E56 at that principle wavelength of 521nm. The material was processed in Agfa-Gevaert Developer G-3p for 2 minutes at 20°C and Fixer G-334 for 2 minutes at 20°C followed by a 5 minuted wash.

KODAK Holographic Film (ESTAR Base) SO-173 is stocked in
two formats:

4 x 5-inch sheets (25 sheets per package). Minimum order
quantity is one package.

35mm x 150-foot rolls, Sp417. (Bell and Howell perforations
on two edges, wound emulsion in on a No. 10 spool with
integral leader and trailer.) Minimum order is one roll.

These products must be ordered through your local photo
dealer for professional photographic materials or through dealers who
specialize in serving the holographic market. When ordering, include
the following information:
1. Number of rolls or packages desired
2. Size (width and length)
3. Product name
4. Specification or Identification Number

Examples: 2 rolls, 35mm x 150 ft, KODAK Holographic Film
(ESTAR Base) SO-173, Sp 417

1 package, 4 x 5-inch, KODAK Holographic Plates,
Type 120-02 (.040-inch glass, unbacked)

SPECTRAL SENSITIVITY

Sensitivity is defined as the reciprocal of exposure, in ergs/cm^2,
required to produce a density (D) above gross fog when the material
is processed as recommended.

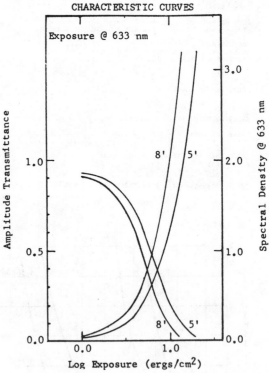

KODAK High Speed Holographic
Film (ESTAR Base) SO-253

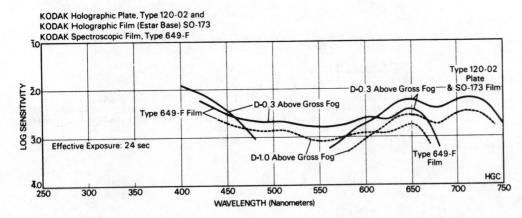

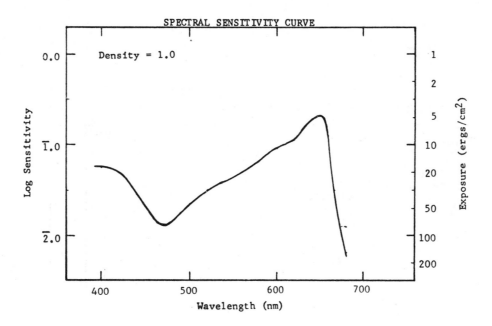

SPECTRAL SENSITIVITY CURVE

Density = 1.0

AMPLITUDE TRANSMITTANCE -- D-LOG E CURVES

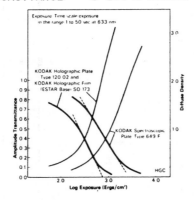

NOTE: The curves shown for the new holographic emulsion were plotted from single-test data; those shown for Type 649-F Film are believed representative of average product. Variations in speed and contrast between film and plate products of the same type, differences between production lots of a particular product, or differences resulting from processing variables may yield an effective speed ratio at any given wavelength greater or less than that depicted. While the curves are intended to illustrate the relative merits of the two emulsion types and to provide a guide for initial exposures of one product based on experience in use of the other, we strongly recommend that you run an exposure series to determine the optimum operating point for a particular product under your exposure and processing conditions.

KODAK PRODUCTS FOR HOLOGRAPHY
AND LASER PHOTOGRAPHY

KODAK PRODUCT	EXPOSURE TO ACHIEVE D = 1.0 (ergs/cm²) ①								RESOLVING ② POWER @ TOC		GRANULARITY ③ @ D = 1.0		CONTRAST γ	EMULSION THICKNESS (µm)	ARU	STANDARD BASE ④	DEVELOPMENT ⑤	SPECIAL NOTES
	HeCd 325	HeCd 442	Ar 488	Ar 515	Nd:YAG 532	HeNe 633	Kr 647	Ruby 694	1000:1	1.6:1	48 µm	6 µm						
High Resolution Plate	(400)	1000	1500	1000	800	-	-	-	2000+		< 5	< 10	8	6		.060 BKD	6-8' HRP/D-19	6
High Resolution Film, SO-343						-	-	-	2000+		< 5	< 10		7		E7B	6-8' D-19	6
Spectroscopic Film, Type 649-GH						-	-	-	2000+		< 5	< 10		7		E4AH	6-8' D-19	6
High Resolution Plate, Type 2	(1000)	3000	2500	2000	2000	-	-	-	2000+		< 5	< 10	5	6		.060 UNB	6-8' HRP/D-19	7
Spectroscopic Plate, Type 649-F									2000+		< 5	< 10	5	17		.040 UNB	6-8' D-19	10
Spectroscopic Film, Type 649-F		500	800	800	1000	900	800	5000+	2000+		< 5	< 10	4	6		E4AH	6-8' D-19	
Holographic Plate, Type 120-02									2000+		< 5	< 10	5	5		.040 UNB	6-8' D-19	10
Holographic Film, SO-173		500	-	-	-	400	400	400	2000+		< 5	< 10	4	6		E4B	6-8' D-19	
Special Plate, Type 125-02						-	-	-						7		.040 UNB	6-8' D-19	8, 10
MINICARD II Film, SO-424	20	80	50	50	100				1250	630	< 5	13	4	< 3	✓	A5CB	6-8' D-19	
Recording Film, SO-141									1250	630	< 5	13	4	< 3	✓	E4C	6-8' D-19	
H.S. Holographic Plate, Type 131-02														7		.040 UNB	6-8' D-19	10
High Speed Holographic Film, SO-253		20/35	40/65	25/35	20/30	5/8	3.5/6	1000+	1250	800	< 5	14	7	9.		E4B	6-8' D-19	
Direct Positive Laser Recording Film, SO-285	-	35	45	55	75	30	50	-	1250	630	< 5	14	-2	< 4	✓	A5C	5' D-19	9
RECORDAK Direct Duplicating Print Film, 5468, 8468	5	100	100	50	40	-	-	-	1000	400	< 5	17	-1.9	3		A5C, A7C	5' D-19	9
High Definition Aerial Film, 3414	.4	.6	3	2	2-	2	2-	5	630	250	9	33	0.8-2.4	< 4		E2.5B	D-19, D-76	
High Contrast Copy Film, 5069	.4	1	4	2	1	6	-	-	630	200	9	43	1.0-2.8	< 4	✓	A5C	D-19, D-76	
Photomicrography Monochrome Film, SO-410	.3	.3	1.5	1.5	1.5	.2	.2	1.0	250	100	7.0 / 5.5	66	4 / 1-3	-		E4AH	D-19 / HC-110 (D,F)	
LINAGRAPH SHELLBURST Film, 2474	.15	.09	.3	.3	.2	.15	.15	2	125	50	24	-	0.5-2	-		E4B	D-19, D-76	
LINAGRAPH SHELLBURST Film, 2476	.15	.09	.3	.3	.2	.15	.15	5	160	63	22		0.5-2	-	✓	E4AH	D-19, D-76	
2479 RAR Film	.1	.05	.2	.3	.2	.08	.08	.5	100	40	24	-	0.6-1.8	-	✓	E4AH	D-19, D-76	
2475 Recording Film	.07	.03	.06	.07	.06	.05	.05	.05	63	22	32	-	0.4-2	-		E4AH	D-19, DK-50	
2485 High Speed Recording Film	.03	.007	.04	.05	.04	.03	.03	.04	50	20	47	-	0.9-1.8	-		E4AH	857, D-19	

HOLOGRAPHIC / PHOTOGRAPHIC

RDA 12-75

other recording materials

Silver halide materials, such as Agfa Gevaert & Kodak plates or film, are by no means the only materials suitable for making holograms. Although less sensitive and usually much more difficult to work with, a number of other substances have been utilized, usually with rather specific commercial or industrial advantages in mind.

photo polymers

Photo Polymers are light-sensitive plastics. Supplied in liquid form, they solidify upon exposure to light. One very successful product, developed by Hughes Research and available presently from Newport Research Corporation, consists of two solutions which are mixed just prior to use. Several drops are placed on a glass plate; a second plate is then hinged down on top of the first to make a sandwich, and, upon exposure to light, the material will partially "set up".

No processing is needed, but the hologram will quickly deteriorate if not temporarily fixed with a flash from a xenon lamp (or suitable substitute). By baking at 150^0 C for 10 minutes, the plate can be permanently fixed.

The advantage of this material is that the hologram is viewable without removing the material to process it. For such uses as holographic nondestructive testing of materials or holographic interferometry, this is indispensible.

The material has excellent resolution, and although less sensitive than silver halide materials (by a couple of orders of magnitude), it compares fairly well with other holographic materials.

The disadvantages should be considered: since the photopolymer is a liquid until cured, the sandwiched plates must be held horizontally until after exposure; (this can involve some creative setups). Besides being difficult to work with in its liquid state, the material is really only suitable for transmission holography. Other photopolymer materials have also been developed*.

*A good description of one by Booth of DuPont can be found in Applied Optics, March 1975 Vol. 14 #3, p.593.

dichromated gelatin

This material offers one of the most exciting mediums for commercial display holography. Dichromated Gelatin (DCG) comes close to being the ideal holographic recording material. Diffraction efficiency can approach the theoretical 100%, or, in other words, almost all of the viewing light will be channeled into making the image or desired effect. It is an especially good material for reflection holograms, which will play back in almost any light and can be adjusted to reconstruct in a variety of colors. These appear to be the most marketable of holograms: hundreds of thousands have already been sold as pendants and other forms of jewelry, or as gift items. Most of these are just simple single beam reflection (Denisyuk) holograms and can be both virtual image (1st generation) masters or image plane (2nd generation) types. A master is simply a first generation hologram placed back into the single beam to generate an image which will be copied with the same beam.

DCG can be used to make extremely bright transmission holograms. Also, ordinary holograms may be contact copied into dichromate material to greatly increase their brilliance. DCG is also nearly ideal for the making of holographic lenses. IBM is now using a dichromate lens in its model 3687 Supermarket Scanner to deflect the laser beam which reads across a product's bar code.

"Cubes" Dichromate hologram by Fred Unterseher and Bob Schlesinger, 1980, Photo by Bob.

Unfortunately, there are a number of characteristics of DCG which render it difficult if not impossible for use by the average home holographer. The material is sensitive only to green or shorter wavelengths of light, with a maximum sensitivity in the ultraviolet. It is normally completely insensitive to the red light of a HeNe laser, requiring the use of an expensive Argon, Krypton, or Helium Cadium laser. (DCG can be made red sensitive* with the addition of a dye, yet needs the enormous energy of 200-600 mJ/cm2 which makes exposures extremely impractical)

*See Applied Optics, vol. 15, #2, Feb. 1976, p. 556

Dichromated Gelatin films are not commercially available. A number of methods for preparing films are discussed in the literature yet much of the information is vague or confusing. With a lot of time and expense, several companies have developed their own methods and formulations for preparing DCG coatings but don't expect them to tell you how.**

Success with DCG is usually elusive for a variety of reasons. First, properly coating an emulsion on a piece of glass is a lot more difficult than one might imagine. It is important to control emulsion thickness, hardness, water content, and sensitivity. The holograms must be made in an environment in which temperature and humidity are carefully controlled. Troubleshooting problems are often difficult as so many ultra-sensitive interdependent variables are involved. In fact, making dichromate holograms is really an art in itself.

Even so, some hard nose experimenters with access to the proper equipment may wish to try it out. We suggest starting with the references listed below.*** The simplest method

**nor us neither!

***

"Dichromated Gelatin Imaging: Guidelines for Beginners" excerpts from "A Cookbok for Dichromated Gelatin Holograms" by C.O. Leonard and B.D. Guenther, published in Holosphere, July/Aug. 1980, P.1

"Color Control in Dichromated Gelatin Reflection Holograms", Stephen McGrew, SPIE proceedings, vol. 215, p.24, Feb. 1980

"Dichromated Gelatin for the Fabrication of Holographic Optical Elements", B.J. Chang and C.D. Leonard, Applied Optics, July 1979, vol. 18, #14, p. 2407

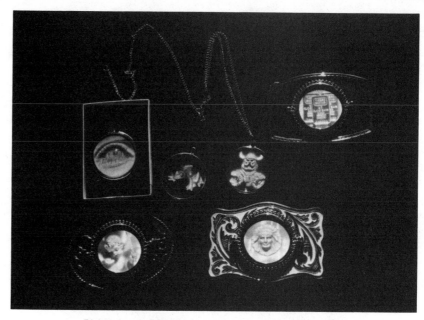

Dichromate holographic jewelry. Hologram masters by Bob Schlesinger. Produced at Dichromate Inc. by Bob, Kent, and Jim, 1980, photo by Bob.

of gelatin preparation (in order to avoid problems with nonuniform coating) is to use an unexposed Kodak or Agfa plate. A fixer is used to remove all of the silver compounds from the emulsion. The plate is then soaked in a solution of ammonium dichromate for sensitization and allowed to dry in the dark.

The plate is then exposed to make a hologram using blue or green laser light. Exposure times must be determined experimentally, as they depend upon such variables as gelatin type and hardness, laser wavelength, sensitizer concentration, and ambient environmental conditions (among others). The plate is processed through a series of temperature controlled baths of water and alcohol, and dried rapidly with a hot air gun.

The finished hologram must be completely sealed to protect it from degrading - the least bit of water will make it "disappear". This will occur even if the hologram is simply left in a humid environment rather than actually gotten wet. Sealing is done by adhering a glass cover plate to the hologram using a good optical cement. A number of materials can be used, including several types of epoxies. Some materials will cure in ultraviolet light, such as one available from Norland Products Inc. of North Brunswick, N.J. (Incidentally, this material can also be used to seal standard silver halide holograms to protect them from becoming scratched). If you've never seen a dichromate hologram - go see one! They are remarkable; their brillance and resolution yield an unbelievable realism. We don't recommend making them, however, unless you've got enormous patience, are a little crazy or (preferably) both.

Chambered Nautilus. Dichromate hologram produced at Dichromate Inc., 1979, photo by Jeannene Hansen.

"Dichromate Jars" by Rich Rallison, 1976, photo © Daniel **Quat.**

thermo plastics

Holograms can be made in deformable thermoplastic film when the thermoplastic is put in contact with a photoconductor. The hologram formed is almost a pure plane phase hologram (no reflection holograms possible) and has the advantages of reusability (the film can be heated to erase the record), good diffraction efficiency, and it isn't messy during use. Its resolution isn't so hot - not more than about 1,000 lines/mm. The preparation of the film is very difficult, and the thermal development is critical. A new commercial thermoplastic system is now offered by Newport Research (pictured below).

A good discussion of thermoplastics for holography, as well as mention of two other materials* (photochromics and ferroelectric crystals) can be found in *Optical Holography*, by Collier, Burckhardt & Lin (Chapter 10).

INSTAVIEW HOLOCAMERA

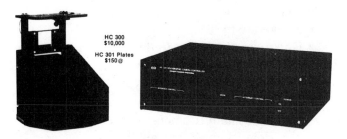

HC 300
$10,000

HC 301 Plates
$150 @

- Hologram viewing in 10 seconds, without chemical processing.
- Reuseable film recycles in one minute.
- High sensitivity and broad spectral response.
- Low noise, high efficiency reconstructions.
- 1/10 the cost of silver halide holograms.
- Automatically compensated, noncritical exposures.
- Compatible with any holographic system, with no special hookup.

photo resist

A surface relief hologram can be formed in photoresist. This material actually becomes etched away during processing, leaving a well-defined series of ridges and valleys. Both positive and negative photoresists are available: the former remove the unexposed areas during processing, and the latter remove the exposed ones.

Photoresist holograms can be used successfully as master holograms for embossing. Photoresist masters have been coated with thin metal layers and used to stamp out copies, in much the same way that phonograph records are made. Although this is only possible with transmission holograms, it does allow the mass replication of copies of everything from white light transmission display holograms, to video discs. Photoresist holograms are well covered in the literature*.

The major disadvantage is that the emulsion is very insensitive, and usually must be exposed in the ultraviolet. For holographic use, photoresist is about 250,000 times less sensitive than 8E75 film.

*For a good overview, see Iwata & Tsujiuchi: Applied Optics, June 1974, vol. 13 #6, p. 1327.

suggested kits

beginner or basic kit

- One 4'x 4' optics table (p.23)
- One laser. Two or more milliwatts recommended (p.96) .
- Three small front surface mirrors (e.g., Edmunds #31,424 or any others - see p. 102)
- One 4"x 5" front surface mirror (or larger, for use as reference mirror - e.g., Edmunds #40,042 - see p.102).
- Two concave lenses of equal focal length for object illumination (e.g., try Edmunds #95453 - See p.103).
- One 50:50 beamsplitter (e.g. Edmunds #31,433 - see p.104).
- One 90:10 beamsplitter (clean ¼ inch plate glass, approximately one to two inches square - from local glass supplier).
- Plastic pipe, masonite, glue, etc., for component construction (p. 66-67)
- One light meter (p. 108 or p.92)

- One 4"x 5" white card.
- One shutter (p.89) simple card o.k.
- Cards or masonite to block scattered light - variety of sizes.
- Translucent diffusing material (optional - for cleaner object beam. Use plastic diffusing sheet).
- Black felt material to surround object (optional).
- Holographic plates - one box each 8E75, 10E75, 4"x 5" glass plates to start (p. 112)
- Darkroom supplies (see list p.54) .
- Squeegee (p. 55)
- Darkroom chemicals (p. 56)
  - Kodak D19 Developer
  - Stop Bath
  - Kodak Rapid Fix
  - Potassium Bromide
  - Potassium Ferrocyanide
  - Mercuric Chloride
  - Kodak 200:1 Photo-Flo

intermediate to advanced kit

Basic Kit, plus the following:

- 4'x 8' optics table. Replace 4'x 4' or build another alongside. (see p. 37)
- Laser of at least 5 mw. Replace your smaller one (see p. 96)
- Variable beamsplitter. This will replace your 90:10 beamsplitter (see p.80 or p.104) .
- Spatial filter with 10X objective and 25μm pinhole. (see p.105-106)
- 8"x 10" plateholder (best to have two - see p.84).
- 8"x 10" white card.
- 4"x 5" insert for 8"x 10" plateholder for test shots (p.84).
- 8"x 10" orlarger front surface mirror (e.g., Edmunds #40,067 - see p.102).

- Collimating mirror. This is optional. You need it if you want to do white light transmission holography (e.g., Coulter 10" f 5.6 - see p.107).
- Film holders - flat and curved (see p.86)
- Three to four additional front surface mirrors for transfer mirrors (see p.102)
- Two each additional lenses of three to four different focal lengths. (see p.103)
- Slit optics. This is optional. You need it if you want to do rainbow holograms (see p.268)
- Mounts for new optics (see p.66)
- 8"x 10" Holographic plates, 4"x 5" or larger holographic film (see p.112)
- Improved shutter (i.e., a mechanical shutter - See p.90 or p.109).
- Index matching holder (this is optional - see p.87).
- Alternate chemicals (see p.58).

professional or super-kit

Intermediate kit, plus:

- Laser of at least 15 mw but preferably 50mw.
- 4'x 8' or larger metal top table (See p.44) or vibration isolation system from Newport Research, Ealing, etc. (See p. 110).
- Magnetic mounts for optical components (see p.91).
- Larger collimating mirror (14" to 17" - see p. 107).

- Additional lenses or mirrors depending on application. (p. 102-103).
- Additional spatial filter for clean master illumination. (p. 105).
- Automatic shutter (See p. 90).
- Film vacuum platen (See p.88).

suppliers' addresses

principal offices (partial listing):

If you are an "Original Equipment Manufacturer" with quality products that fall in one of the following categories, we will be happy to consider listing your organization in one of our future editions. Send your literature to the publisher.

lasers (helium neon):

Aerotech Inc.
Electro Optical Division
101 Zeta Drive
Pittsburgh, Pa. 15238
Tel. (412) 963-7470

Coherent Inc.
3210 Porter Drive
Palo Alto, Calif. 94304
Tel. (415) 493-2111

Ealing Corporation
Pleasant Street
South Natick, Mass. 01760
Tel. (617) 655-7000

Edmund Scientific Co.
101 E. Gloucester Pike
Barrington, N.J. 08007
Tel. (609) 547-8900

Hughes Aircraft Co.
Laser Products
6155 El Camino Real
Carlsbad, Calif. 92008
Tel. (714) 438-9191 ext. 584

Jodon Engineering Assoc.
145 Enterprise Drive
Ann Arbor, Michigan 48103
Tel. (313) 761-4044

Melles Griot
1770 Kettering St.
Irvine, Calif. 92714
Tel. (714) 556-8200

Metrologic Instruments, Inc.
P.O.Box 307
143 Harding Ave.
Bellmawr, N.J. 08031
Tel. (609) 933-0100

Spectra-Physics
Laser Products Division
1250 West Middlefield Road
Mountain View, Calif. 94042
Tel. (415) 961-2550

Laser Lines Ltd.
19 West Bar
Banbury, Oxon
OX16 95A
England
Tel. (0295) 57581

Rolfin Ltd.
Winslade House
Egham Hill, Egham Surrey
TW20 OAZ
England
Tel. (07843) 7541

Nippon Electric Co. Ltd.
252 Humbolt Court
Sunnyvale, Calif. 94086
Tel. (408) 745-6520

laser kits:

Metrologic Instruments (see above)

Maplin Electronics Supplies Ltd.
P.O. Box 3
Rayleigh, Essex
SS6 8LR
England
Tel. (0702) 554000

laser safety:

Mr. Robert G. Stohl
U.S. Food & Drug Adm.
Bureau of Radiological Health
Room 108
586 North First St.
San Jose, Calif. 95112
Tel. (408) 275-7670

optics:

Edmund Scientific* (see above)

Ealing Corporation (see above)

Rolyn Optics Co.
P.O. Box 148
300 Rolyn Pl.
Arcadia, Ca. 91006
Tel. 213-445-6550
Telex 67-5312

Offord Scientific
113 Lavender Hill
Tonbridge, Kent
TN9 2AY
England
Tel. (0732) 364002

variable beamsplitters:

Ealing Corporation (see above)

Newport Research Corporation
18235 Mt. Baldy Circle
Fountain Valley, Calif. 92708
Tel. (714) 963-9811

Jodon Engineering (see above)

spatial filters:

Ealing Corporation (see above)

Jodon Engineering (see above)

J.A. Noll
Box 427
Cumberland, Maryland 21502
Tel. (301) 777-7887

Newport Research Corp. (see above)

* Best prices.

microscope objectives:

Ealing Corp. (see above)

Edmund Scientific (see above)

Laboratory Equipment Co.*
532 McAllister
San Francisco, Calif.
Tel. (415) 863-1185

Newport Research Corp. (see above)

Rolyn Optics (see above)

Swift Instruments, Inc.
1190 North 4th Street
San Jose, Calif.
Tel. (408) 293-2380

pinholes:

Edmund Scientific (see above)

Ealing Corp. (see above)

Melles Griot* (see above)

Newport Research (see above)

Optimation*
P.O. Box 310
Windham, New Hampshire 03087
Tel. (603) 434-2346

collimating (telescope) mirrors:

Coulter Optical Company*
P.O. Box K, 54121 Pinecrest Road
Idyllwild, Ca. 92349
Tel. (714) 659-2991

Edmund Scientific (see above)

light meters:

Davis Publications
Science & Mechanics Division
380 Lexington Ave.
New York, New York
10017
Tel. (212) 557-9100

Radio Shack (for parts)
Tandy Corporation
Fort Worth, Texas 76102
Tel. (817) 390-3011

Spectra Lumicon
% Simon & Assoc.
20 Sunnyside
Mill Valley, Calif. 94941
Tel. (415) 381-0835

shutters:

Jodon Engineering (see above)

Newport Research Corp. (see above)

Ealing Corp. (see above)

Melles Griot (see above)

Vincent Associates Inc.
1255 University Ave.
Rochester, N.Y. 14607
Tel. (716) 473-2232 or
(800) 828-6972

plate holders:

Jodon Engineering (see above)

Newport Research Corp. (see above)

vibration isolation tables:

Ealing Corp. (see above)

Newport Research Corp. (see above)

optical component mounts:

Mitutoyo Measuring Instruments
18 Essex Road
Paramus, N.J. 07652
Tel. 201-368-0525
Telex 134317

Newport Research Corp. (see above)

Jodon Engineering (see above)

Optimark Corp. (see above)

Ealing Corp. (see above)

Aerotech Inc. (see above)

recording materials (holographic plates and film):

Agfa-Gevaert Products:

Newport Research Corp. (see above)

or

Intergraf
Box 586
Lake Forest, Ill. 60045
Tel. (312) 234-3756

Note: In addition to the above dealers, some of the larger photo supply stores may special order holographic materials, Plates or film cannot be ordered directly from Agfa-Gevaert, however, the company may be contacted for technical information or the location of the nearest distributor. In the U.S., the main office is:

Agfa Gevaert
275 North Street
Teterboro, N.J. 07608
Tel. (201) 288-4100

Eastman Kodak Products: (SO 173, SO 253, Type 120 plates, Type 131 plates, 649 F). Kodak products are available through Jodon Engineering (address above), some photo-supply stores, or may be ordered directly from Kodak in some cases. Sales or technical information can be obtained by contacting:

Mr. Robert D. Anwyl
Scientific & Technical Photography
Eastman Kodak Co.
343 State Street
Rochester, N.Y. 14650
Tel. (716) 724-4000 ext. 443-445

optical adhesive:

Norland Products Inc.
695 Joyce Kilmer Ave.
New Brunswick, N.J.
08902
Tel. (201) 545-7828

cylindrical mirrors for rainbow holograms:

G.M. Vacuum Coating Lab, Inc.
882 Production Place
Newport Beach, Ca.
92663
Tel. (714) 642-5446

comprehensive directories of suppliers:

Optical Purchasing Directory
Optical Publishing Co. Inc.
P.O. Box 1146, Berkshire Common
Pittsfield, Ma.
01201
Tel. (413) 499-0514

Laser Focus Buyers Guide
1001 Watertown St.
Newton, Ma.
02165
Tel. (617) 244-2939

suppliers - additional branch offices:

A number of the suppliers listed above operate sales offices in many parts of the world. Some of them appear below:

AEROTECH, INC.
ELECTRO OPTICAL DIVISION
101 Zeta Drive, Pittsburgh, PA 15238
(412) 963-7470 • TWX 710-795-3125

Sales Offices

**Aerotech Midwest
Leetonia, OH
(216) 482-9957**
Illinois, Indiana, Kansas, Kentucky, Missouri, Ohio, West Virginia, Western New York, Wisconsin

**Aerotech Northeast
Hartford, CT
(203) 673-3330**
Delaware, Eastern New York, Eastern Pennsylvania, Long Island, New England, New Jersey

**Aerotech West
Sunnyvale, CA
(408) 738-1220**
Northern California, Oregon, Washington

**Aerotech West
Los Angeles, CA
(213) 376-9698**
Southern California

**Aerotech West
Phoenix, AZ
(602) 941-5341**
Arizona

**CW Radiation GmbH
(Subsidiary of Aerotech, Inc.)
Munich, Germany
(089) 311-1061**
Austria, Germany, Switzerland

Representatives

Spectro Systems
Silver Spring, MD
(301) 949-8261
Maryland, Virginia, Washington, D.C.

Moore Associates
Dallas, TX
(214) 690-4929
Arkansas, Louisiana, Oklahoma, Texas

Kim Controls, Inc.
Minneapolis, MN
(612) 636-9594
Minnesota, North Dakota, South Dakota

McMahan Associates
Orlando, FL
(305) 645-0463
Alabama, Florida, Georgia, North Carolina,
South Carolina, Tennessee

Grady Moore Associates
Albuquerque, NM
(505) 262-1416
New Mexico

Grady Moore Associates
Denver, CO
(303) 494-3351
Colorado

Optikon, Ltd.
Waterloo, Ontario
(519) 885-2551
Canada

Digital Motor Sales
Detroit, MI
(313) 474-0666
Michigan

Laser Lines Ltd.
Banbury, England
(0295) 57581
United Kingdom

Laser Optronic
Versailles, France
953-23-00
France

Laser Optronic
Brussels, Belgium
(02) 771-2166
Belgium, Holland

Quentron Optics Ltd.
Allenby Gardens,
South Australia
23 2345
Australia

Laser Optronic
Milan, Italy
(02) 34-90-265
Italy

Holtek Optronics AB
Uppsala, Sweden
(018) 301200 or 302354
Denmark, Finland,
Norway, Sweden

Y. Ben Moshe
Tel Aviv, Israel
03-739535
Israel

Engineering Equipment Co.
Tokyo, Japan
03-572-7071
Japan

Agfa Gevaert

SALES ORGANIZATIONS
OF THE HOME MARKETS:

Benelux: Mortsel, Rijswijk,
Luxembourg
General Management:
Mr. J. Davignon
F. R. of Germany:
Leverkusen
General Management:
Mr. J. Gehrt

SUBSIDIARIES

Argentina: Agfa-Gevaert
Argentina S.A.C.I.,
Buenos Aires
General Management:
Mr. R. Rozas
Australia: Agfa-Gevaert Ltd.,
Nunawading, Victoria
General Management:
Mr. P. Hennessy
Austria: Agfa-Gevaert
G.m.b.H., Vienna
General Management:
Mr. H. Kaminski
Chile: Agfa-Gevaert
Chilena de Productos
Fotográficos Ltda.,
Santiago de Chile
General Management:
Mr. P. Berger

Denmark: Agfa-Gevaert A/S,
Glostrup, Copenhagen
General Management:
Mr. J. Dau
Finland: OY Agfa-Gevaert AB,
Espoo-Suomenoja
General Management:
Mr. W. Staudinger
France: Agfa-Gevaert S.A.,
Rueil-Malmaison
General Management:
Mr. A. Carpentier
Great Britain:
Agfa-Gevaert Ltd., Brentford
General Management:
Mr. G. Ahrens
Greece:
Agfa-Gevaert A.E., Athens
General Management:
Mr. A. Nothdurft
Iran: Agfa-Gevaert Iran
S.S.K., Teheran
General Management:
Mr. K. Filtzinger
Ireland (Eire):
Agfa-Gevaert Ltd., Dublin
General Management:
Mr. R. Vervoort
Italy: Agfa-Gevaert S.p.A.,
Milan
General Management:
Mr. B. Chiucchini
Japan: Agfa-Gevaert
Japan Ltd., Tokyo
General Management:
Mr. R. Paus
Mexico: Agfa-Gevaert de
México S.A., Mexico

New Zealand:
Agfa-Gevaert NZ Ltd.,
Auckland
General Management:
Mr. H. R. Thieme
Norway: Agfa-Gevaert A/S,
Oslo
General Management:
Mr. A. Naper
Portugal: Agfa-Gevaert Lda.,
Linda-A-Velha
General Management:
Mr. B. Worthmann
Spain: Agfa-Gevaert S.A.
Barcelona
General Management:
Mr. E. Carton
Sweden: Agfa-Gevaert AB,
Stockholm
General Management:
Mr. R. Van Roosbroeck
Switzerland:
Agfa-Gevaert AG/S.A.,
Dübendorf
General Management:
Mr. R. Klett
United States of America:
Agfa-Gevaert Inc.,
Teterboro N.J.
General Management:
Mr. R. A. M. Coppenrath
Venezuela: Agfa-Gevaert de
Venezuela S.A., Caracas
General Management:
Mr. M. Brenzel

PRODUCTION
CENTRES

Argentina: Fabricación
Industrial Fotográfica
Argentina S.A.C.I.F. (FIFA),
Buenos Aires
General Management:
Mr. H. Vandenbroele
France: Agfa-Gevaert S.A.,
Rueil-Malmaison
Factories in Pont-à-Marcq
General Management:
Mr. N. Everaert
India: The New India
Industries Ltd., Baroda
Factories in Baroda and
Mulund
General Management:
Mr. D. M. Ghia
Portugal: Agfa-Gevaert
Indústrias Fotográficas
Portuguesas Lda., Coimbra
General Management:
Mr. J. Heinen
Spain: Manufacturas
Fotográficas Españolas S.A.
(MAFE), Madrid
Factories in Aranjuez
General Management:
Mr. E. Garriga
United States of America:
Metacomet Inc.,
Teterboro N.J.
General Management:
Mr. D. Collura

Peerless Photo Products Inc.,
Shoreham, N. Y.
General Management:
Mr. D. Collura

Coherent, Inc. Offices

Corporate Headquarters
3210 Porter Drive
Palo Alto, CA 94304
(415) 493-2111
Telex: 34-8304

Eastern United States
1000 West Ninth Ave.
Suite A
King of Prussia, PA 19406
Sales: (215) 337-3035
Service: (215) 337-1750
Telex: 84-6179

Midwestern United States
870 East Higgins Road
Suite 131
Schaumburg, IL 60195
Sales: (312) 843-1650
Service: (312) 843-1652
Telex: 28-3592

West Germany – East Europe
Coherent GmbH
Hermanstrasse 54-56
D-6078 Neu Eisenburg
West Germany
6102-27073-77
Telex: 4-185656

France
Coherent, SARL
55 Rue Boussingault
75013 Paris, France
589-89-39
Telex: 204-909

United Kingdom
Coherent, Ltd.
Science Park
Milton Road
Cambridge, CB4-4BH U.K.
0223-68501
Telex: 817466

Benelux
Coherent, B.V.
Meenthof 15
1241 C.P. Kortenhoef
The Netherlands
35-62504
Telex: 844-43514

Italy
Coherent S.r.l.
Residenza Mestieri
Milano 2
20090 Segrate
(02) 2138905
I 2138910

South America
Coherent S.A.
Casilla 2876 C.C.
1000 Buenos Aires
Argentina
392-9666

Japan
Coherent Japan
I.P.O. Box 5330
Tokyo 100-31/Japan
(03) 355-0341
Telex: 781-2322712

Ealing

UNITED STATES
Eastern Region (Corporate Headquarters)
The Ealing Corporation
Pleasant St., South Natick, MA 01760
Tel: (617) 655-7000, Telex: 948339

UNITED STATES
Western Region
The Ealing Corporation
(Western Regional Office)
Suite 203
3900 Birch St.
Commerce Park
Newport Beach, CA 92660
Tel: (714) 833-9826, Telex: 181570

MEXICO, CENTRAL, SOUTH AMERICA
The Ealing Corporation
Pleasant St., South Natick, MA 01760
Tel: (617) 655-7000, Telex: 948339

CANADA
Ealing Scientific Ltd.
P.O. Box 238
Pointe Claire-Dorval, Quebec H9R 4N9
Tel: (514) 631-1807, (514) 631-5171,
Telex: 05821557

UNITED KINGDOM
Ealing Beck Ltd.
Greycaine Road, Watford WD2 4PW, England
Tel: Watford 42261, Telex: 935726

FRANCE
Ealing S.A.R.L.
1030 Boulevard Jeanne D'Arc,
59500 Douai, France
Tel: (27) 88 48 65, Telex: 842160330

GERMANY
Ealing GmbH
Bahnhofstr. 8, Postfach 1226, Höchst,
West Germany
Tel: 06163-3909, Telex: 04191945

ITALY
Ealing Italia
Via Mario Greppi, 5, 28100 Novara, Italy
Tel: (0321) 28065 Telex: 200622

ALL OTHER COUNTRIES
Ealing Beck Ltd.
Greycaine Road, Watford WD2 4PW, England
Tel: (0923) 42261, Telex: 935726

HUGHES
HUGHES AIRCRAFT COMPANY

International

CANADA
Aptec Engineering Limited
4251 Steeles Avenue West
Downsview, Ontario M3N 1V7
TEL: (416) 661-9722 TLX: 06-965819

UNITED KINGDOM
Barr & Stroud, Ltd.
Caxton Street, Anniesland
Glasgow G13, 1HZ, Scotland
TEL: (041) 9549601 TLX: 778114

NETHERLANDS
Koning en Hartman
Elektrotechniek B.V.
P.O. Box 43220
2504 AE The Hague
TEL: (070) 210101 TLX: 31528

ISRAEL
Solgood, Ltd.
P.O. Box 33432
Tel Aviv 64928
TEL: (03) 252216 TLX: 033430

JAPAN
C. Itoh & Company, Ltd.
Central P.O. Box 136
Tokyo 100-91
TEL: (03) 6392946 TLX: 22295

SCANDINAVIA
Marwell AB
Kyrkbacken 27
S-171 50 Solna, Sweden
TEL: (08) 838281 TLX: 17015

GERMANY, AUSTRIA, SWITZERLAND
Atomika Technische Physik GmbH
D-8000 München 19
Kuglmuellerstrasse 6
Federal Republic of Germany
TEL: (089) 152031 TLX: 5215129

FRANCE
Soro Electro-Optics, S.A.
26 Rue Berthollet
94110 Arcueil
TEL: (01) 6571283 TLX: 260879

ITALY
GSG Laser Ltd.
Via Garibaldi 7
10122 Torino
TEL: (011) 555075 TLX: 210310 PPTO

AUSTRALIA
Laser Electronics Pty., Ltd.
P.O. Box 359
Southport, Queensland 4215
TEL: 321699 TLX: 41225

newport

AUSTRALIA
QUENTRON OPTICS PTY. LTD.
576-578 Port Road
Allenby Gardens,
South Australia 5009
Telephone: (08) 46 6121
Contact: Boris Balin or
Alex Stanco

EGYPT
SCIENTIFIC & TRADING CO.
P.O. Box 114
Heliopolis West
Cairo, Egypt
Telephone: 831025
Contact: Hosny Nooh

ENGLAND
METAX, LTD.
P.O. Box 315,
Gladstone Road,
Croydon, Surrey,
England CR9 2BL
Telephone: 01 689 6821
Contact: Bob Roberts

ISRAEL
ISRAMEX COMPANY, LTD.
25, Arlozorov Street
Tel-Aviv, 62-488, Israel
Telephone: 248213-4-5
Contact: Nissan Gazenfeld

ITALY
DB ELECTRONIC INSTRUMENTS S.R.L.
Via Torino 5
20032 Cormano
Milano, Italy
Telephone: 92.32.313
92.97.201
Contact: Aurelio Pessina

JAPAN
KYOKUTO BOEKI KAISHA LTD.
7th Floor, New Otemachi Bldg.
2-1, 2-Chome, Otemachi
Chiyoda-ku, Tokyo 100-91, Japan
Telphone: 244-3511
Contact: S. Komatsu

NETHERLANDS
KONING EN HARTMAN ELEKTROTECHNIEK B.V.
30 Koperwerf
P.O. Box 43220
2504 AE The Hague, Netherlands
Telephone: (70\ 678380
Contact: Hans Siedsma

SWITZERLAND AND LIECHTENSTEIN
W. STOLZ, AG.
Industrievretungen
Bellikonerstrasse 218
8968 Mutschellen, Switzerland
Telephone: 057 54655
Contact: J. Goessi

U.S.S.R.
MULTIC GMBH
Schleissheimerstr. 371
D-8000 Munchen 45
West Germany
Telephone: 089-35 17 001
Contact: Peter Brandstetter

WEST GERMANY AND AUSTRIA
CARL BAASEL LASERTECHNIK KG.
Sandstrasse 21
8000 Munchen 2
West Germany
Telephone: 089/521100
Contact: Carl Baasel or
Friedrich Meyer

RADIO SHACK

TANDY CORPORATION

AUSTRALIA

280-316 VICTORIA ROAD
RYDALMERE, N.S.W. 2116

BELGIUM

PARC INDUSTRIEL DE NANINNE
5140 NANINNE

U. K.

BILSTON ROAD, WEDNESBURY
WEST MIDLANDS WS10 7JN

 Spectra-Physics

Laser Instruments Division
1250 W. Middlefield Road
Mountain View, CA 94042, USA
Telephone: (415) 961-2550
TWX: (910) 379-6941
Telex: 348488

Albuquerque, New Mexico
(505) 881-7577

Boston, Massachusetts
(800) 631-5693

Los Angeles, California
(714) 770-8545

Piscataway, New Jersey
(800) 631-5693
In N.J. (201) 981-0390

Strongsville, Ohio
(800) 631-5693

Washington, D.C.
(800) 631-5693

Spectra-Physics France
3, rue Leon Blum
Zone Industrielle de Glaises
91120 Palaiseau, France
Tel: 920 25 00
Telex: (842) 691183

Spectra-Physics GmbH
Alsfelder Strasse 12
D-6100 Darmstadt
West Germany
Tel: 06151-708-1
Telex: (841) 419471

Spectra-Physics, Ltd.
17 Brick Knoll Park
St. Albans, Herts, AL1 5UF
England
Tel: (727) 30131
Telex: (851) 23578

Spectra-Physics B.V.
Kanaaldijk Noord 61
5642 JA Eindhoven
The Netherlands
Tel: 040-81 45 55
Telex: 51668

Canada, Latin America, Pacific Areas
Spectra-Physics International
2905 Stender Way, Santa Clara, California 95051, USA
Telephone: (415) 249-5200
TWX: (910) 338-0220, Telex: 357-460

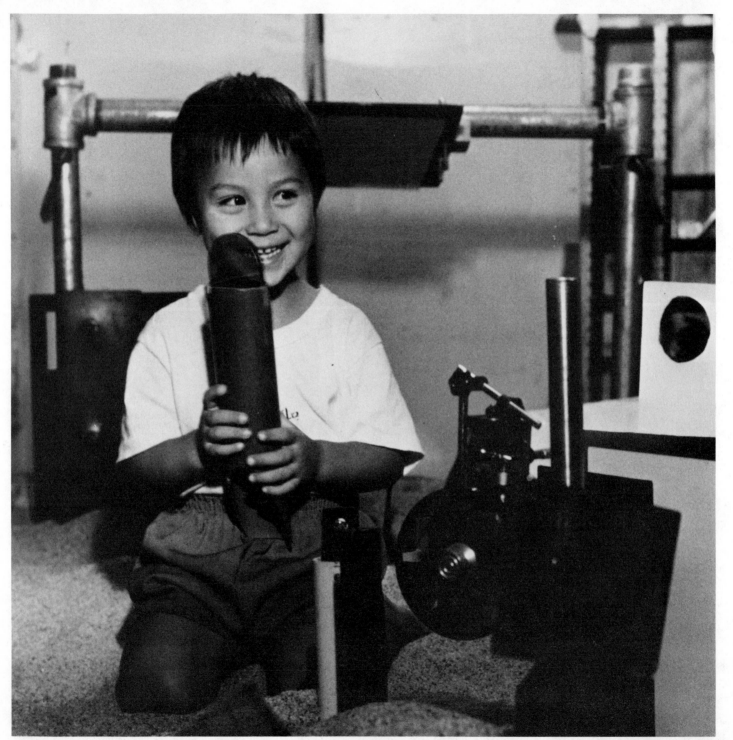

Photo by Bob Schlesinger

basic
holography

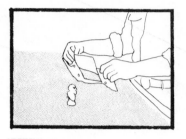

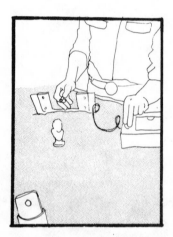

making holograms

In this part of the book, we will demonstrate how easy it is to make a hologram, and how satisfying it is to use your own home-built equipment to do so.

First, we'll discuss some holographic setups which can be made and with a minimum of time and effort, and are especially good for use with low power lasers (in that only very short exposures are usually required). Then we'll move to some arrangements which are a little more complex, but which allow a greater degree of control over the results. All of these setups can be done on a small sand table using a minimum of simple optics.

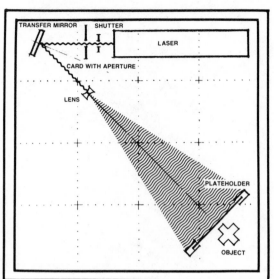

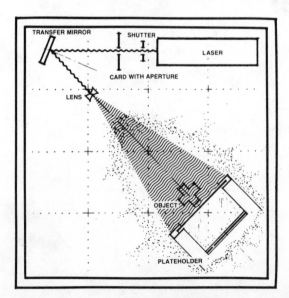

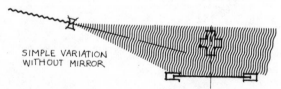

SIMPLE VARIATION WITHOUT MIRROR

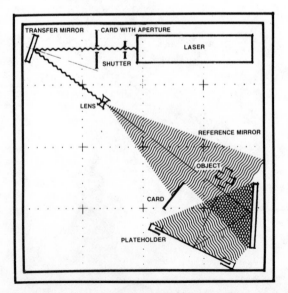

transmission holography

What do you think of when you hear the word "hologram"? What kinds of holograms have you seen? Although there are a great number now available to view or make (most of which will be discussed later in this book), there is one type, the simple transmission hologram, which is really the granddaddy of them all.

We hold a plate of glass with a hologram on it up to a light from a laser or a filtered light. Peering into the plate, we view an image that magically appears to focus in space behind the holographic plate. These are called transmission holograms since light is transmitted through the hologram to the observer. When properly viewed, a "virtual" image is formed appearing on the same side of the plate as the light source. We shall deal with this in greater detail in the theory section. We can say at this point, however, that with the many hybrid types of holograms which have been perfected, it is still the regular transmission hologram which yields an image with the greatest depth and clarity. We think it's the best place to start.

This was the type of transmission hologram first made by Leith and Upatnieks in their early experiments. Their invention was labeled "off-axis" holography (so called because light struck the holographic plate from a different angle or along a different axis from the object light). This allowed for the first time the making of transmission holograms of real 3-D objects (all prior transmission holograms were of flat objects, usually transparencies). The single beam setup we describe here will duplicate their earliest experiments.

Special note: It is important that you do not skip this section because we will be discussing the processing of transmission holograms, among other things, and all transmission holograms are essentially processed alike. It makes no difference what kind of holographic setup is used. Therefore, the processing described in this section will not be repeated again but will be referred to in later sections covering split beam transmission holograms, image planing, white-light transmissions, and diffraction grating.. Therefore any "experts" who will be tempted to skip this chapter, will be confused later.

single beam transmission

As we discuss in detail later, a hologram is formed by the interference caused by reflected light from an "object" called the object beam encountering a pure beam from the laser (called the reference beam). There are numerous ways of causing this to happen but, perhaps the simplest method involves the use of the same beam of light from the laser both to illuminate the object, and also to act as the reference beam. This type of hologram was first made by Emmett Leith and Juris Upatnieks at the University of Michigan between 1962 and 1964.

Leith and Upatnieks later improved upon their invention by splitting the beam into two distinct parts which could be manipulated separately for maximum control over results.

Generally speaking, the quality of an image from a single beam hologram is usually not as good as one obtained from a split beam setup. They are, however, recommended as an ideal way of quickly becoming acquainted with the use of basic optics to form a hologram with a minimum of time and effort.

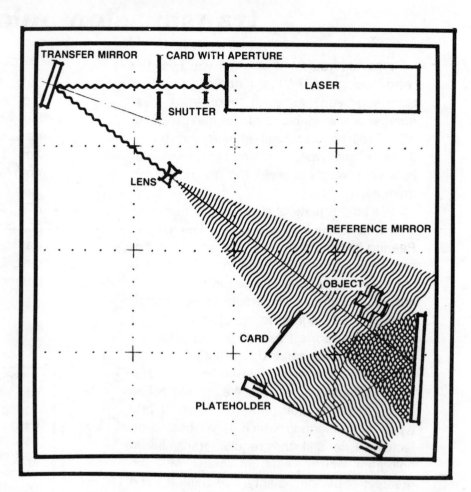

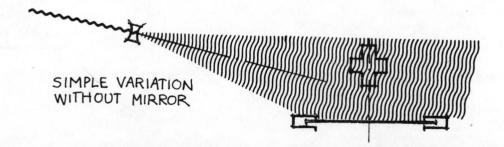

SIMPLE VARIATION
WITHOUT MIRROR

the essentials:

- **stability**
- **light ratios:** 4:1 (ref:obj).
- **exposure:** 8E75NAH or 10E75
- **development:** 15% dark by eye.

what's needed:

- laser (of course).
- negative lens.
- a mirror.
- a plateholder.
- holographic plates: 10E75 or 8E75.
- object (bright & solid).
- white 4''x 5'' card.
- black felt and/or stick for objects (optional).
- transmission holographic processing chemicals.
- several black cardboard scatter cards.

exposure chart:

exposure charts for Science & Mechanics meter:

8E75 nah
photometer scale

reference beam reading	2	3	4
3	6$_s$	60$_s$	600$_s$
5	4$_s$	40$_s$	400$_s$
10	2$_s$	20$_s$	200$_s$
20	1$_s$	10$_s$	100$_s$
30	2/3$_s$	6$_s$	60$_s$
50	1/2$_s$	4$_s$	40$_s$

10E75 ah
photometer scale

reference beam reading	2	3	4
3	3$_s$	30$_s$	300$_s$
5	2$_s$	20$_s$	200$_s$
10	1$_s$	10$_s$	100$_s$
20	1/2$_s$	5$_s$	50$_s$
30	1/3$_s$	3$_s$	30$_s$
50	1/5$_s$	2$_s$	20$_s$

Time in seconds.

the set-up:

1. place laser
2. position transfer mirror
3. place object
4. place plateholder
5. position reference mirror
6. place lens
7. put card in plateholder
8. place reference mirror
9. card off unwanted light
10. take light meter readings
11. load plate
12. allow to settle
13. expose

processing:

1. develop - D19 . . . . . . .15% dark by eye
2. stop . . . . .1 minute
3. fix . . . . .3 - 5 minutes
4. rinse . . . . .3 - 5 minutes
5. hypo clear . . . . .2 - 5 minutes
6. rinse . . . . .2 - 5 minutes
7. bleach* . . . . .until pink
8. rinse . . . . .3 - 5 minutes
9. photo flo
10. squeegee

* Potassium ferrocyanide preferred. See page 156 for alternate processing.

single beam transmission

yes objects no

Try starting with a fairly bright object. Something that is white, yellow, or a light red is good. The best way to tell if something is bright is to spread out the laser beam with your lens, turn off the lights, and see what it looks like. Try to avoid something that is too shiny (it will act more like a mirror than a object and will reflect direct light from the laser back at the plate). It should also be something fairly solid so that it will not move during the exposure (remember your stability requirements) stay away from flowers, pieces of paper, living things, etc. Also, try starting with something small at first (no larger than 2"x 2"x2").

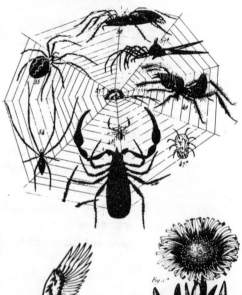

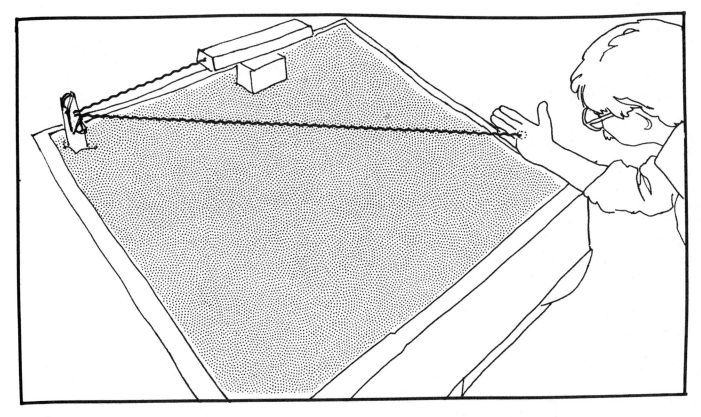

1. laser

The actual holographic setup is quite simple. You may either position the laser on the sand table (or on another surface along side the sand table), or in another area if you made provisions for this when you designed your system. If you are "doing it in the sand", carefully place your laser on a board on the sand surface with the beam traveling along the edge as shown. The beam should travel parallel to the surface and about 3 to 4 inches above it. You can help this situation along by smoothing out top of the sand off with your hand or a card (you'll also have a much happier laser if you take care not to let any sand into or near the casing).

2. transfer mirror

A transfer mirror placed as shown will direct the beam diagonally across the table.

A transfer mirror is simply a mirror that reflects and redirects the laser beam.

3. object

Place the object at a point near the center of the table and in the beam path. Position the front of the object as shown, allowing the dot of red light to appear on its surface.

If you do not wish to have sand in your final hologram, you can place the object on a piece of black velvet material. The area around the object will turn out black, with the effect of the object floating in space.

Some objects may be best mounted on a stick (a ¼" or less dowel, painted black, is good for this). The object may be mounted on the stick in such a way that it won't be visible through the plate holder, but will be securely set into the sand behind the object.

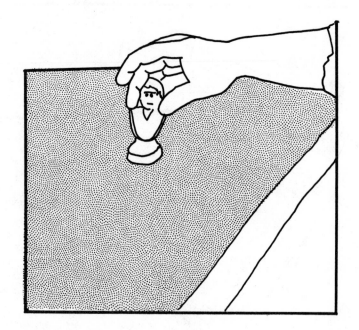

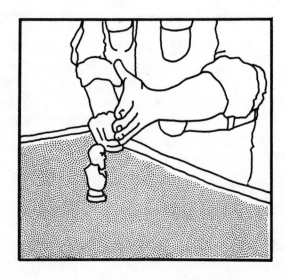

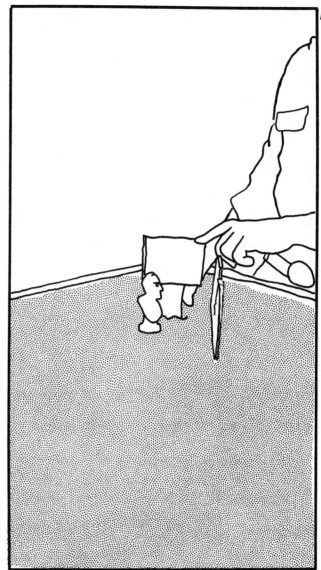

4. plateholder

Now place the plateholder, as shown, about *6 to 8 inches in front of the object*. Look through the plateholder window and position the object to achieve the desired view. Be careful not to look directly into the beam. Start getting used to the idea of the plateholder opening being a "window" which you look through to see a scene beyond. This is the gateway to the holographic space you will design.

5. reference mirror

No more than 3 to 4 inches beyond and to the right of the object, place a 4"x 5" mirror as shown. Known as the "reference mirror", it will, when the setup is complete, deflect some of the light back to the plateholder.

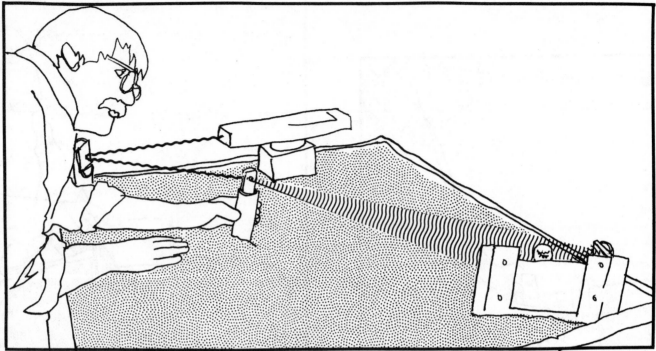

6. lens placement

We now want to place a lens in the system to spread the beam so that it evenly covers the mirror and evenly illuminates the object. The best way to adjust the spread of the light is to move the lens back and forth in the beam path. The focal length of the lens will determine the point at which it must be placed to accomplish this. The more negative the focal length, the less the light will spread, and the further back from the plate the lens will have to be placed. Ideally, you will want a lens with a relatively long focal length which can illuminate both the plate and the object evenly without wasting too much light around them. Lenses with extremely short focal lengths can sometimes yield results which distort things slightly.

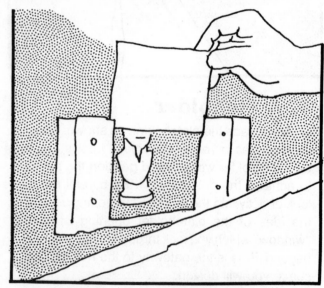

7. card in plateholder

Place a white card in the plate holder and go back to the laser.

8. reference beam

Position the mirror in such a way as to direct light towards the card in the plateholder. As we stated, an even coverage of light is necessary; however, the object should not cast a shadow onto the card. The object can be moved to remedy that situation but the lens and/or beam may have to be rechecked and repositioned to compensate for any lost light due to the change.

9. carding off

Place a card in the sand to block any light going *directly* from the laser to the plate. Make sure light is still illuminating the object and the mirror. See illustration.

In addition, a card with a small aperture may be placed directly after the shutter, allowing the beam to pass through while blocking any stray light coming from the laser housing.*

*This is a wise step to include in all your setups.

10. light meter

Pile additional sand up around the plate holder and lens (and mirror, if present). A light tapping will insure that they are solidly placed and will not move.

A light reading is necessary to determine the proper exposure time. Since we highly recommend the Science and Mechanics model No. A-3, an exposure chart has been included for it.

After turning off all multi-frequency lights, place the meter's probe in the plate holder window pointed directly towards the reference mirror. When making this reading, keep the probe in the plane of the opening of the plate holder (make believe it is resting against a pane of glass), move it to various spots in the window, and record an average of the readings obtained. This, of course, will be your "reference beam reading" You will use your average reference beam reading when you refer to the chart for suggested exposure time.

If you are not using the Science and Mechanics model, it is not difficult to construct your own chart, which may be referred to time and time again. To do this, it is necessary to make a test strip, similar to the type used for determining exposures for printing photographs. Simply place a card with a slit approx. ¾" wide cut out in front of the plateholder*. Different exposures across the same plate can be made (e.g., 3, 5, 10, 20, 50 sec.) by moving the card and reshooting for the next desired time. Remember to move the card in the dark, and to allow a

settling time for each new exposure. You should end up with a plate with a number of strip holograms of the same scene. After processing, a scan across the plate will make it apparent which exposure time was best. To develop a chart, simply match this exposure to the reading obtained on *your* meter. In the future, a reading of half as much (or in the case of the volt-ohmmeter type, twice as much) will require twice the exposure, and so on.

11. loading plate

Block the beam with the shutter. Place the box of plates near the plateholder and turn off the lights. In total darkness, remove a plate from the box and locate the emulsion side of the glass. An easy trick is to place a corner tip of the plate between your lips to the point where they are wet; the emulsion side of the plate should stick to one lip more than the other.

Slide the plate into the holder so the emulsion sees the object (*this is a must when using backed plates*). Be sure the plate is in the holder securely.

If, after 4 or 5 hours (or whatever your breaking point is), you still can't get the plate in properly, put it back in the box, and turn on the lights. Try loading just a card in the holder with your eyes closed, until you get it.

Another great problem that occurs at this point, is knocking over the carefully placed components while you try to find the plateholder in the dark.

If this is a problem, you might try placing some cards in the sand. Have them stretching from the edge of the table to one edge of your plateholder. It will be easy to find the plateholder by simply running your hand along the edge of the table until you find your card, and then following it to the holder.

Note: Cost conscious (or broke) holographers may wish to do test shots with smaller plates to insure all is working properly (to prevent wasting a full plate). This can be done by carefully cutting a plate into strips (e.g. 1"x 5") in the dark or under a green safelight. Then expose one at a time until any problems are worked out. Practice cutting with regular glass first to get the hang of it.

Making A Test Plate

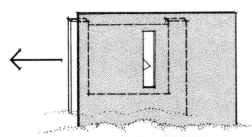

Leave plate holder still.

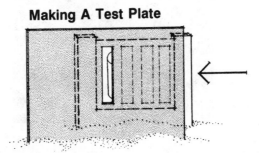

Move mask with single slit to left and increase exposure with each shot.

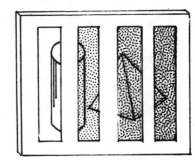

Result, when developed, is holographic test strip.

12. settling time

When you are all set and the plate is loaded, leave the room and let everything settle for at least 5 or 10 minutes. Holograms are so sensitive, even to minute motion, that we need to have a period of quiet or settling time before the exposure in order to allow ambient vibrations time to die out; (note: these are caused by physical motion on or around the table, and can be due to influences as small as the heat from one's fingers placing the plate in the holder).

You might wish to take a friend, the dog, or baby, or all three, for a walk; turn off the stereo, phone, TV; tell your neighbors to stop playing drums for a few minutes, etc. You should be able to get a good exposure even if there is a little noise in the area, but (being a perfectionist) you'll want to make an extra effort to insure the best possible shot.

During the settling period, you may wish to check your darkroom to see that all chemicals are prepared and accessible. You will need Developer, Stop, Fix, Hypo clear, Bleach for transmission holograms and Photo Flo. Make sure your safelight is working well.

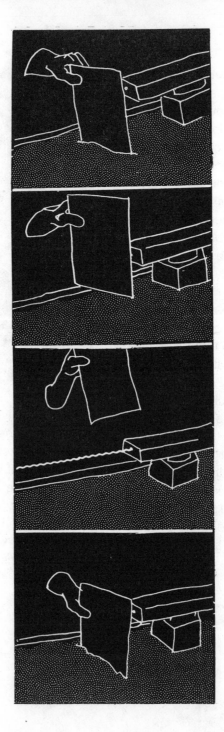

13. expose

And now the moment of truth. Simply open the shutter for the required amount of time. If the shutter is on or near the table, you will have to be extremely careful about making the exposure (i.e., you must add as few vibrations to the process as you can). The most difficult shutter to operate with this in mind is the simple card-in-the-sand variety. If you must use this type, walk very quitely to the table and stand still near the card for a few minutes. Then, delicately and slowly, pull the card out of the sand while making sure it still blocks the beam. You should be holding the card in midair above the table. Now, after a wait of a few more moments, you can pull the card out of the beam path for the required exposure time, and then replace it in the sand.

Remember that each of these steps is to insure that the system is as stable as possible after each motion and absolutely so during the exposure.

After the exposure is over, find your way back to the plateholder; remove the plate carefully (always holding it securely by the edges), and head for the developing room. If your darkroom is not adjacent to the optics table room, the exposed plate should be placed in a transfer box (any light tight box). Place it, emulsion up, in the box so as not to scratch it.

Transmission Processing Summary

(see p.56 for mixtures)

safelight
{
1) Develop (D-19, Neofin, PAAP, etc.) to 15% density.
2) Stop Bath (30 sec. - 1 min.)
3) Fixer (Rapid fix 2 min or regular fixer for 5 min.)
}

regular light
{
4) Rinse (briefly)
5) Hypo Clear (2 min)
6) Rinse (5 min.)
7) Bleach till clear (optional)
8) Rinse (5 min.)
9) Photo Flo (30 sec.)
10) Squeegee
11) Dry.
}

transmission processing

The developing room should be dark except for the green safelight. You may wish to wear rubber gloves throughout the processing (some people discover that their skin is sensitive to some of the chemicals).

The processing sequence must be followed in the order given. If you reverse the order of developer, stop, and fix, you will not get a hologram, and you will ruin the chemicals too; (if fixer goes into the developer, it renders it inactive). The two chemicals which you should be wary of mixing, are fixer and the bleach (we will be using potassium bromide and potassium ferrocyanide). Beware of getting any fixer in the bleach or any washes thereafter. The two together will clear the plate and all you'll have is a nice clean piece of glass to show for it.

Before actually beginning the development process, you must allow your eyes to become accustomed to the dark. This is extremely important since we will be determining visually, with the aid of the safelight only, when the plate has reached the proper degree of development. Try to avoid going directly from a bright room into the darkroom and immediately plunging your plate into the developer.

When your eyes are ready, and this should take only a minute or two, hold the exposed plate by the edges (while remaining aware of which side the emulsion is on), and place it in the developer. If you are using trays, make sure the emulsion is facing up. Agitate the plate slowly for about 10 seconds or so.

You may wish to check the plate at this time. You can quickly look through it at the safelight or at a white card on the wall.

The plate may be relatively clear, with no change in density across it. Place the plate back in the developer and agitate. Keep checking the plate often. Proper development is usually in the range of a density of 15% to 20%, and is usually determined by trial and error. A simple method that usually gives good results is as follows: as the plate develops, the area struck by the light becomes increasingly dense. Since the edges of the plate were covered by the grooves of the plateholder, they were not exposed to the light. Thus, distinct lines should begin to form along the two ends. As a rough guide, the plate will visually be 15% to 20% more dense when the plateholder lines begin to become clearly visable.

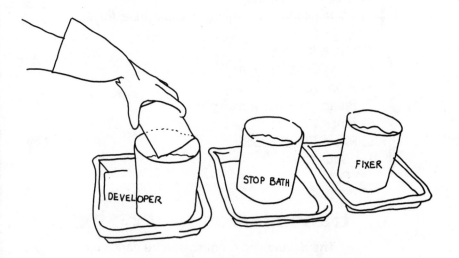

If the plateholder lines become visible right away, it means the plate was probably overexposed or the developer was too strong. It may still be salvageable. Get it quickly into the stop bath. More than likely, however, the plate will be transparent when you first check it and you will have to develop it longer. We don't advise going more than 30 to 45 seconds without checking it. On the other hand, you don't want to be checking the plate all the time, or the plate may develop unevenly. The entire development process may take anywhere from 15 seconds to five minutes, depending on the degree of exposure, the developer strength, and temperature. It will usually develop in one to two minutes.

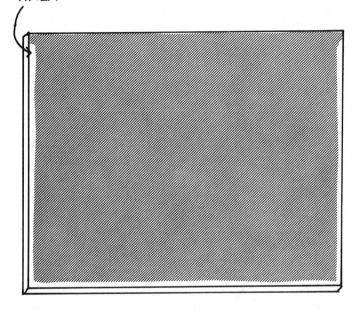

UNEXPOSED AREA

When the edge line is clearly visible, place the plate in the stop bath and agitate for about 30 seconds. Then place it in the fixer where it should remain for 5 to 10 minutes, if you are using regular fixer, and 2 to 4 minutes for rapid-fix. The plate may be agitated occasionally but this is optional.

By now you are probably going out of your mind with wild anticipation, wondering if you've got a hologram. Well, it's about time you found out. The plate can be exposed to regular light at this point, so quickly rinse it off with some water and take a look at it.

You should have a hazy, slightly darkened, piece of glass (big deal, right?). If you hold the hologram up to the light as in the illustration, you should see some color flashes. The light will have to make the same angle with the plate as the light from the laser did originally in order for this image to appear. For this to happen, you may have to fiddle around a bit. Make sure the light bulb for white light in your darkroom is a clear, unfrosted, bulb. *You will not see your original object in this test, but rather, a fuzzy colored blur.* Believe it or not, this is a holographic image. Certainly not the clear crisp floating red image we've been breathlessly waiting for, but we'll look at that a little later.

If, by chance, you didn't see a color burst, don't give up! We've got some more processing to do and there may be a hologram there yet.

Put the plate in the hypo clear for about 2 minutes. You will want to make absolutely sure that all of the fixer is removed before you go into the bleach. The plate can then be rinsed in gently running water for 3 to 5 minutes. The next step involving bleach is optional. Under most circumstances, the bleach will greatly improve the qualities of the hologram by making the image considerably brighter. It will also add a certain degree of "noise" or graininess to the image. The noise increases as the density increases. So, if your plate came out fairly dark, it might be best to skip the bleach step. If it is on the light side, bleaching will be almost essential (see trouble shooting).

Place the plate into the bleach. You should be wearing rubber gloves for this step. The plate will clear fairly rapidly, especially if it is gently agitated.

Rinse the plate for about 5 minutes, then place in the photo-flo solution. The darkness in the emulsion will go away fairly rapidly, especially if it is gently agitated.

BLEACH

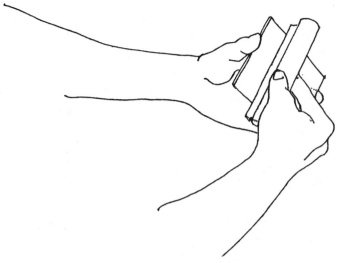

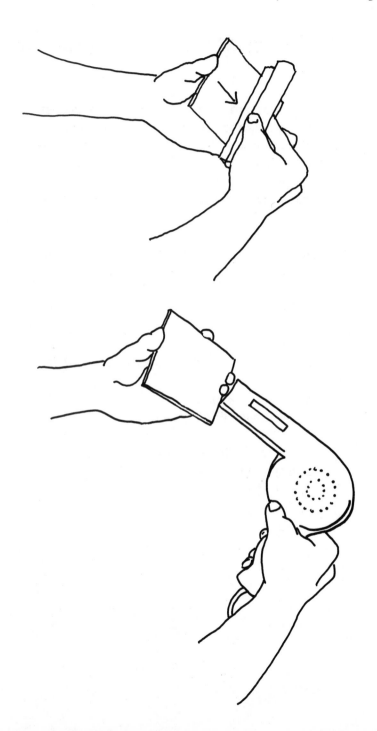

Now for the gentle art of squeegeeing. Good squeegeeing is a real talent, but anyone can learn the hidden secrets and pleasures of this art with a minimum of effort. Following the drawings, simply wipe across the plate in a continuous direction with a relatively even pressure. You may wish to go back over it again lightly, but don't do this too many times on the gelatin side, or you will undoubtedly scratch it. If you get a good deal of streaking, you can always resoak the plate in Photo Flo and try again. Don't be discouraged if you don't get perfect results the first time.

Now for drying: the plate can be left out to air dry or, if you are as impatient as we usually are, dry it with a hair dryer.

Congratulations! Your first hologram is done. You've joined the enigmatic electromagnetic elite, and you're probably the first one on your block.

alternate transmission processing

There are more than one way to process a hologram, and with a virtual (no pun intended) revolution in holo-chemistry taking place (see Darkroom Chemicals), variations in methodology are prevalent. Some which are used for transmission holograms are included below:

A) This sequence is currently recommended by Agfa for use with their HD emulsions, especially for white light transmission holograms:

1) Develop in GP;61 (to 15% density) approx. 2 min.
2) Rinse 1 - 2 minutes
3) Fixer (non hardening such as F-24) for 2 minutes.
4) Rinse 1 - 2 minutes.
5) Bleach in GP-431 till clear (usually about 30 seconds).
6) Rinse 5 minutes.
7) Photo-flo.
8) Squeegee.
9) Dry.

B) When using Neofin Blue developer, some prefer to use it undiluted and add 0.3 grams/liter of benzotriazole and 120 grams/liter of Sodium Metaborate.

C) Reversal Bleaching: when making a bleached hologram, the fixer step may be omitted, with the bleaching occuring just after development, i.e.:

1) Develop as usual.
2) Stop Bath or Rinse.
3) Bleach 1 - 2 minutes.
4) Rinse 5 minutes.
5) Photo-flo, squeegee and dry.

viewing

We imagine you will probably wish to see what you've just made.

Go back to your sand-table and put the hologram in the plateholder. The hologram must be placed in the same position relative to the laser light as it was when it was exposed. To do this, turn or flip the plate around until you see the image appear.

Looking into the plate, you should be able to see your object from different angles just as though it were really there. You can remove your object from the sand for all these tests.

Now flip the hologram over (top to bottom) and place it back in the holder. If you step back a bit and move around, you will see an image standing off in front of the plate toward you (between you and the plate! Not behind the plate). Since the plate is now being viewed from the back side, the image appears inside out and backwards (pseudoscopic). This projected image is formed by the light actually focusing in space in front of the holographic plate. You don't believe it? Run your finger through it! Or try this: take a piece of translucent material (e.g., a piece of frosted glass or plastic). Using this as a rear projection screen, you should see various parts of the three dimensional image going in and out of focus. This is a good way of getting a sense of how the image actually occupies a three-dimensional space. If you don't have a piece of translucent material to use as a screen, try using a white card viewed from the other (plateholder) direction.

Another interesting trick can be performed as follows: remove the lens from the beam path. There should be simply a dot of light on the plate. Now move the card or screen out to where the image is focusing in space. You should see a crisp image of the full object projecting out of that one spot. Now, if you move the plate around so that the dot hits the plate at different places (take it out of the plateholder to do this), you will always see the full object projected but the angle of view will change. That is, you will see your object tumble around in space, and you will see it from different perspectives (just as if you were able to fly around it). This is an example of how the whole image becomes stored on all areas of the plate. We can imagine that when the whole plate is illuminated, it is tantamount to all of the dots projecting all of the angles of view simultaneously. The hologram really acts as a super-sophisticated lens, focusing or spreading light that passes through it to such a highly controlled degree, that it replaces the *way in which* light came from the original object. Sound confusing? More on this later.

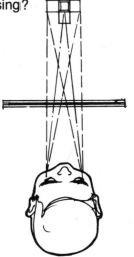

troubleshooting single beam transmission

Problem	Cause	Solution
The image has bands of light and dark, or parts of the object are missing while others are there.	The object was moving during the exposure.	Secure the object more firmly, or use a more solid object.
The object is visible in its entirety, but there are bands of light and dark across the surface of the plate. The object is visible as you look through some areas of the plate, yet is very dim or not visible at all through other areas.	The plate was moving during exposure.	Check to see that the plateholder sits securely in the sand, and that the plate fits securely.
Entire image is dim or not visable.	Entire system unstable.	Check with interferometer.
	Object or component motion.	Resecure all components.
		Replace with rigid object which appears bright under laser light.
	Plate not being viewed properly.	Make sure you try all possible orientations of plate with respect to viewing light before you give up. Plate must make same angle with light as it did during exposure.
		Check light readings.
	Over/under exposed or over/under developed.	Make certain processed correctly. Also plate should be processed soon after exposure to prevent decay of latent image.
	Laser not in TEM$_{00}$ mode.	Use proper equipment.

Problem	Cause	Solution
Plate is entirely dark.	Plate overexposed or overdeveloped or developer improperly mixed, or plate fogged (opened to ordinary light).	Correct accordingly
Plate is entirely light (clear) or it appears that part or all of plate not exposed at all.	Under exposed or developed.	Expose/develop more or heat developer slightly (to 90-100 degrees).
	An obstruction blocked some or all or the beam. Shutter not working properly.	Check to see plate and object properly lit and shutter is working.
	If using AH plate, emulsion facing wrong direction.	Check AH backed plates.
	Fixer not sufficiently washed out before bleaching.	Use longer wash time after fixer or make sure hypo clear is good.
Plate is very grainy; image is "noisy".	Bleaching of plate that is too dark.	Expose/develop less or don't bleach.
	Optics dirty or of poor quality	Clean or replace or use spatial filter in place of lens.
Plate "spotty" or "mottled" in appearance.	Chemical contamination or improper procedure (also due to omission of photo flo, poor squeegeeing or drying).	Replace chemicals. Review processing instructions. Make certain photo-flo is good and squeegee doesn't leave large streaks.
	Object surface is too reflective, or light from some other surface is inadvertently reaching plate.	

Problem	**Cause**	**Solution**
Undesirable flashes of color/light visible while viewing image.	Object surface is too reflective, or light from some other surface inadvertantly reaching plate.	Replace w/object having diffuse (matt) finish. Check to see that laser light only lights object and plateholder opening. Make sure edge of glass plate is covered by plateholder to prevent internal reflections.

The following may be the cause of some of the problems:

If you are near the table: moving around, breathing, or talking during the exposure.

Table located in "noisy" area, or near radiator, appliances, or unstable floor or higher story of a building.

Components not sufficiently immobilized - mound up sand around bases. Make certain they can't fall over.

Laser not warmed up. Exposures should not be made unless laser has been on for at least ½ hour.

alternative single beam transmission

1. place laser
2. position plateholder
3. dig hole
4. place object
5. place lens
6. take light readings
7. load plate
8. allow to settle
9. expose

Here's another method of making this type of hologram. The advantage to it is the ease and short time for setup. The disadvantage is that the object is backlit only.

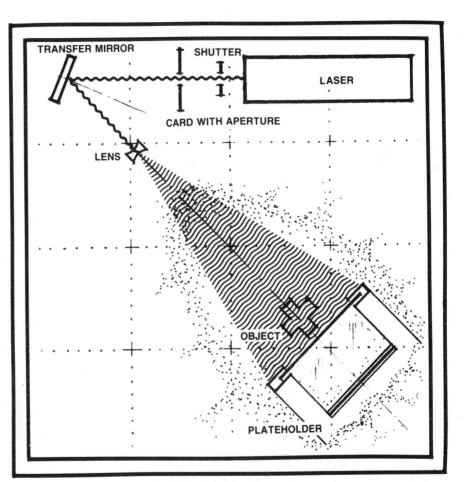

1. laser

Clear the table except for laser, transfer mirror, and plateholder. Set up so you are sending a beam diagonally across the table.

2. plateholder

At the end of the table, opposite the laser and in line with the beam, place the plateholder in the sand. Tilt the top of the plateholder forward towards the beam. Your plateholder should appear as in the diagram, with the plate tilted from about 40 to 70 degrees from the sand.

Now get behind the plateholder, and look through the opening. If your line of sight is perpendicular to the plateholder, you should be looking through the opening down into the sand.

3. digging hole

About 2 feet behind the plate (going toward the transfer mirror), scoop out a depression in the sand 3 - 4 inches deep. This will allow the object to be placed so that it does not interfere with the reference beam.

4. object placement

The object can be temporarily placed in this depression. You will determine its final position in a moment.

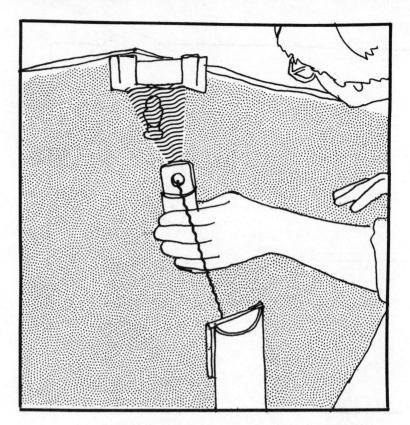

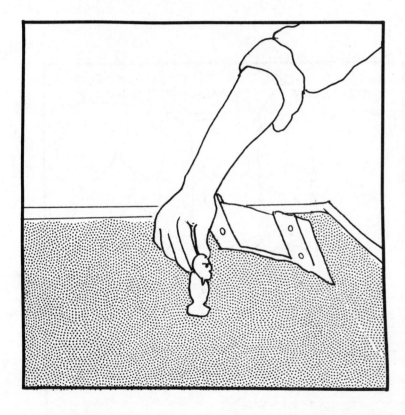

5. lens

Turn down the overhead lights. Now move the lens back and forth in the beam path, until the light is spread evenly across the card in the plateholder. This same light should afford the best illumination possible for the object (as seen through the plateholder window). One of the chief drawbacks to this type of setup becomes apparent at this point. Since the light is coming from behind the object, as seen through the plate holder, it appears to be predominantly lit from behind. With this in mind, a little more time is well spent re-positioning the object to achieve the best vantage point for efficient or satisfactory illumination.

7. loading plate

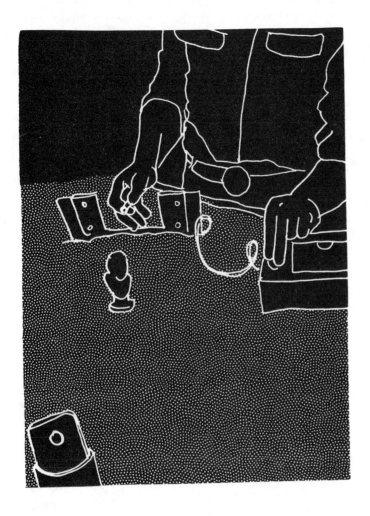

6. light reading

Using a photometer, point the probe directly towards the lens. Obtain an average reading throughout the plateholder window.

8. as in previous setup:

check exposure time
settle
shoot

single beam reflection

It's true that regular transmission holograms (when properly made) can yield the best depth, parallax and resolution obtainable, but they are not viewable in ordinary light. This can often be very inconvenient. How does one show holograms away from special-lighting situations?

Fortunately, there *is* a hologram available for display in such circumstances. Reflection holograms (known as Braggs angle holograms) can be viewed in sunlight or with a point source of light. We might imagine these holograms acting as extremely complex mirrors (as opposed to the transmission hologram's lens analogy), where the reflected light is very precisely manipulated by the hologram. These holograms are thus viewed with the observer and the light source on the same side of the plate. The depth obtainable in a reflection hologram is much less than that of a transmission, typically only a few inches, depending upon the circumstances. Even so, the results when observed properly, are quite impressive.

The first reflection holograms were developed by Y.N. Denisyuk of the Soviet Union in 1961 (see Theory section). The setup for these differs from that of transmissions, in that light reflected from the object hits the opposite side of the plate from the reference light. Again, there are countless ways to achieve this, and the simplest involves using a single beam from the laser. The original Denisyuk holograms were of this type.

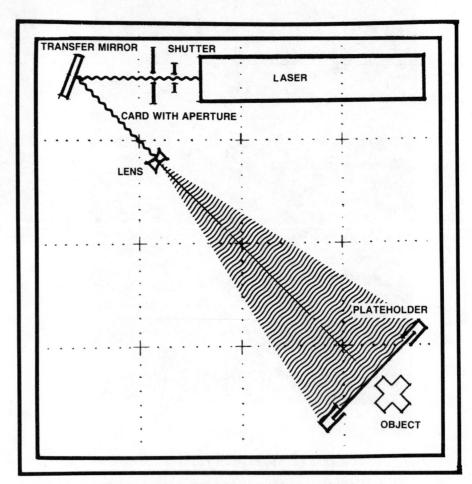

The concept behind this setup is very simple. We want a spread out beam to travel through the plate to the object. Some of the light from this beam will be reflected by the object back to the plate.

the essentials:

- **stability**
- **light ratios:** 2:1 (ref:obj).
- **exposure:** 8E75NAH only
- **development:** 70-90% dark by eye.

what's needed:

- laser (of course).
- negative lens.
- plateholder.
- object (bright & solid).
- black felt (optional).
- shutter.
- white card.
- Agfa Gevaert 8E75 NAH (unbacked!) plates.

the set-up:

1. place laser
2. position transfer mirror
3. position plate holder
4. position object
5. place lens
6. take light readings
7. load plate
8. allow to settle (mix bleach now!)
9. expose

exposure chart:

exposure charts for Science & Mechanics meter:

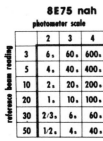

8E75 nah
photometer scale

reference beam reading	2	3	4
3	6s	60s	600s
5	4s	40s	400s
10	2s	20s	200s
20	1s	10s	100s
30	2/3s	6s	60s
50	1/2s	4s	40s

Time in seconds.

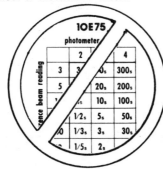

10E75
photometer

reference beam reading	2		4
3	3..	..0s	300s
5		20s	200s
		10s	100s
	1/2s	5s	50s
	1/3s	3s	30s
	1/5s	2s	

processing:

1. **develop** . . . . 70 - 80% dark
2. **stop** . . . . .1 minute
3. **rinse** . . . . .1 - 2 minutes
4. **bleach*** . . . . .P.B.Q.
5. **rinse** . . . . .5 min.
6. **photo-flo** . . . . .30 sec.
7. **squeegee**
8. **dry**

* See page 173 for alternate processing.

1. laser   2. transfer mirror

Either by using a transfer mirror or just the laser, run the beam diagonally across the table.

3. plate holder

Set the plateholder straight up and down in the sand near one end of the table (allow room behind the plateholder for the placement of your object). Place a card in the holder. Now aim the beam directly at the center of the card in the plateholder. The beam should hit the card at a 60 to 90 degree angle.

For best results, however, the angle should not be much less than about 70 degrees. This angle can be either up and down (achieved by moving the top of the plateholder forwards or back) or left and right (by turning the plateholder left or right in the sand).

Those of you with protractors can check this exactly; (otherwise, if the beam appears to be *nearly* perpendicular to the card, it should be fine - don't worry about it.)

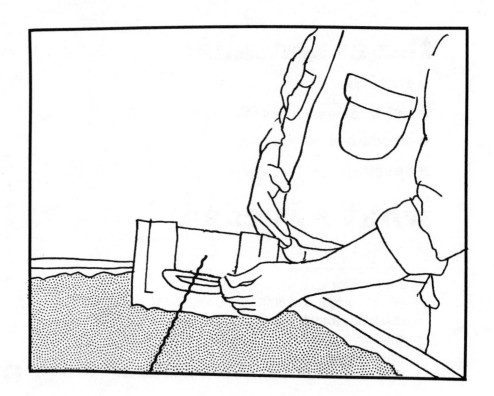

4. object

The object you will want to use should be very bright and very solid. Stability for reflection holograms is about 10 times as critical as for transmission holograms. Reflection holograms are also much less efficient in the playback of light than transmissions.* We thus need something which will reflect back a lot of light, and which will definitely not move. Something metallic is good (as long as it is not highly polished or shiny). White ceramics or ivory, etc. are also o.k.

Try to use something with no more than 1"-2" of depth to start.

The object should be placed as close to the empty plateholder as physically possible, on the side opposite to that from which the beam originates. (the object can even rest against the glass for the exposure, although this is sometimes difficult to arrange in the dark without scratching the emulsion or knocking over the object while inserting the plate).

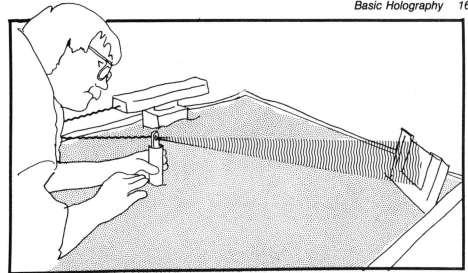

5. lens placement

Center the object in the window. You should have a red dot in the middle of your object at this point.

Place the white card back in the holder. The beam should now be centered on it. Now place a negative lens in the beam path so that you get a nice even coverage of light on the card, yet don't waste too much around the edges (you will lose some, so don't worry). This should be checked with all overhead lights off.

Now, if you remove the card, your object should also be well-lit. Get yourself around to the side of the table opposite the object, and look at it through the plateholder to check this. If you don't like what you see, re-position the object until you are satisfied. Once again - what you see is what you'll get, so make sure you're happy with it.

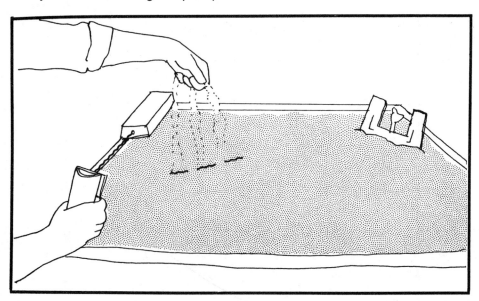

*True for silver halide emulsions, which most readers will be using (Agfa & Kodak plates).

6. light reading

To determine the exposure, take the light-meter probe and, taking care not to knock into the object, point it into the beam. Since the object will be in the way, you will not be able to take an average from all areas of the window (unless you use a little ingenuity), but this is not all that important (especially if you had a fairly even coverage of light over the area to begin with). Check the readings you get against the exposure chart for 8E75 plates to find the proper time (a common error at this point is to look at the wrong chart for the plate being used).

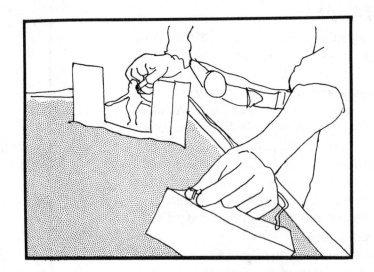

7. loading plate

Recheck all of the components to make sure they sit securely in the sand. You should now be ready to make your first reflection hologram. Set the shutter and, in the dark, pull out a new 8E75 plate (make sure they are *unbacked* or 8E75 NAH).

Using the wet lip test (remember, from your transmission hologram days?), find the emulsion (sticky) side and place it toward the object*. Now get yourself out of the room and let it all settle for about 15 minutes or so.

*Actually, the emulsion can face either way. We recommend facing the object so that the front of the final hologram is protected by glass from fingerprints, scratching, etc. A reflection hologram with the emulsion towards the beam will tend to be slightly more red than one with the emulsion towards the object.

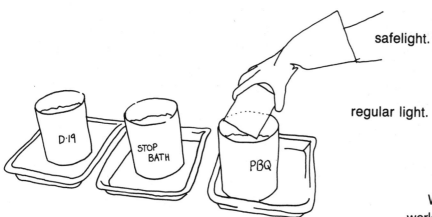

Reflection Processing Summary
(see p.56 for mixtures)

safelight.

1) Develop (70%)
2) Stop 30 sec
3) Rinse 1 - 2 min.
4) Bleach (PBQ)

regular light.

5) Rinse 5 min
6) Photo Flo 30 sec
7) Squeegee
8) Dry

reflection processing

And while you wait...

The darkroom setup will be a little different from that for transmission holograms. You will be using a different bleach - if using P.B.Q., you must use the sequence described for it. Under no circumstances should the plate be fixed after bleaching. The Mercuric bleach is used in a completely different processing sequence (see "alternate processing" section). the plate.

These plates are developed to a much greater density than transmissions. You will want these to go to 70 to 90% percent dark or let only a third to a tenth of the light pass through (density of 1.0 to 2.0 for you photo people).

When you feel that things are stable and the world is fairly quiet, expose for the required amount of time, and begin developing.

You will want to monitor the plate's progress with the safelight as you did with the transmission hologram (only allow it to go much darker). It can take anywhere from 1 to 5 minutes for the plate to become 70 percent dark.

Many holographers become very superstitious about the relationship between exposure, developer strength, development, temperature, and time. All are integrally intertwined, as are such factors as the age of the developer, the type of water used at the time (its PH, or acid-alkaline state), water hardness, the tides, your horoscope for the day, etc.

Since these conditions may vary from time to time and from locale to locale, the best mix for you will probably be determined by experience. Your magic formula may or may not turn out to be very much different from what we've suggested. Try it, you may like it.

Our best results have occurred with exposures close to, or slightly longer (up to 2x) than, those recommended by the chart. The best images usually develop in about 1 to 3 minutes in developer diluted 4:1 as previously described. Long development times may be reduced, with good results, by heating the developer slightly (to 80 or 90 degrees). This can be done with an aquarium heater, or by placing the bucket of developer into a hot pan of water. When the plate has reached the proper development, it should head for the stop bath. Next, the plate is rinsed for 1 to 2 minutes and then placed carefully into the P.B.Q. bleach which was just recently prepared. After a minute or two in the bleach, the regular light can be turned on and the plate removed and rinsed for at least 5 minutes.

After the final wash, the fun begins. First, however, a short bout with the Photo Flo. Gently rub with a flo-soaked paper towel and squeegee. This is one place where good squeegeeing is important.

So far you haven't seen any image, right? Well, here's something to make them all clap and squeal! With a light from a good high-intensity lamp or unfrosted bulb coming over your shoulder, turn on the hair dryer and start drying the emulsion side of the plate. As you dry and tilt the plate around to find the proper angle - the image will magically begin to appear. If you don't see anything, continue with the hair dryer until the plate is totally dry. Sometimes the moment of truth doesn't come until the very end.

The hologram can be viewed in any bright point source of light. The sun is best but light from a slide projector or high intensity lamp will also do fine. Keep flipping the plate over and turning it around, if you don't see an image yet. Remember, the hologram must make the same angle to the light as the original reference beam used to form it.

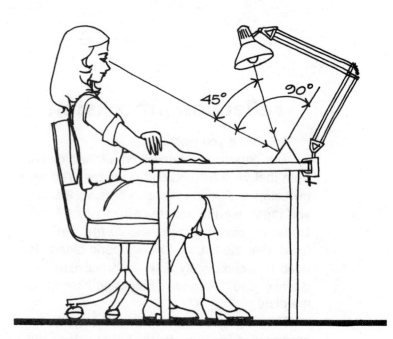

alternative reflection processing

under safelight. { 1) Develop (D-19, Neofin, PAAP, etc.) to 70% dark
2) Stop Bath 30 sec.
3) Fixer 2 - 5 min.

under ordinary light. {
4) Rinse 1 -2 min.
5) Bleach (Mercuric Chloride) until all of darkness turns pink.
6) Rinse until color shifts blue.
7) Return to fixer (until dark)
8) Rinse briefly (30 sec.)
9) Hypo clear 2 min
10) Rinse 5 min
11) Photo Flo 30 sec.
12) Squeege
13) Dry

The plate is developed as before, to a density of approximately 70%. Next is the stop bath for about 30 seconds. The plate then goes into the fixer (as with the transmission hologram sequence) for 2 - 5 minutes. At this point, the regular light may be switched on. Rinse the plate for about 5 minutes. Now, place it carefully into the Mercuric Chloride bleach. Be sure to wear gloves and keep from splattering the stuff around. In the bleach, the plate will turn a pinkish color. Wait until all of the original darkness disappears before removing the plate.

Rinse the plate again until the pink turns a bluish color (usually 3 to 5 minutes).

Place the plate back into the fixer. It will magically turn dark again after a few seconds; (some holographers prefer to use 2 different fixer trays to prevent chemical contamination between the bleach and fixer; this is optional in this case).

Then comes hypo clear for about 2 minutes, followed by a water wash of about 5 minutes. The hypo clear step may be omitted in favor of a longer wash time, but this step usually insures that no residual fixer will remain in the emulsion to oxidize and cause degradation of the hologram over time.

Next, the hologram is placed into the photo-flo and, as with the other sequences, squeegeed and then dried.

B) Other variations:

Other developers and bleaches can be substituted into the basic P.B.Q. reversal sequence. Best results have been obtained using P.A.A.P. or Neofin developer (undiluted) with one of the P.B.Q. bleaches. Benton adds 0.5 grams/liter of Ammonium Triocyanate to the P.A.A.P. developer when using 8E75 plates and grossly overexposes (2100 ergs/cm^2). These amounts may have to be adjusted for different chemical batches, however.

C) Alternate drying techniques:

Streaking can often be a problem in reflection holograms. It is usually due to non-uniform drying after squeegeeing. Obtaining a good squeegee stroke can at times be difficult and, to avoid undue frustration, you may wish to try something different. Benton suggests the following method:

After the final water rinse:
1) Soak in 50% alcohol 50% water.
2) Soak in 75% alcohol 25% water.
3) 95% alcohol 5% water.
4) Rinse with wash bottle containing 100% very pure, clean alcohol.
5) Squeegee dry (suggests using air, rubber, or vacuum squeegees).

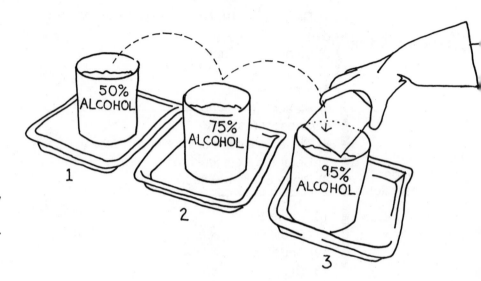

The soak times and results will vary with the type and quality of the alcohol used.

Isopropyl or methanol are usually used. Isopropyl is much slower, but does not have the toxicity of methanol. The final alcohol bath may need to be relatively long. If uneven drying still persists, the plate should go through an extra hardening step before the final water rinse; (see Darkroom Chemicals - Alternate Fixers).

If you use a red laser to record the reflection hologram, the image will reconstruct with a greenish color. This is due to shrinkage in the emulsion from processing.

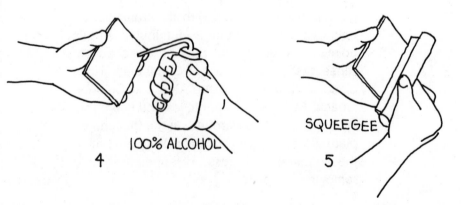

It is possible to adjust chemically the emulsion thickness so that the hologram will play in almost any single color, from red through violet. The chemical used is triethanolomine and the result can be achieved in either of two ways:

1) **Post Soaking -** The finished hologram is placed in a 5 to 10% trietanolomine solution. (for those of you who are not chemists, start with 50 ml. of triethanolomine, and dissolve it in a liter of water).*

The hologram is soaked for a couple of minutes, carefully squeegeed (with a different squeegee from the one ordinarily used so as to prevent contamination), and dried. The image color will depend on the precent solution, soak time, solution temperature, drying time and temperature of drying air. One should experiment with all of the variables in order to develop a preferred personal technique, as well as to gain some understanding of the sensitive relationship between emulsion structure and image characteristics.

Often times this method will result in non-uniform "color streaking" in the image. This is due to certain areas of the emulsion expanding more than others. It is often caused by poor squeegee work. To help, a small amount of photo-flo may be added to the triethanolomine solution and the squeegee step may be eliminated in favor of a longer "drip-dry".

The photo-flo may, however, allow better squeegeeing so that the original procedure might be followed. The post soak may be done any number of times, so if you're not pleased with the results, try again. If you decide you want the green image back, simply wash in water, photo-flo, squeegee and dry.

2) **Pre - Soaking -** The emulsion thickness may also be adjusted *prior to exposure* of the plate. The plate is soaked in triethanolomine solution, then dried and exposed; it is then processed in the usual manner. John Kaufman reports excellent results using this method, along with reversal bleaching, to achieve very bright red-orange images, as well as images in any single desired color of the spectrum. A hybrid system of both post and presoaking can be used to add greater flexibility to the "search for color".

Once the desired color has been achieved, image contrast may be heightened by spraying the back of the plate with quick-dry flat black paint. If the plate has the emulsion on the back, the image will disappear as the paint hits it, but will reappear after drying. This step also provides a protective coat to the emulsion.

*Triethanolomine and equipment it comes into contact with should be isolated from the other materials to prevent contamination.

troubleshooting single beam reflections

In addition to problems listed in transmission troubleshooting (p. 158), difficulties may arise from the following:

Problem	Cause	Solution
No motion problems were encountered with transmission holograms, yet in the reflection hologram, the object or plate appears to have moved.	Stability is much more critical with reflection holograms.	Resecure all components. Use a more solid object. Move table to quieter area if necessary (we hope you don't have to do this).
Image is visible, yet extremely dim, even in bright sunlight. Plate density seems o.k.	Object not reflective enough. Single beam holograms are often by nature not very bright. Problem also may be caused by using improper film of insufficient resolution. Also remember that reflection holograms must be bleached.	Use more reflective object. Problem often solved by using split beam methods. Use only Agfa 8E75! Review reflection processing sequence.
Plate won't turn dark while developing.	In addition to causes listed in transmission troubleshooting, it can be result of using a plate with AH backing.	Use NAH plates.
Plate turned dark while developing yet came out clear at end of process.	Wrong bleach used.	Mercuric Chloride bleach is the only one which should be used in the sequence involving a second fix step.

You may wish to photocopy a bunch of
these to use to design all of your setups.

NAME :	TYPE :	REMARKS :
PLATE/FILM :	READINGS :	
EXPOSURE :	RATIO :	
DEVELOPER :	DISTANCES :	
BLEACH :	LASER :	

Birds-eye views of setups in this section:

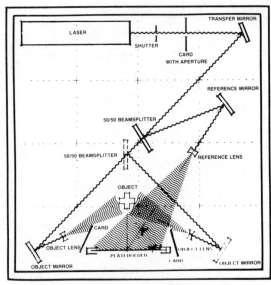

Split beam transmission

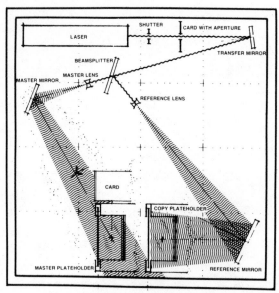

Alternate image plane reflection

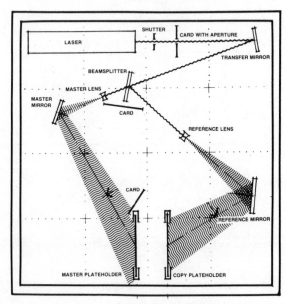

Image plane reflection

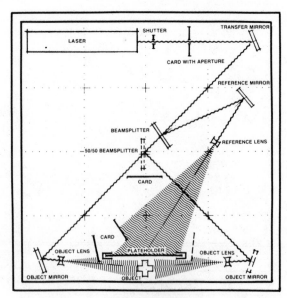

Split beam reflection

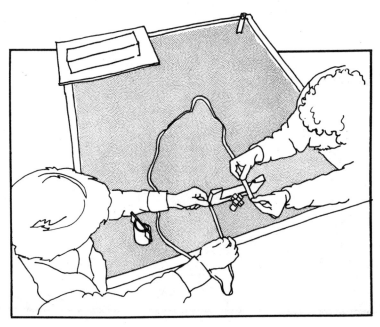

split beam holography

Although simple to do, there are some limitations to the single beam setups previously described. For one thing, there is no way of controlling the ratio of intensity of reference beam to object light - an important factor in obtaining a bright, efficient hologram. In addition, there is no way of controlling the lighting of more complex objects (e.g., of achieving side or back lighting, etc). By splitting the beam into two or more distinct parts, and by separately manipulating the object illumination and reference light, we can add the flexibility necessary to make holograms of a wide variety of objects. They can be lit in different ways, and with consistently better results.

Split beam holography is considered *real* holography. There are countless ways to devise a setup to achieve a given result. Each new arrangement may become a unique expression of the individual holographer. Those shown here are well-tried, save time, and ultimately cause fewer headaches.

split beam transmission

This is really *the* basic setup. Most of the others which follow use what are essentially modifications of the concept which is presented here. The most sensible approach, we feel, is to maintain a symmetry in the placement of the components. This is important to insure that all beam path distances from the beamsplitter to the plate are equalized, to guard against undesirable shadows or stray light, and to allow enough working room for adjustments. With this in mind, variations on our schemes are encouraged, yet if you do, remember K.I.S.S.

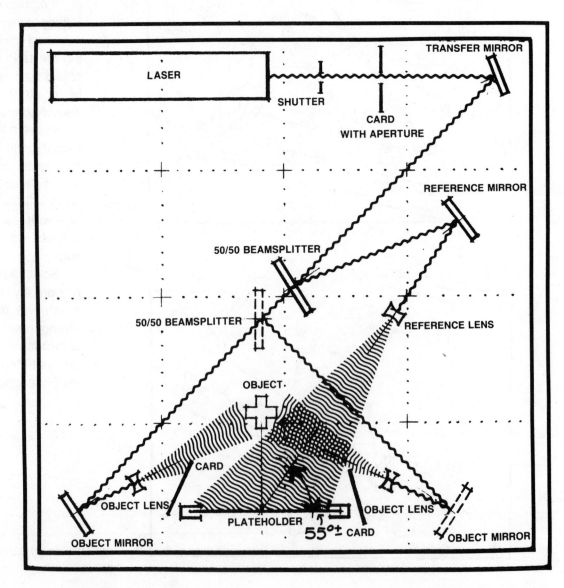

the essentials:

- **stability**
- **distances:** equal distances from first beamsplitter
- **light ratios:** 4:1 (ref:obj).
- **exposure:** 8E75NAH or 10E75
- **development:** 15% dark by eye.

what's needed:

- laser (of course).
- 3 mounted front surface mirrors (two 2''x 2'' and one 4''x 5'').
- 3 mounted negative lenses.
- 4''x 5'' plateholder.
- one or two beamsplitters (90:10 or variable beamsplitter; 50:50).
- object (bright & solid).
- white 4''x 5'' card.
- black felt and/or stick for objects (optional).
- protractor. • tape measure (cm/in).
- several black cardboard scatter cards.
- light meter. • shutter.
- transmission holographic chemicals.
- Agfa Gevaert 8E75 or 10 E75 plates.

exposure chart:

exposure charts for Science & Mechanics meter:

8E75 nah

photometer scale

reference beam reading	2	3	4
3	6 s	60 s	600 s
5	4 s	40 s	400 s
10	2 s	20 s	200 s
20	1 s	10 s	100 s
30	2/3 s	6 s	60 s
50	1/2 s	4 s	40 s

10E75 ah

photometer scale

reference beam reading	2	3	4
3	3 s	30 s	300 s
5	2 s	20 s	200 s
10	1 s	10 s	100 s
20	1/2 s	5 s	50 s
30	1/3 s	3 s	30 s
50	1/5 s	2 s	20 s

Time in seconds.

the set-up:

1. place laser
2. position plate holder
3. place object
4. place transfer mirror
5. place object mirror
6. place beamsplitter
7. measure distances
8. place reference mirror
9. place reference lens
10. position object lens
11. card off light
12. measure light ratios
13. load plate
14. allow to settle
15. calculate exposure time
16. shoot

FOR SECOND OBJECT BEAM:

6a. place beamsplitter #2
6b. measure distances
6c. place object mirror #2
6d. place object lens #2

processing:

1. develop - D19 . . . . .15% dark by eye
2. stop .1 minute
3. fix .3 - 5 minutes
4 rinse . . . . .3 - 5 minutes
5. hypo clear . . . . .2 - 5 minutes
6. rinse . . . . .2 - 5 minutes
7. bleach* . . . . .until pink
8. rinse . . . . .3 - 5 minutes
9. photo flo
10. squeegee

* Potassium ferrocyanide preferred. See page 156 for alternate processing.

1. laser

Place the laser as shown.

2. plate holder

Position the plateholder at the opposite side of the table. It should be about 6 inches in from the edge, and in the center of the table. Make this as stable as possible by mounding up sand at the base.

3. object

Set the object about 10 - 15 cm (4" - 6") in the sand just beyond the plate (see diagram). To do this properly, get your tail end down and look through the plate holder window. Position the object so it is well centered and affords the view you want as you move your head from side to side. If the object is too low, mound up the sand underneath it or secure it on a stick to help center it in the window.

4. transfer mirror

Place the first transfer mirror in the corner across from the laser and direct the beam diagonally across the table. An easy way to see the beam is to sprinkle some sand lightly along the path, but *gently*, so that you do not sand blast your optics.

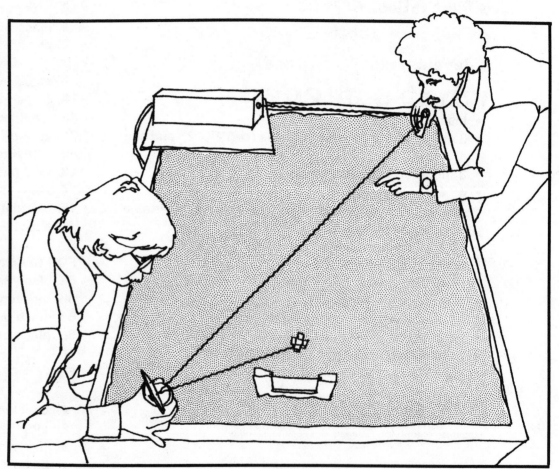

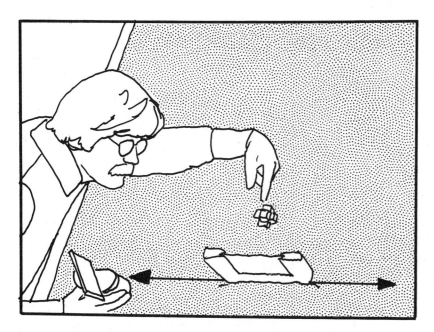

5. object mirror

Draw a line in the sand that follows the plane of the plate from both sides of the plateholder. Place a mirror (called the first object mirror) securely in the sand behind the line. Then direct the beam, by adjusting the first transfer mirror, to the center of the object mirror.

Direct the beam towards the center of the object by wiggling the mirror in the sand (you should see a red spot on your object). This is the first object beam.

6. beamsplitter

Now place a reference beam mirror in the setup in the same manner used for split beam transmissions. For a reflection hologram, however, the reference beam should hit the plate at an angle of about 60 to 80 degrees.

A line drawn in the sand out from the plateholder at this angle may be beneficial.

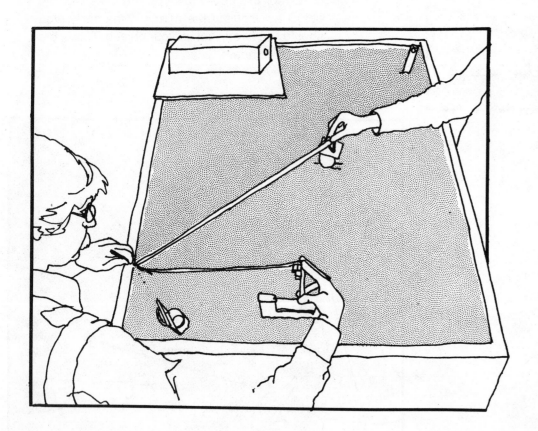

7. measure distance

In split beam setups, *it is essential that the distance travelled by each beam be equal.* Each should be measured following its setup. Using a friend's assistance for this can be very helpful.

With a tape measure or string, measure the distance from the beamsplitter to the first object mirror, then to the center of the object, and *then to the center of the plate*. Be sure to measure the *total* distance (a common mistake is forgetting to measure the additional object to plate distance). The total measurement to the plate is the object beam distance. Record the distance.

8. reference mirror

Now set up the reference beam and arrange its path to be equal in distance to the object beam path.

The reference beam should strike the plate at about a 30 to 45 degree angle. A good way of achieving this, with the use of a protractor, is to draw lines in the sand out to the right of the plateholder at 30 degrees and 45 degrees. Now, using the same length of tape or string used for the object beam (remember your recorded distance), place one end at the center of the plateholder while holding the other end of the tape at the beamsplitter.

Enlist a friend's help to pull the tape out to the right and take up the slack with a finger or pencil. By moving the pencil (or finger) back and forth and letting the taut tape slide around it, an arc can be described in the sand. Any point along this arc will equal the proper distance, from the beamsplitter to the plate, for the reference beam.

At some point on this arc, between the 30 degree and 45 degree lines you previously drew in the sand, place a mirror (reference mirror). Slowly pivot the beamsplitter to direct the reflected beam towards this mirror. Now, place a white card in the plateholder and direct the beam from the mirror to the center of the card. This reference beam distance should now be equal to that of the object beam.

The checklist that follows will serve as a quick roadmap.

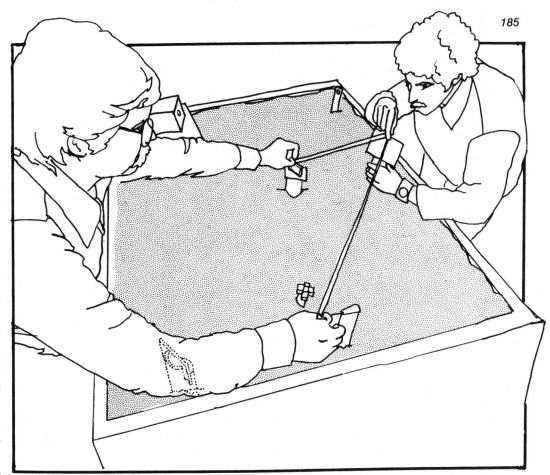

Object Beam Path: The sum total of the distances from beamsplitter to object beam mirror to object to center of plate.

Reference Beam Path: Equal to the object beam path and is the sum total of the distances from beamsplitter to reference beam mirror to center of the plate.

9. reference lens

Reference Beam: place a lens in the beam path so that a fairly even coverage of light appears on the white card (The lens may have to be moved back and forth in the beam path or possibly the choice of a lens with a different focal length will help to achieve this). It's best to place the lens between the reference mirror and the card (not too close to the plateholder, however). If better coverage can be obtained with the lens between the beamsplitter and the mirror (and the mirror is fairly clean), do it that way.

10. object lens

Object Beam: remove the card from the plateholder. Place a lens in the object beam path so that the object is evenly bathed in red; (make sure the lens is behind the line of the plateholder). The lens should *not* be placed where it can be viewed through the window (plateholder).

By playing around with the object as well as the lens, you can achieve the most suitable lighting.

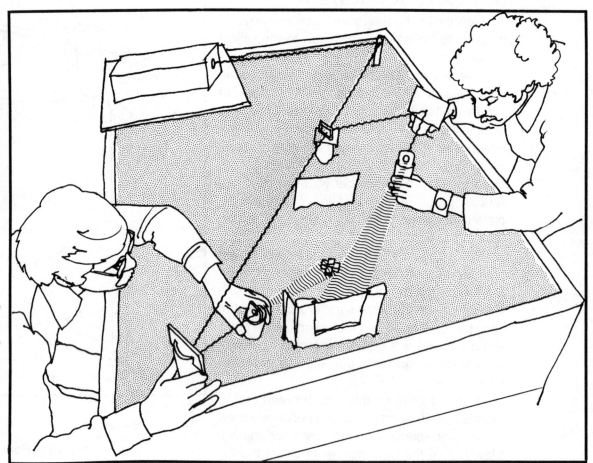

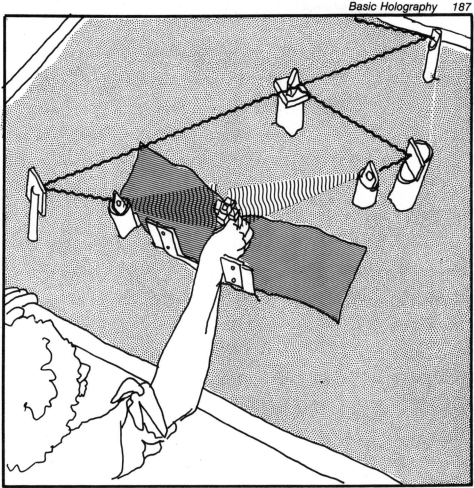

While moving the object around, make sure it does not cast a shadow in the reference beam and onto the plate area. You will be removing and replacing the white card *frequently* during this procedure. Remember, too, that what you see *through the plateholder* is what you'll get in the final hologram. So, get down to the level of the plateholder and make sure the object looks good from this position. Black felt may be placed around the object at your option to give it the effect of "floating" in space.

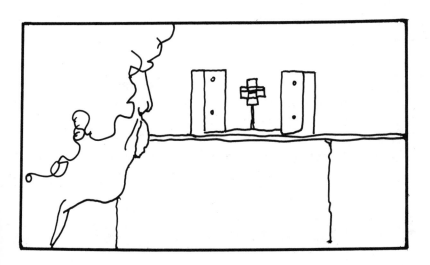

Don't worry too much about the exact placement of the lenses now unless you are using a variable beamsplitter. More than likely, their positions will be changed when the light readings are taken. Remember, too, that once the beam paths have been set, the placement of the lenses will have no effect on distances.

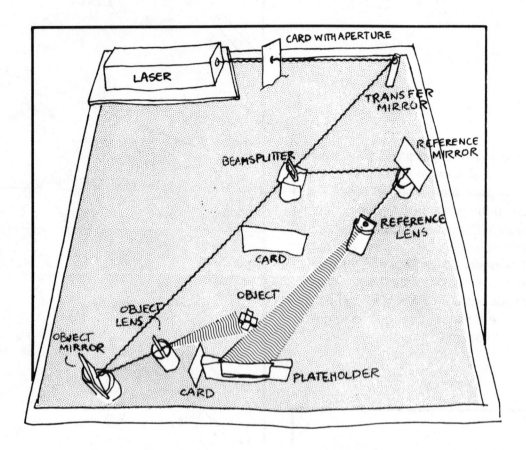

After a fairly thorough job of tracking and tracing scattered light, go back and check to see that a nice even coverage of reference light still falls across the plateholder opening and the object is well lit. A final check is to look through the window. What you see is what you will get. Therefore, you should *only* see your object and the reference beam.

At this point, you may wish to consider an optional object illumination setup. Up to now, the object has been lit from only one side. This is adequate for a number of objects, but for many an undesirable shadow will fall to one side.

options

Adding an additional object illumination beam by splitting the first one into two parts will correct this. However, it also adds a whole new level of complexity. This new beam's distance must be matched to that of each other beam (from the first beamsplitter to the plate). The new optics become additional sources of scattered light which must be blocked off. The fact that the object beam is split in two means a bit more light will be "lost", probably resulting in a longer exposure. When adjusting ratios, both object beams must be considered together; (this makes the movement of lenses a little more tedious.)

The advantage, though, is that the lighting can be controlled much more precisely and the light can find its way into those harder-to-reach places. We suggest developing expertise with a single object beam the first time around, before adding this option. Ultimately, you may wish to experiment with a number of object beams.

a. beam splitter

To add another object beam, place a 50:50 beamsplitter in the path of the first object beam a short distance after it passes through the first beamsplitter.

b. measure distance

To place the second object beam mirror properly, draw a line in the sand extending the plane of the plate out to the right side of the plateholder. Now find the distance recorded for the original object beam (and reference beam) on the tape measure or length of string. Place one end at the *first* beamsplitter and the other at the center of the plateholder. We wish to place the mirror at a point where the distance required covers the following path:

1st (90:10) beamsplitter to 2nd (50:50) beamsplitter to new object beam mirror to object to plate.

c. object mirror #2

A friend can be very useful in helping to stretch the tape on this path. Place the second object beam mirror slightly behind the line in the sand just as you did for the first object beam.

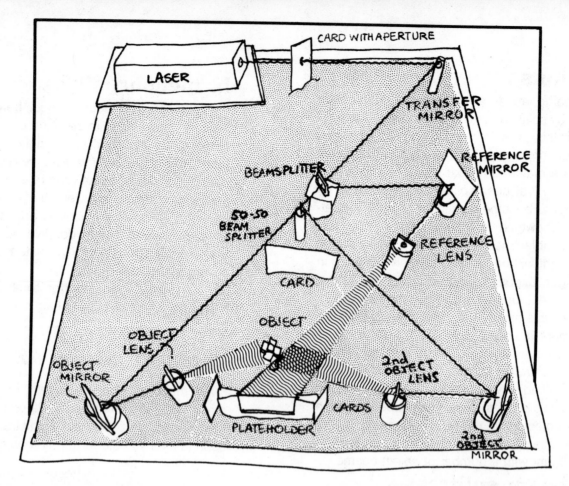

d. object lens #2

Using the new beamsplitter, direct the reflected beam to this new mirror, and in turn direct the beam to the object. A lens can be placed in this path to illuminate the object from this new direction. The lens can be placed either before the object mirror (as shown) or after, depending upon which gives the best object illumination. Try to adjust the lighting with both object beam lenses, so that both sides of the object appear lit with approximately the same intensity (as viewed through the plateholder). Card off excess light as you did for the other object beam.

The object should now be evenly illuminated from both sides. From now on, the term "object illumination" will refer to the total effect of *all* object light.

Recheck the reference beam at this time to insure that none of the new components casts a shadow on the plate.

11. carding off

Carding Off - This is an essential step in all split beam setups and one often overlooked in guides to holography. The step is actually simple to remember if one remains aware of what actually goes on when a hologram is made. During the exposure, *only* the light from the reference beam and light reflected form the object should reach the plate.

We cannot stress this enough!

Any other scattered light will interfere with the formation of a good clean hologram.

Results of light scatter often show up as distracting patterns or flashes of light. If severe enough, scattered light can almost completely obscure an image.

Causes and cures of scattered light:

Perhaps the most common cause of scattered light is direct light from the object illuminating beam hitting the plate. *This beam is for lighting the object only.* To insure this, place a card at the edge of the plateholder to block light coming from the direction of the object beam.

Take care not to place the card too far out in front, or light will be blocked off from the object. Angle it slightly out to the left as you look through the plateholder so it will be out of view. A shadow from the card may be visible in the sand between the plateholder and the object, letting you see the area blocked. If you will be using plates without an antihalation backing, be sure also that the card extends back far enough to prevent any light from striking the other side of the plate.

Additional scattered light may come from other optics or directly from the laser itself. It may be wise to place a card between the beamsplitter and object, as shown. Be careful not to block off either of the beams. Another good practice is to take a card with a hole a little larger than the beam diameter, and place it in the beam path before the beamsplitter (so that the beam can pass through easily).

One of the real advantages of split beam holography is that the efficiency of the hologram can be improved by controlling the *ratio* of reference-to-object light intensity. This makes it possible to make holograms of objects with a wide variety of reflective qualities by merely reading the two beams with a light meter, and adjusting their ratio to compensate for "brighter" or "dimmer" objects.

Transmission holograms will work with ratios ranging from 2:1 to 10:1 (reference to object; the reference beam should always be brighter than the object light). Best results are obtained at a ratio of 4:1, or when the reference beam is about 4 times brighter than the object light. These intensities must be measured "at the plate" where the hologram will be formed.

12. light ratios

There are two ways to determine ratios. The first is a visual test to gauge the situation approximately. With the white card in the plateholder, place a finger in the reference beam so as to cast a shadow on the card. The area in the shadow should still be receiving light from the object (make sure your finger is not in line between the object and card), and one can visually approximate how much brighter one source is than the other. The eye, with a bit of training, can become a very precise indicator - some holographers swear by this method. We do not recommend it for beginners, however.

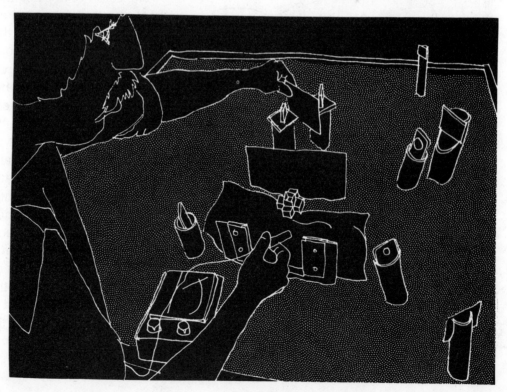

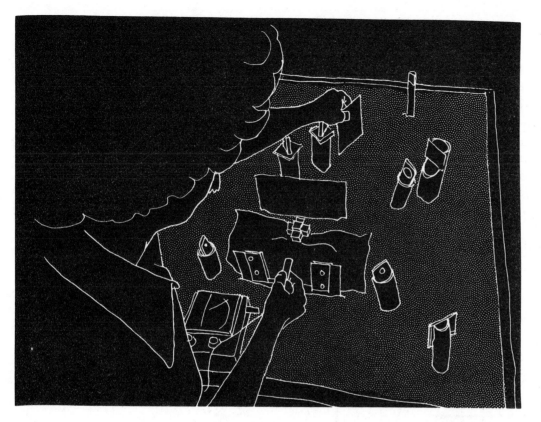

Now, remove the card blocking the object beam. Place it to block off the reference beam, and again place the probe in the plane of the plateholder window, this time pointing towards the object (which should be beautifully lit). Record an average reading.

To change the ratio, simply move either of the two lenses back or forward in the beam path. Moving the lens closer to the plateholder will brighten that beam's effect, while moving it further away will dim it (the illuminated area on the plate/card or object may also change). It will take some playing around with this to feel comfortable with it. The goal is to achieve the ratios needed with a minimum of light lost (not forgetting the balanced lighting across the plate and object).

As you may have guessed by now, it can be extremely tedious to move lenses slowly back and forth, constantly rechecking the setup, the scatter cards, etc. simply to achieve the desired ratio. The holographers who have already invested in variable beamsplitters are probably smiling about now. The beauty of this optical component is that with only a simple motion, the ratios may be adjusted without disturbing the setup; (see advanced components for further discussion).

Make a note of the average *reference* beam reading after the 4:1 ratio has been found.

The proper exposure can be determined by consulting the chart already established for single beam holograms for the type of plate being used. The reference reading is needed for this.

If you wish to be more precise, separate measurements of the two beams can be obtained with a light meter. First, place a card in the object beam path to block it off. Now, point the probe of the light meter into the plateholder window towards the reference beam. Be careful to keep the probe *in the plane* of the plateholder, as if it were resting against a glass plate. Move the probe around in the window while keeping it aimed at the reference beam, and record an average reading.

A gentle tapping down of the components and plate holder will insure that they settle quickly in the sand. Stability requirements are more stringent for split beam setups, and the greater the number of optics used, the more chance there is that something may move or even fall over. Make a habit of this final check.

The final step, in all holographic setups will be the same. Check for uniform reference coverage and object illumination. Check and secure the scatter cards if they are out of place.

Now set the shutter, turn off the lights, load a plate (remembering that anti-halation plates are loaded emulsion side toward the object), let the system settle for 5 or 10 minutes, expose it for the required amount of time, and process it according to the transmission hologram instructions in the single beam section.

14. load plate

15. settle

13. shutter

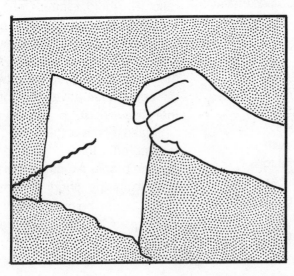

16. check exposure time

17. shoot

10E75 ah
photometer scale

reference beam reading	2	3	4
3	3s	30s	300s
5	2s	20s	200s
10	1s	10s	100s
20	1/2s	5s	50s
30	1/3s	3s	30s
50	1/5s	2s	20s

8E75 nah
photometer scale

reference beam reading	2	3	4
3	6s	60s	600s
5	4s	40s	400s
10	2s	20s	200s
20	1s	10s	100s
30	2/3s	6s	60s
50	1/2s	4s	40s

18. process as per transmission instructions (p. 151)

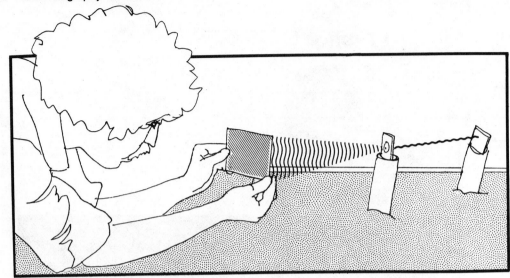

viewing

The split beam setup just completed utilizes a "side" reference beam. Viewing this hologram with the laser light intercepting the plate at the same angle as when it was made. If the setup hasn't been disturbed, simply place the plate back into the plateholder properly (after flipping it around a few times to find the proper side), and the holographic image will concide exactly with that of the real object (although it will be dimmer because of the split beam).

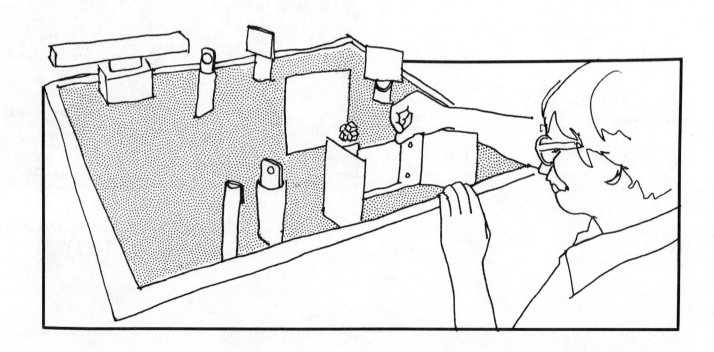

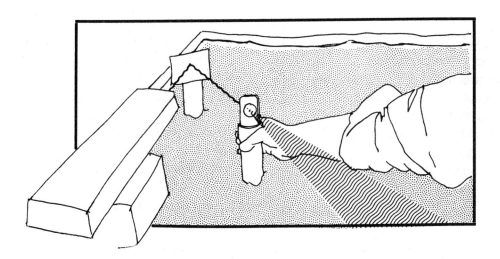

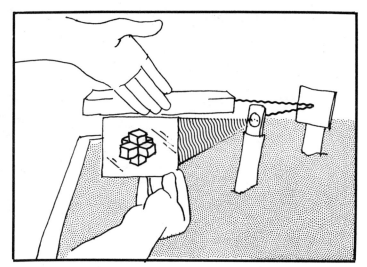

To view the hologram in full power laser light without disturbing the setup, simply place an unused mirror in the beam path before the 90:10 beamsplitter, and "steal" the beam by directing it away from the other components. Another lens can be placed in line to spread out the light in order to illuminate the hologram. When this new mirror and lens are removed, the beam will follow its original path.

splitbeam transmission troubleshooting

Problem	Cause	Solution
Object dim or not there	1) Object or optics motion. 2) Unequal beam path lengths. 3) Exposure/development off. 4) Improper light ratios. 5) One of beams inadvertantly left blocked during exposure. 6) Poor object positioning.	Secure object and optics. Check object and reference beam distances to assure they are equal. Adjust if necessary. Check plate density (15% dark). Check light readings and adjust for 4:1 (reference:object). Measure *at the plate*. Check to see cards not blocking reference or object light. Make certain object does not cast a shadow on plate (in reference light).
Plate has "dirty" or mottled apperance, or flashes of spectrum sometimes visable and obscuring image.	1) Unwanted light scatter reaching plate. 2) Optics dirty. 3) Chemicals gone bad. 4) drooling over image. (also review single beam troubleshooting p. 158)	Check to see *only* object and reference light reaches plate, all else is blocked off. If problem persists, try less reflective object or spray object with matt-spray. Clean or replace dirty optics. Use better quality componants. Replace chemicals. Remember to photo-flo, and back off while viewing.

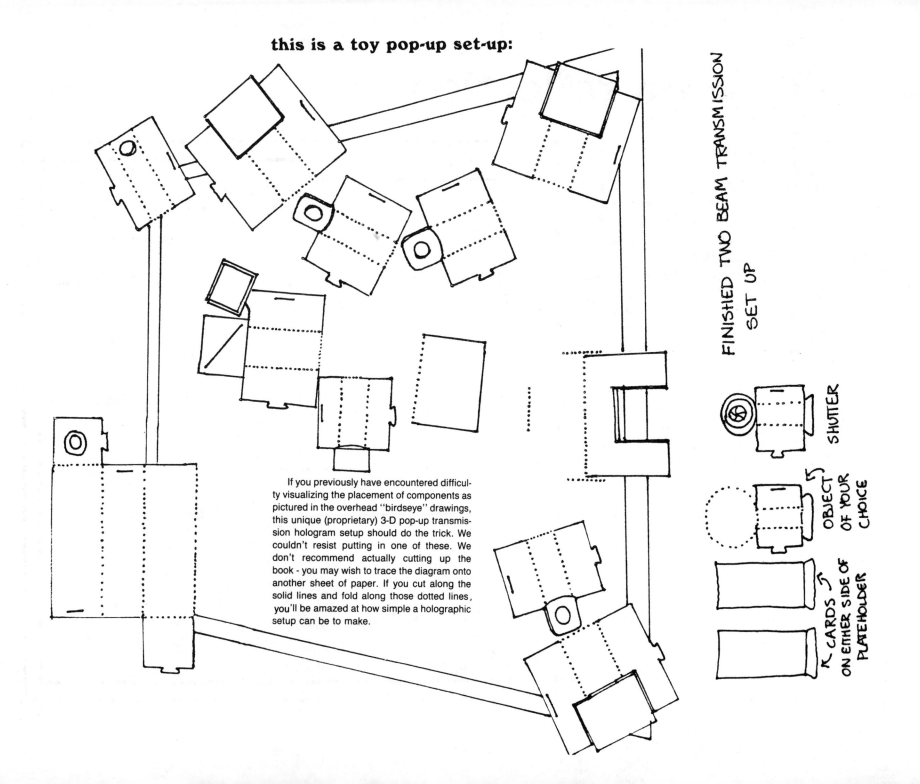

this is a toy pop-up set-up:

If you previously have encountered difficulty visualizing the placement of components as pictured in the overhead ''birdseye'' drawings, this unique (proprietary) 3-D pop-up transmission hologram setup should do the trick. We couldn't resist putting in one of these. We don't recommend actually cutting up the book - you may wish to trace the diagram onto another sheet of paper. If you cut along the solid lines and fold along those dotted lines, you'll be amazed at how simple a holographic setup can be to make.

FINISHED TWO BEAM TRANSMISSION SET UP

SHUTTER

OBJECT OF YOUR CHOICE

CARDS ON EITHER SIDE OF PLATEHOLDER

split beam reflection

The setup for this hologram is almost identical to that of the split beam transmission type, except the object is on the opposite side of the plate. In addition, the efficiency of this hologram will invariably be much greater than that of the single beam reflection variety (also *much* nicer - the difference is unbelievable).

Reflection holograms are very dependent upon a proper reference to object ratio for good results (more so than the transmission kind). This type of setup will yield a dynamic hologram when viewed with a point source of ordinary white light.

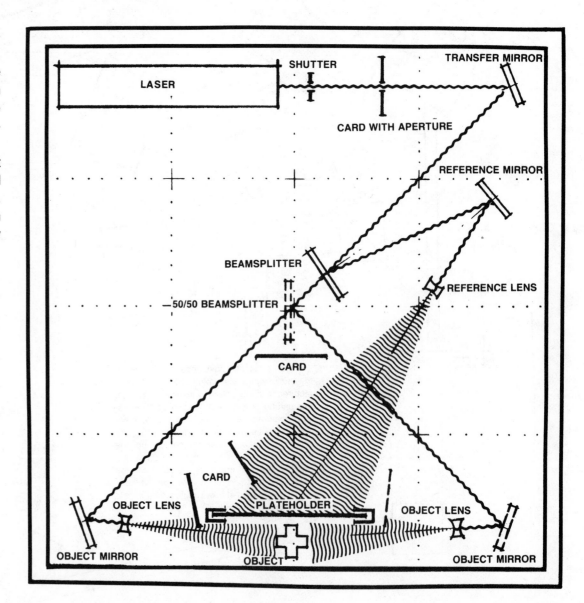

the essentials:

- **stability**
- **distances:** equal distances from first beamsplitter
- **light ratios:** 2:1 (ref:obj).
- **exposure:** 8E75NAH only
- **development:** 70-90% dark by eye.

what's needed:

- laser (of course).
- 3 negative lenses.
- plateholder.
- one or two beamsplitters (90:10 or variable beamsplitter; 50:50).
- object (bright & solid).
- 4''x 5'' card.
- black felt and/or stick for objects (optional).
- protractor.
- several black cardboard scatter cards.
- light meter.
- shutter.
- reflection holographic chemicals.
- Agfa Gevaert 8E75 NAH (unbacked!) plates.

exposure chart:

exposure charts for Science & Mechanics meter:

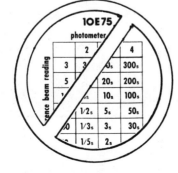

8E75 nah

photometer scale

reference beam reading	2	3	4
3	6 s	60 s	600 s
5	4 s	40 s	400 s
10	2 s	20 s	200 s
20	1 s	10 s	100 s
30	2/3 s	6 s	60 s
50	1/2 s	4 s	40 s

IOE75

photometer

reference beam reading	2		4
3	3 s	0 s	300 s
5		20 s	200 s
		10 s	100 s
	1/2 s	5 s	50 s
0	1/3 s	3 s	30 s
	1/5 s	2 s	

Time in seconds.

the set-up:

1. place laser
2. position transfer mirror
3. position plate holder
4. position object
5. place object mirror
6. place beamsplitter
7. position reference mirror
8. position reference lens
9. position object lens
10. card off unwanted light
11. measure light ratios
12. load plate
13. allow to settle (mix bleach now!)
14. calculate exposure time
15. shoot

FOR SECOND OBJECT BEAM:

6a. place 50:50 beamsplitter
6b. place object mirror #2
6c. measure distances
6d. place object lens #2

processing:

1. develop . . . . 70 - 90% dark
2. stop . . . . 1 minute
3. rinse . . . . 1 - 2 minutes
4. bleach* . . . . P.B.Q.
5. rinse . . . . 5 min.
6. photo-flo . . . . 30 sec.
7. squeegee
8. dry

*See page 173 for alternate processing.

Split beam reflection

1. laser
2. transfer mirror
3. plateholder
4. object
5. object mirror

Position the beam so that it travels diagonally across the table (as with the transmission setup). Set the plateholder about 6 inches from the edge opposite the laser. Place the object on this side of the plateholder as close to the window opening as possible (see illus.). Go around to the laser side of the table to view the object through the plateholder. Draw lines in the sand as if to extend the plateholder out on both sides. Now place a mirror (first object mirror) on the line so that it intercepts the beam. Placing the mirror in the same plane as the plateholder with this kind of setup will yield the best object illumination with the least possibility of light scatter. Rotate the mirror to direct a dot of light onto the object.

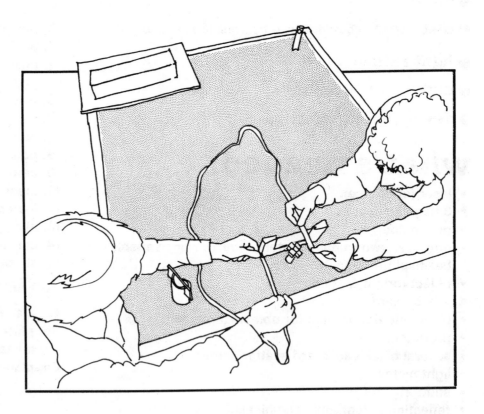

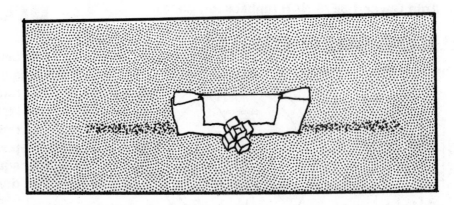

6. beamsplitter

Place the beamsplitter (90:10 or variable) near the center of the table intercepting the beam. Measure the distance from the beamsplitter to the first object mirror to the object to the center of the plateholder. Record this total object distance.

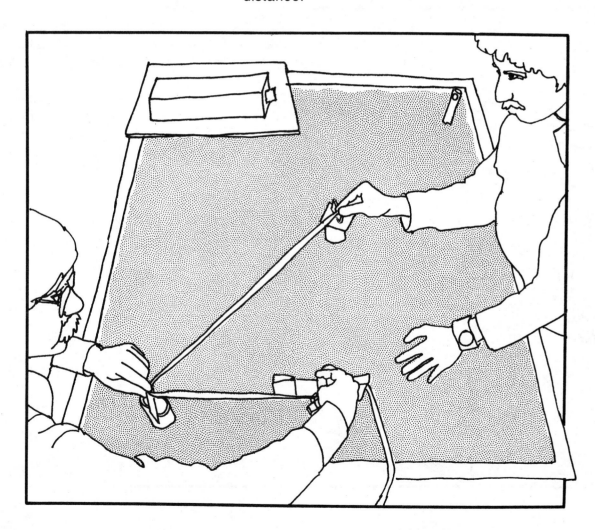

7. reference mirror

Now place a reference beam mirror in the set-up in the same manner used for splitbeam transmissions. For a reflection hologram, however, the reference beam should hit the plate at an angle of about 60 to 80 degrees.

A line drawn in the sand out from the plateholder at this angle may be beneficial

Find the distance on the tape previously recorded for the object beam and, with a little help from a friend, stretch the tape out tight. Swing it back and forth in the sand (an arc will be formed). Place the reference mirror at the point at which the arc intercepts the line from the plateholder. Measure the distance from the beamsplitter to the reference mirror to the center of the plateholder, to insure that it still agrees with the object beam measurement.

Rotate the beamsplitter to direct the light to the center of the reference mirror and, in turn, direct the beam to the center of the plateholder. A card placed in the plateholder should now have a dot of light in its center and, on the opposite side, there should be a dot near the center of the object.

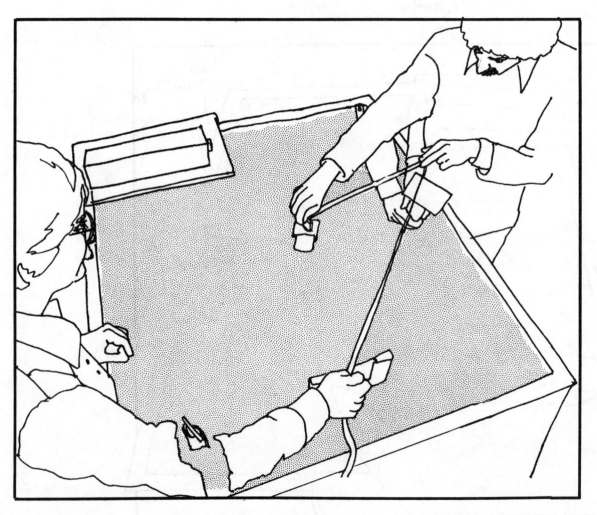

8. reference lens
9. object lens

To illuminate the object properly and provide a reference beam, place lenses into the system, as shown. Make sure the reference beam lens is placed so as to illuminate the white card evenly in the plateholder. Likewise, the object beam lens should evenly illuminate the object *as viewed through the plateholder from the laser side of the table.*

10. carding off

Scatter cards are essential to split beam setups. It is difficult to illuminate an object without some of this light also striking the plate directly. With the white card in the plateholder, so that you may see what is happening, place a scatter card at the edge of the plateholder. Allow it to protrude back only far enough to place the plateholder card in shadow (without cutting off light from the object). With some objects, it is almost impossible to do this without either moving the piece further away from the plate, or cutting off light from the very front. A compromise is usually best- remember, reflection holograms require objects to be as close to the plate as possible (because of the limited depth obtainable), without loss of uniform lighting. If it is too close, (you will find it impossible to light without also lighting the plate. Fortunately, in these cases, the very front of the object will probably be well illuminated by the reference beam (as with single beam reflection holog-

rams). The effect of this light will often be different from that of the regular object light, but with many objects it is not really noticeable.

If the object should have to be moved very slightly, take into consideration the possible effect on the equality of the beam paths. To be certain, check the distances again at this time.

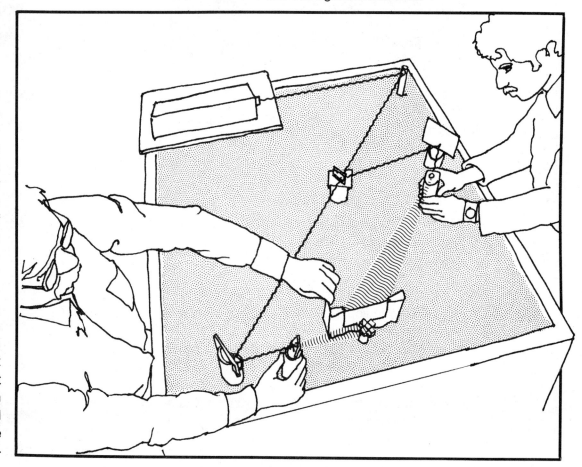

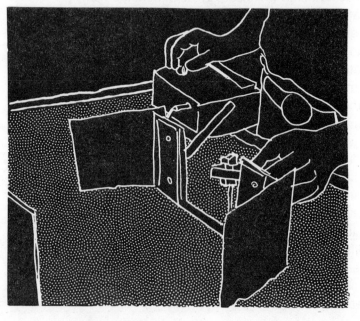

11. light ratios

The light readings will be read from opposing sides of the plateholder. To achieve a ratio of 2:1, point the probe towards the reference beam, making sure it remains in the plane of the plateholder, and record an average reading. This reading is a bit difficult to obtain because the probe must be moved around the object. Now, from the other side of the window, point the probe towards the object and record an average reading as the probe "sees" the object from a number of points of view. Adjust the lenses (or variable beamsplitter) until the proper ratio is achieved.

12. shutter

13. load plate

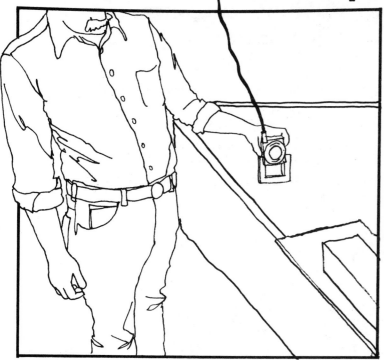

Check the set-up to make sure the object is properly lit, even reference illumination is intact, scatter cards are in place, and optics are secure with sand piled around them. Set the shutter, turn off the lights, and load an 8E75 NAH plate (only NAH plates!) into the holder. Let everything settle for 10-15 minutes while you determine the proper exposure by consulting your chart. Cross your fingers, hold your breath, and then open the shutter (if you have any fingers left) for the required amount of time.

Process the hologram in the same manner as the single beam reflection hologram. When dry, view it in the sun or a bright point source of light (slide projector light, etc.).

14. settle

15. check exposure time

16. shoot

17. process as per reflection instructions (p. 171)

An optional second object beam can be introduced into the system if you have become familiar with the use of one object beam.

a. 50:50 beam splitter
b. object mirror # 2

Place a 50:50 beamsplitter in the beam path a short distance beyond the original beamsplitter. Draw a line in the sand extending the plateholder out to the right (this sort of thing should be old hat by now). Place the second object beam mirror at the proper point on this line so that the new beam path distance equals the other two.

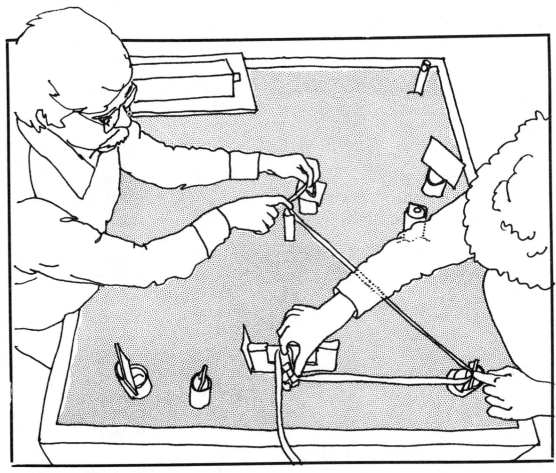

c measure distance
d. object lens # 2

This path is measured from the first beamsplitter to the 50:50 beamsplitter to the new object mirror to the object to the plate. Once the optics are properly arranged, direct the beam along this path, and center the red spot on the object. Place a lens in the system to illuminate the object from this side, balancing the intensity with the other object beam, to yield an even illumination across the whole object. Place another scatter card by the right side of the plateholder, and position it carefully to prevent unwanted light from the second object beam from reaching the plate (exactly as the first object beam was carded). Again, care must be taken not to block light from reaching the object, nor to block the reference beam.

Light readings may now be taken and the exposure made.

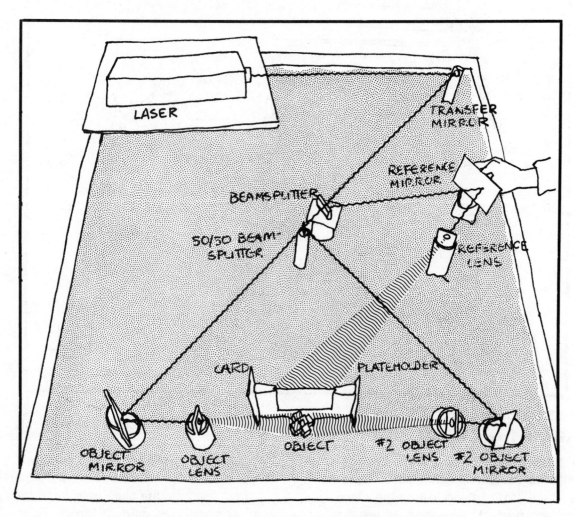

completed split beam reflection

troubleshooting

By this time, familiarity with some of the common problems associated with both transmission and reflection holograms should be understood. Some of the problems which may be encountered with split beam reflection holograms are now quickly summarized:

Problem	Cause	Solution
Object dim or not visable.	1) Object or optics in motion.	Resecure all components.
	2) Unequal beam path lengths.	Measure distances and equalize.
	3) Improper exposure or development.	Develop to 70% density.
	4) Improper light ratio.	Ratio 2:1 (reference:object).
	5) One beam inadvertantly left blocked during exposure.	Check all light before shutter set in place.
	6) Plate fogged by light leak.	Get new plates & fix darkroom.
Plate "mottled" in apperance.	Chemical contaminaton, poor photo-flo or squeegee work.	New chemicals. Remember to photo-flo.
Flashes of color visable. Patterns obscure image.	1) Unwanted light scatter reaching plate.	Block off all undesired light.
	2) Object too reflective (mirror-like surface).	Use object with matt finish.
Image not "clean" looking.	Optics probably not too clean.	Clean or replace. Use spatial filter for reference beam.

image plane holograms

making a hologram from a hologram

One of the ways of demonstrating a completed transmission hologram, you may remember, involved viewing the projected, or real, image. By flipping the plate around and allowing light to travel through it, the image actually focused in front of the hologram.

Of course, this real image is three dimensional, just as was the original object, and may be said to occupy a certain space. By moving a piece of translucent material back and forth through the image, different areas come into focus and then "fuzz out". If we hold the screen still for a moment, we can imagine it bisecting the image at one plane. This plane is visible as the area in focus, with part of the image in front of and the remainder behind the screen.

By replacing the screen with a light-sensitive plate, and utilizing a new reference beam, it's possible to make a hologram at this plane of the image. The resultant holographic image will appear to form with the glass plate cutting it in half. Part will be seen behind and the rest will project out from the plate.

This hologram of a holographic image, or image plane hologram, can very often display more interesting characteristics than the original master hologram from which it is made. It becomes possible to take a regular transmission hologram and generate reflection copies viewable in white light. These holograms are usually much brighter than ordinary reflection holograms, with the added option of having part of the image magically floating out in front. This also allows the making of holograms of objects with greater depth than those of the ordinary reflection type. Of course, the same idea can be used to make transmission image plane holograms.

Most advanced hybrid holograms utilize the image plane concept. This includes white light transmission varieties such as rainbow holograms, full or open aperture holograms, and white light stereograms displaying moving images of live subjects. These and more are discussed later in the book.

THE MASTER
(usually referred to as H-1)

To make this hologram, a transmission hologram master must first be made. A second setup is then constructed, using the master to make the reflection image plane.

The master is simply a transmission hologram with the following special qualities:

a. Object is small relative to the plate size (no more than 2"x 2" for a 4"x 5" copy plate).

b. Object is positioned in the center of the plate, or slightly above.

c. The object has been placed 12-15 cm away from the plate.

d. The reference beam approaches the plate at a thirty degree angle.

e. The image is very bright and clean, and not excessively grainy (due to careful exposure and processing).

You may wish to review the section on split beam transmission holograms for the finer details on this type before making the master. We will run through the steps here briefly, incorporating the above special conditions, but we are assuming some familiarity in setting up by this time.

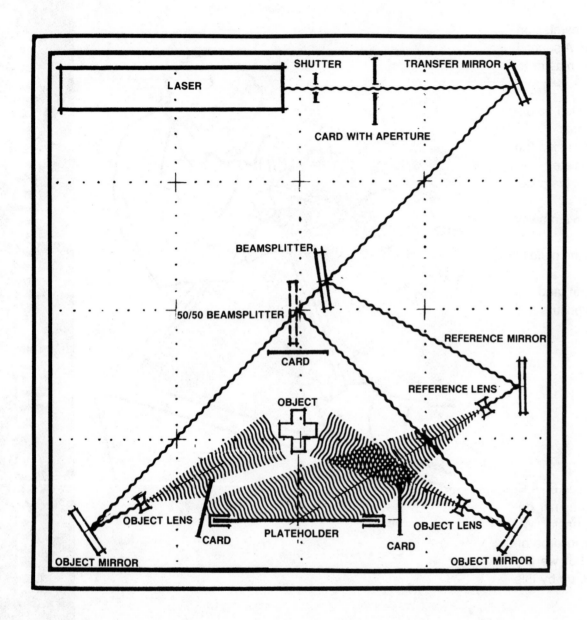

1) Laser placed on table as shown.

2) Plateholder as shown, at least 6 inches from table edge.

3) Object placement: 12-15 cm in front of plate. When looking through plateholder, the object should appear in the center, or slightly above the center of window. If not, mound up sand, or glue object to stick and place properly.

4) Transfer mirror to send beam across table.

5) Object mirror placed as shown to direct beam to object.

6) Beamsplitter in center of table intersecting beam.

7) Measure object beam distance (beamsplitter to object mirror to object to plate).

8) Reference mirror placement. A line from the reference mirror should intersect the plate at a 30 degree angle, and the mirror should be placed so that the distance from the beamsplitter to mirror to plate equals the object beam distance.

Note: There is only one spot on the table where the mirror can be placed to meet both these conditions.

9) Lens placement - place so as to obtain good object illumination and a good reference beam coverage at plateholder.

10) You will probably wish to add another object beam as shown (beamsplitter 1 to beamsplitter 2 to object mirror 2 to object to plate).

11) All unwanted light should be carded off.

12) Light readings adjusted for 4:1 (reference:object).

13) Expose and develop.

Additional explanations:

The position of the object and the angle of reference beam are very important in master making. The object must be far enough away from the plate so that when the finished master is illuminated, the illumination beam can not intercept the copy plate where the image reconstructs (see diagrams). To insure this condition, the object is best placed 12-15 cm away from the plate. Under this condition, a reference beam striking the plate at a thirty degree angle will translate into an illumination beam that will clear the copy plate and reconstructed image.

It's possible to check for correct object placement in a "ready" setup. Place scatter cards on either side of the plateholder. Next, plant a second plateholder (with white card inserted) in the sand directly across from the object, on the opposite side of the original plateholder. Move the second plateholder away from the first, out of the path of the reference beam light. No light should reach the card in the plateholder. Measure the distance between the parallel holders.

The object should be the same distance away from the first plateholder (on the other side). This quick check will insure that the image will eventually reconstruct in the proper plane for copying.

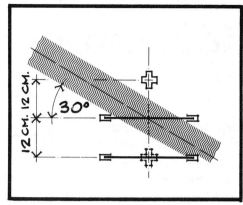

OBJECT PLACEMENT FOR MASTER MAKING

A NOTE ON PROCESSING:

Proper exposure and development are crucial to good master making. Low noise levels are important, so we recommend guarding against over-development. When using D-19 processing sequences, the hologram should not generally be bleached, unless it is considerably underexposed or underdeveloped. The best masters are those which have least density (maximum transmissiveness) before losing their image brightness while maintaining low noise.

ANOTHER NOTE ON QUALITY:

Good quality masters also result from good optics. Make sure yours are clean. In addition, although these setups will work with simple inexpensive optics, the use of a spatial filter on the reference beam will make a dramatic difference in the results. The making of the master with better optical components is discussed in the advanced section.

reflection image plane

(usually referred to as H-2)

This reflection image plane (IP) setup is fairly simple. The beamsplitter is positioned near the center of the table. The reference beam is deflected toward the copy plate at a 70 degree to 80 degree angle. The master (which replaces the object of previous setups) is positioned parallel to the copy hologram.

The master illumination beam (object beam) is deflected toward the master at the angle which causes it to produce the brightest image. The master and copy holders are separated by the proper amount to cause the image to be located in the desired plane. A reflection hologram is produced since the reference light approaches the copy plate from the side opposite that of the master image. The first setup outlined will produce a hologram which will reconstruct with side illumination. An alternate method, used to obtain a hologram viewed with overhead or underneath lighting, is then described.

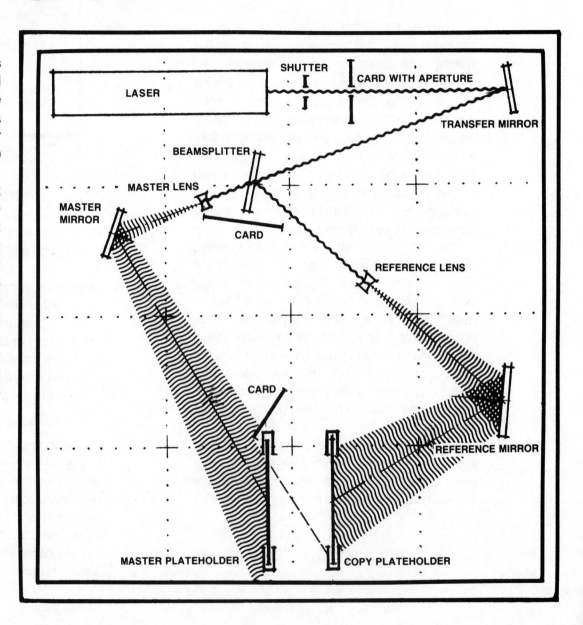

the essentials:

- **stability**
- **distances:** equal distances from first beamsplitter
- **light ratios:** 1 ½ :1 (ref:obj).
- **exposure:** 8E75NAH only
- **development:** 70-90% dark by eye.

what's needed:

- laser (of course).
- transfer mirror.
- master.
- master plateholder.
- beamsplitter.
- reference mirror.
- copy plateholder & white card.
- lenses
- several black scatter cards.
- shutter.
- Agfa Gevaert 8E75 NAH or HD plates.
- chemistry for reflection processing.

exposure chart:

exposure charts for Science & Mechanics meter:

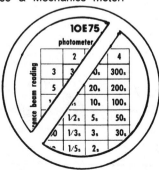

8E75 nah

photometer scale

reference beam reading	2	3	4
3	6.	60.	600.
5	4.	40.	400.
10	2.	20.	200.
20	1.	10.	100.
30	2/3.	6.	60.
50	1/2.	4.	40.

10E75

photometer

reference beam reading	2		4
3	3	0.	300.
5		20.	200.
		10.	100.
	1/2.	5.	50.
0	1/3.	3.	30.
	1/5.	2.	

the set-up:

1. place laser
2. position master plateholder
3. position master illumination mirror
4. setup transfer mirror
5. position master illumination beam lens
6. reverse master plate
7. find image plane
8. set up copy plate
9. place variable beamsplitter
10. measure distances
11. place reference mirror
12. position reference beam lens
13. card light off
14. measure light ratios
15. load plate
16. calculate exposure time
17. allow to settle (mix bleach now!)
18. expose

processing:

1. develop . . . . 70 - 80% dark
2. stop . . . . .1 minute
3. rinse . . . . .1 - 2 minutes
4. bleach* . . . . .P.B.Q.
5. rinse . . . . .5 min.
6. photo-flo . . . . .30 sec.
7. squeegee
8. dry

*See page 173 for alternate processing.

reflection image plane

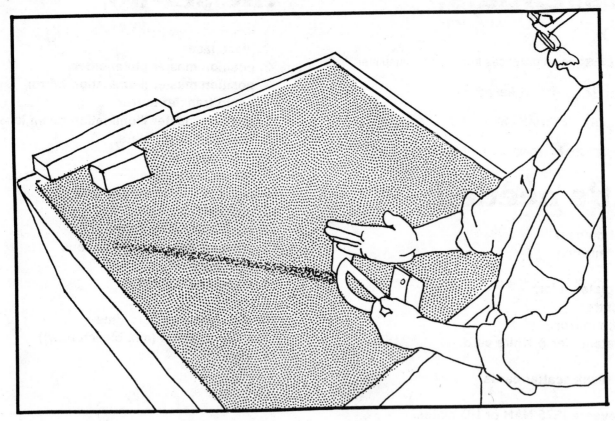

1. laser   2. master plate holder

The master plateholder should be placed near the center, and perpendicular to the edge, of the table opposite the laser. Using a protractor, scribe a line from the center of the plateholder, out into the sand, at an angle of 30 degrees.

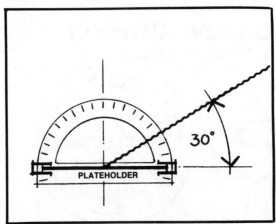

3. master illumination beam mirror

Set the master illumination mirror at a point
on this line, as shown.

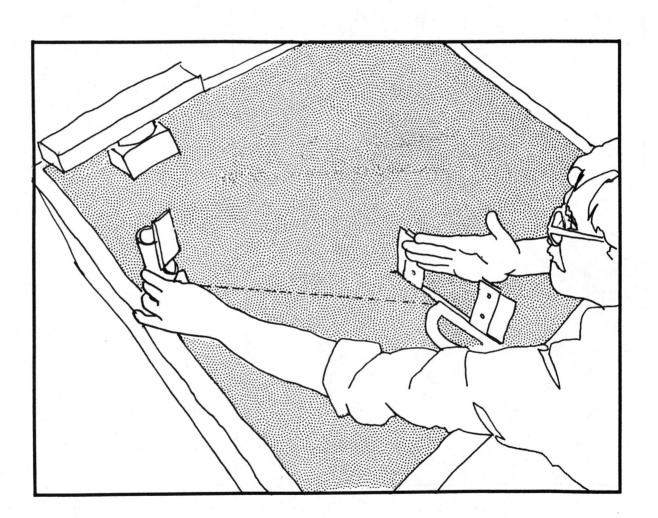

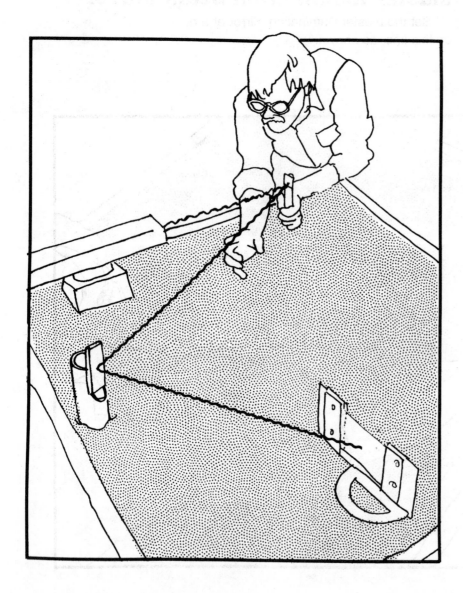

4. transfer mirrors

Using the transfer mirror, direct light from the laser to the master illumination mirror. In turn, twist the master mirror to direct a dot of light in the center of a card placed in the master plateholder. Using the protractor, you may wish to check to see that this beam mades a 30 degree angle with the plate.

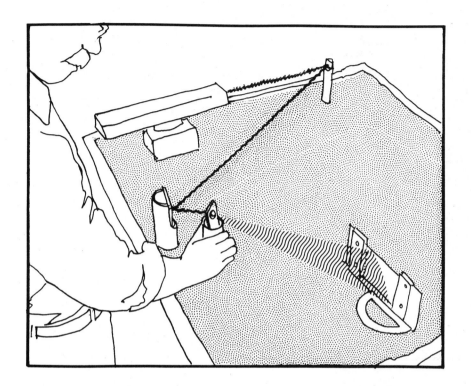

5. master illumination beam lens

Place a lens in the beam to spread light out onto the card.

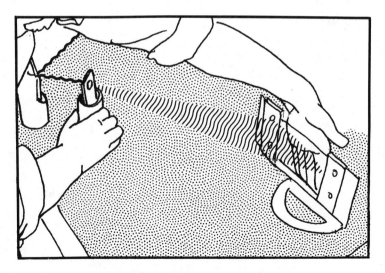

6. proper master placement

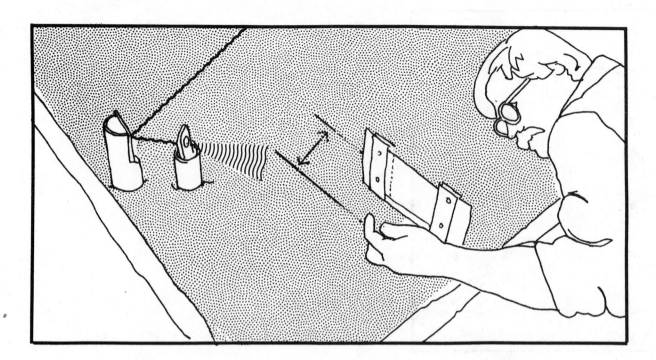

Place the master in the holder so that as you look through it, the image appears behind it (virtual image). You may wish to check and see that the image is playing back at its brightest by removing the plate and playing with the angle it makes with the light. Compare other angles to the one made with the plate in the holder and adjust the holder if necessary.

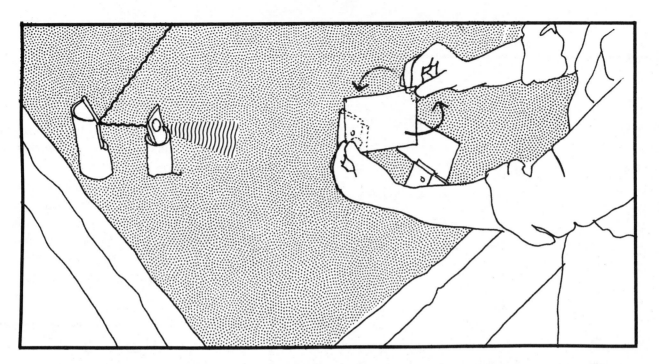

Looking through the plate, you can actually try to reach out and touch the image. Choose a point on the image you wish the copy to "bisect," and mark its position temporarily with a card in the sand. Measure the distance between the card and the plate. Now move the card to the same distance on the *opposite* side of the plateholder. Turn the master around 180 degrees so that the image now projects in front, onto the card.

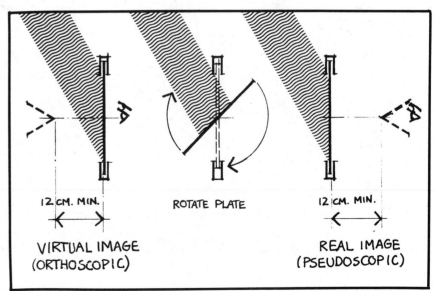

12 CM. MIN.   ROTATE PLATE   12 CM. MIN.

VIRTUAL IMAGE
(ORTHOSCOPIC)   REAL IMAGE
                                   (PSEUDOSCOPIC)

8. copy plate placement

Replace the card with the copy holder. This holder must be centered with respect to the master, and parallel to it.

You can test the proper placement of the copy holder using a white card. First place a scatter card on the side of the master holder towards the laser, to block any stray light that may be going by the master. You should observe a fuzzy image on the white card.

Another technique can be used. Temporarily remove the lens from the beam so only a dot of light hits the master. A crisp two dimensional image of your object should be visible on the card.

Either of these previsualization methods will assure the holographer that the image is well centered (if not, the copyholder must be repositioned). Also, by moving the card closer or further away from the master, various parts of the image will come into focus (yet another method to determine which image plane to copy).

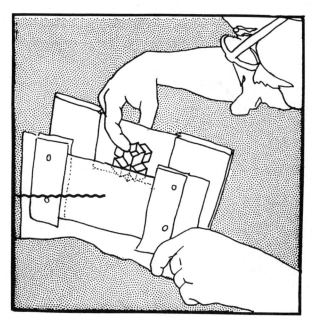

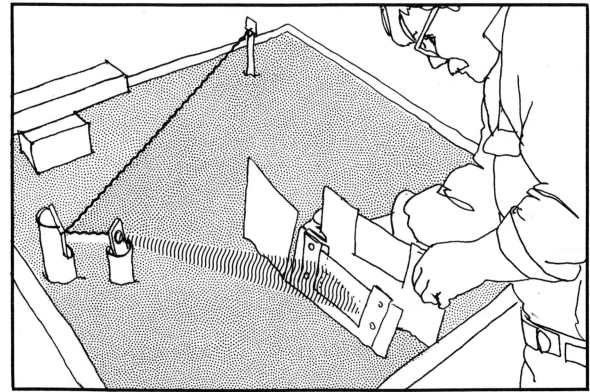

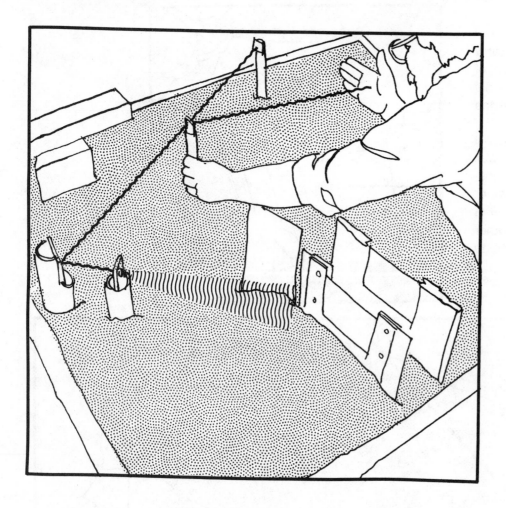

9. beamsplitter

Place a beamsplitter between the transfer mirror and master illumination mirror. It should be situated almost in line with the copy holder.

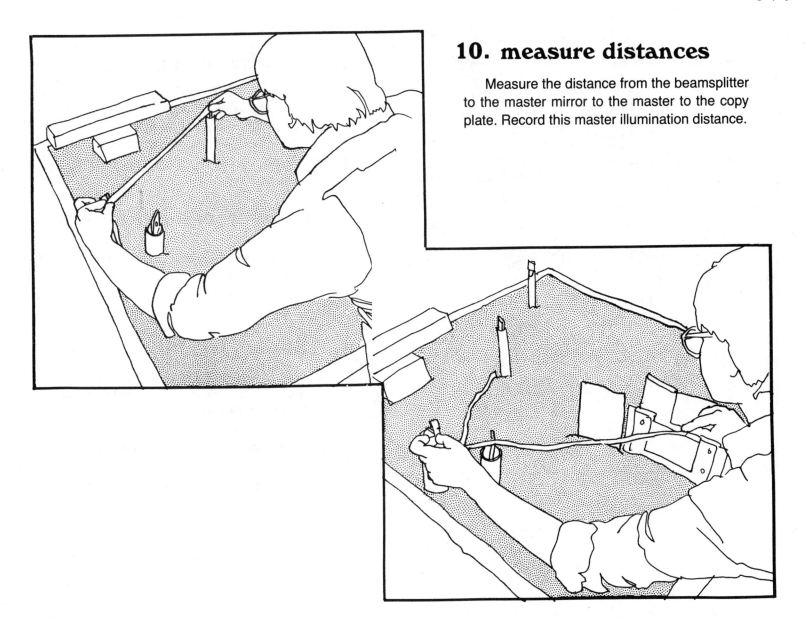

10. measure distances

Measure the distance from the beamsplitter to the master mirror to the master to the copy plate. Record this master illumination distance.

11. reference mirror

Draw a line from the copy plateholder out to the right at a 70-80 degree angle. Plant the reference mirror at a point on this line so that the distance from the beamsplitter to that point to the copy plate equals the master illumination distance.

12. reference beam lens

Direct the beam, now reflecting from the beamsplitter, to the copy holder. A lens in this beam will spread the reference light evenly onto a white card in the copyholder.

13. carding off

Use cards to block stray light (especially useful is one placed alongside the master holder, on the side the illumination light arrives from, to keep the beam from striking the copy directly. In addition, the placement of some black felt material on the sand between and on either side of the two plateholders will help hold down scatter.

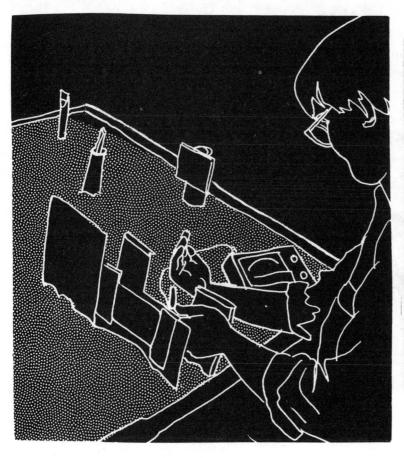

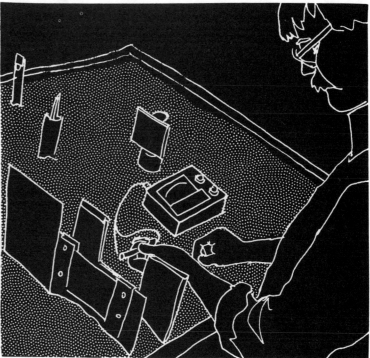

14. light ratios

The ratio desired for an image plane reflection hologram is approximately 3:2 (or 1½:1 reference:object). The reference reading is taken in the same manner as for previous holograms. Make sure the probe remains in the plane of the copy holder, pointed towards the reference mirror, while you move it about, recording an average reading.

While recording the object reading, keep the probe in the plane of the copy holder (directed towards the master). Be careful, an image may concentrate light in only a small area of the image plane, leaving little or no light to be detected elsewhere in the copy holder opening. Locate the image in the space, and take into account these radical areas of light and dark when obtaining the light readings.

The brightest spots should register no more than 1:1 (image/object:reference beam), with 1½:1 as an average. To compensate for too much or too little light, move the reference beam lens (or if using variable beamsplitter, adjust it instead).

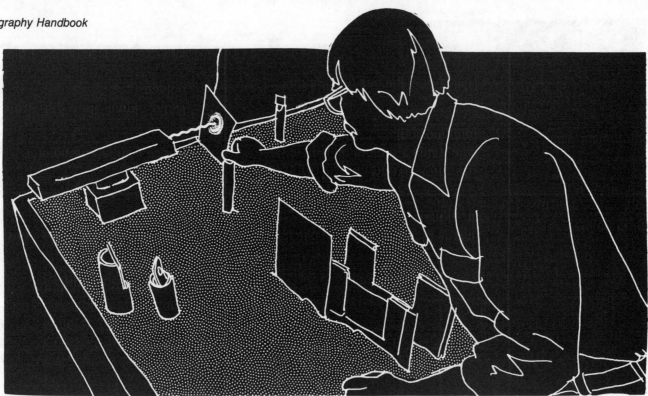

Satisfied with your reading? Optics securely placed? Distances correct? Reference *and* object illumination reaching the copy holder? Stray light blocked off? If so, it is time to tap things down, ready the shutter, turn off the lights and load an 8E75 NAH or HD plate into the copy holder (NAH is the only choice for the 8E75 plates).

Expose according to the chart and develop the hologram as you would a regular reflection hologram. This hologram will be viewable with a point source of white light.

15. shutter

16. load plate

17. check exposure time

18. settle

19. exposure

20. process as per reflection instructions (p. 171)

setup modification

You may modify the existing setup as follows or you may start from scratch and set up as shown on page 234.

Up to this point, all of the setups have utilized side reference and illumination beams, which facilitate building a simple setup. However, the final hologram must be viewed with side lighting. In some instances, it may be desirable to display the hologram with lighting from above or below. Surprisingly, the changes in the setup are simple. We offer the following procedure, which will yield a reflection hologram lit from above, using the optics already positioned in the sand. 3 Tilt the plateholders back about 15 - 30 degrees away from the reference beam. The plates must remain parallel, and the image from the master must be centered in the copy plateholder opening. So, mound up sand to support a somewhat elevated copy holder.

The reference beam can now approach the plates straight on. The reference mirror must be positioned so the reference distance remains the same (beamsplitter to reference mirror to master plate to copy plate). This mirror also transfers light from the beamsplitter to the plate at an angle of 90 degrees) (due to the tilt of the plate, the light also approaches it at an angle of 70 - 80 degrees.)

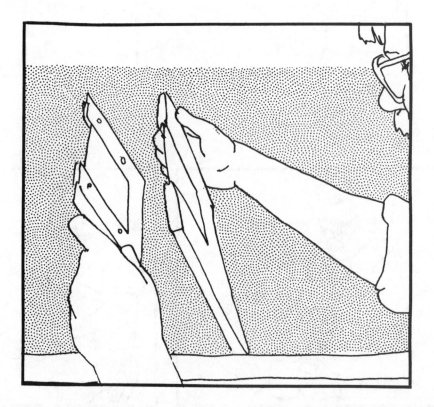

Replace the lens at some point in the reference beam path to evenly illuminate a white card in the copy holder. Place a scatter card as shown.

To obtain maximum brightness, it may be necessary to reposition the master illumination mirror. Push it deeper into the sand, and tilt it back to the same angle as the plateholders. For the beam to reach this mirror now, the transfer mirror and beamsplitter must be adjusted.

Of course this is all much easier to build from scratch:

1. Tilt master and copy plate back.

2. Tilt master mirror back and push lower into the sand.

3. Direct beam from transfer mirror to master mirror to master. The hologram should play back at its brightest.

4. Set reference mirror in line with both holders. Equalize the reference distance to the object distance.

If underneath lighting of the final image plane reflection copy is desired, simply tilt the plates in the opposite direction (toward the reference beam), and adjust optics accordingly.

trouble shooting

Problem	Cause	Solution
Image fuzzy. Object too far in front of plate.	Image placement. Copy plate too far from master.	Reposition copy plate and check distances.
Image dim even though master was bright.	Stability (remember reflection holograms are more difficult to keep from moving). Ratios or distances may be off.	Check with interferometer. Ratio must be 3:2 (reference: object). Remeasure distances carefully.
Only part of image fits on plate.	Object too large for image planing into size used.	Replace with master of smaller object or use larger copy plate.
Large color flashes or dark bar on part of plate.	Master illumination light hitting copy plate directly.	Card off this excess light.
Image plane correct, yet image has poor resolution, is fuzzy or dim at points away from image plane.	Reference beam for copy at too oblique of an angle. Incorrect ratios/exposure/processing. Used wrong type of plate (Use 8E75). Distances off.	Correct accordingly.

advanced
holography

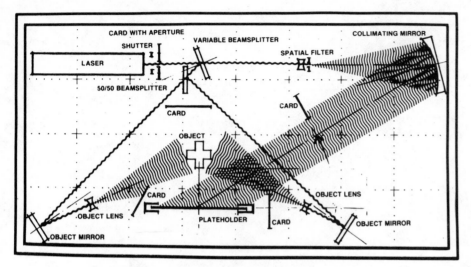

Split beam transmission master.

Birds Eye View of Advanced Setups

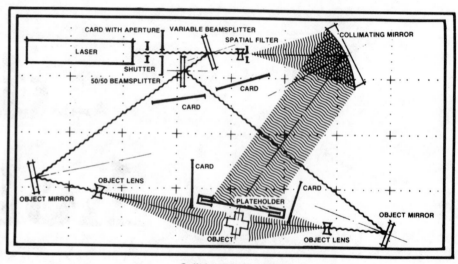

Split beam reflection.

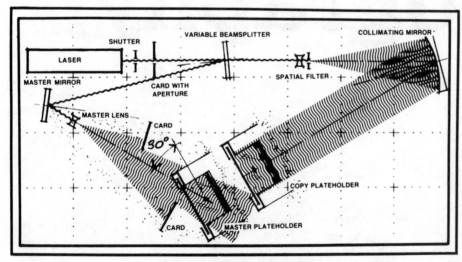

Split beam reflection image plane.

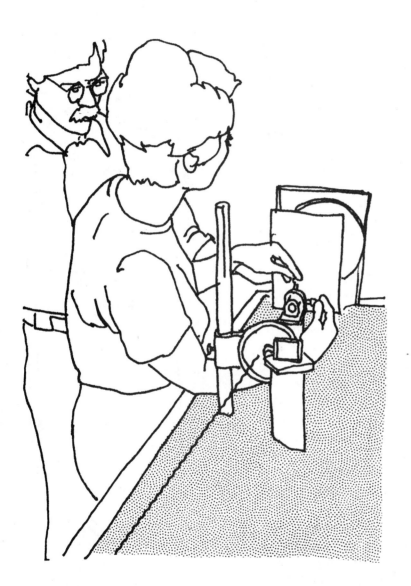

overview

In this section, we'll discuss the making of better quality holograms by utilizing three important new optical components, the spatial filter, variable beamsplitter and collimating mirror.

We begin by making a large (8"x 10") transmission master, an 8"x 10" reflection hologram, and a reflection image plane from the large master.

We then cover white light transmission holograms, including rainbow and achromatic-open aperture varieties. The making of curved as well as integral holograms follows. Additional advanced techniques such as double exposures, matte screening, and double channels, will be discussed in the art section.

Although we begin with the use of large plates, we recommend that the reader gain familiarity with the new optics by using 4"x 5" plates first. Some beginning holographers, who acquired these optics at the start, may even wish to do their first split beam holograms from notes in this section. If you do begin here, the earlier setup descriptions should be carefully reviewed.

advanced equipment needed

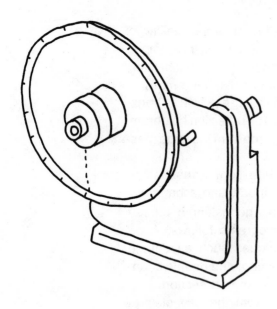

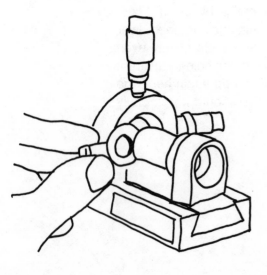

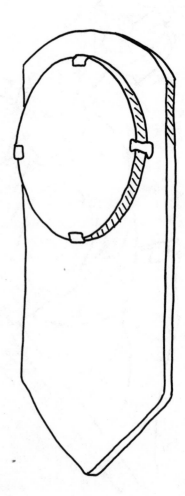

All of the setups described up to this point have utilized only simple, inexpensive optics, arranged on a relatively small (4'x 4') table. This is fine for getting a sense of the basics, yet is somewhat inadequate for serious work. It is true that good quality holograms are largely a function of technique; they also result from the use of the proper equipment. In addition, more room is needed for placement of optics to make larger or more advanced hybrid holograms. Even simple holograms become easier to do when there is more room to spread out. For these reasons, the descriptions that follow assume the use of a table at least 4'x 8', as well as the use of a variable beamsplitter, spatial filter, and collimating mirror. (Note: if you had decided earlier to build a 4'x 4' table and now wish to do advanced work, simply build another one alongside it and bolt or clamp the two at the center. The common edge of the two tables will probably only occasionally be inconvenient.).

It might be wise at this time to review the basic descriptions of the variable beamsplitter, spatial filter and collimating mirror in the equipment section. In the setups that follow, we used a Jodon LPSF-100 spatial filter, Jodon VBA-200 variable beamsplitter, and Coulter Optics 10'' f5.6 parabolic mirror for collimation. You, of course, may use other varieties; as long as they're in good condition - that's what counts. If your collimation mirror has a different focal length, some of your setups may vary slightly from ours.

spatial filters

The spatial filter is used to obtain a very clean and even reference beam over the whole holographic plate. This component can take a little getting used to, but once you get the hang of it, you'll notice the most dramatic improvement in the quality of your holograms. You may wish to practice aligning the beam through it first, before actually using it in a setup.

The spatial filter consists of a microscope objective and a precision pinhole. The beam must go through the objective, which focuses it to a tiny point at its precise focal length. The pinhole must be located exactly at this point. This is done by manipulation of the pinhole over an x - y axis, using precise micrometer heads or ultra-fine threaded screws, along with movement of the objective carefully over a z axis to bring it slowly to its focal point.

The following is a description of how this is done using the Jodon model; (the Noll spatial filter is very similar as are a number of others.) If you have any trouble adjusting other types, check with the manufacturer for directions.

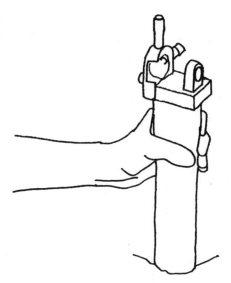

1. placing spatial filter

With the objective and pinhole removed, place the spatial filter in the sand so that it intercepts the beam. Move it around until the beam travels through both holes (where the objective and pinhole will go).

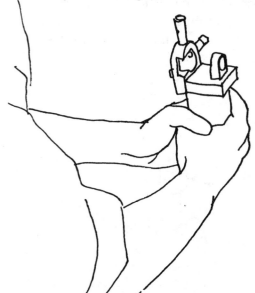

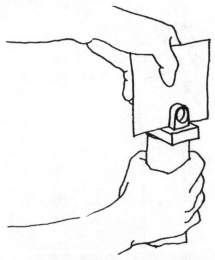

2. positioning the beam

Using a card, center the beam through *both* holes by carefully moving the spatial filter around, and noting the beam's path by looking through the hole at the card. You will be going back and forth from one hole to the other, since as soon as you think the back one is aligned, the front will be off, or vice versa. Have patience, and they'll both line up.

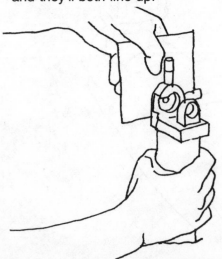

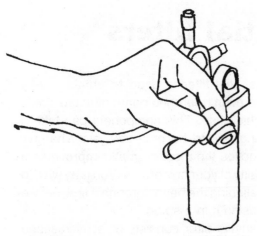

3. insert objective

Screw the objective into the rear opening. Make sure that the two openings are far enough apart so that the objective can fit easily between them. To insure trouble-free alignment, you will also want the objective to be as far back as possible when you start the focusing process with the pinhole. Adjust with the Z axis screw (located in the front of the Jodon and at the rear of the Noll model) if necessary.

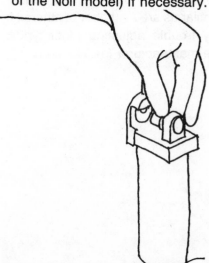

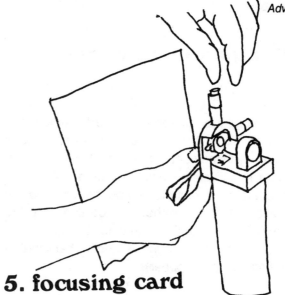

4. placing pinhole

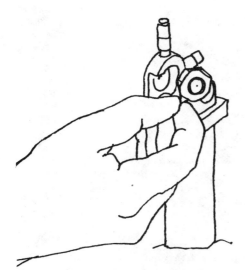

Now place the pinhole in the front opening (the Jodon design holds the pinhole mount magnetically, the Noll mount screws in.) You will notice that the objective focuses a spot on the pinhole.

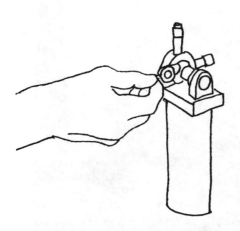

5. focusing card

Place a white card in the sand a short distance in front of the spatial filter. This card will be used to observe the progress of focusing. Again, make sure the objective is located about as far from the pinhole as it can go.

6. adjusting pinhole

Turn down any room lights.

Using the micrometer heads, adjust the pinhole so that the focused beam spot hits it squarely in the center. Looking at the card, continue to adjust the pinhole until a dot of light appears. The pinhole is now centered to the point that some of the focused light is going through. If you can't find that dot, keep at it - you will need it; (try moving the objective back further if it still doesn't appear).

Now slowly work the Z axis screw to bring the objective closer to the pinhole, while watching the spot on the card. As you do this, you will also have to recenter the pinhole with the micrometer screws. Work all three screws together as part of an ongoing process. If you lose the spot, back up the Z axis until it is visible, and try again.

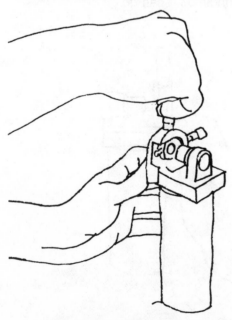

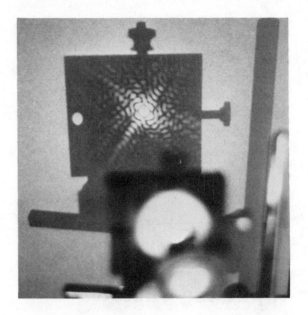

1) Light is first visible through the pinhole.

2) As the objective moves closer, the spot on the card will get brighter. You will notice that it is surrounded by a series of light and dark concentric rings.

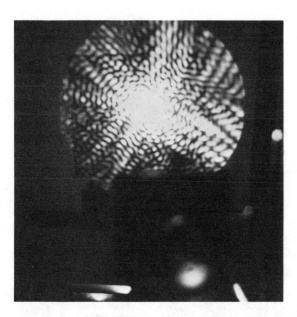

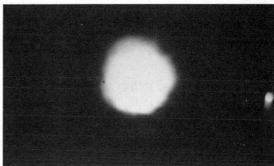

3) These also become brighter as the objective moves in. The closer it gets to the focal point, the more delicate the adjustments become to keep the light centered and visible.

4) Finally, you will reach a point where the rings will start to merge into a single blob of light.

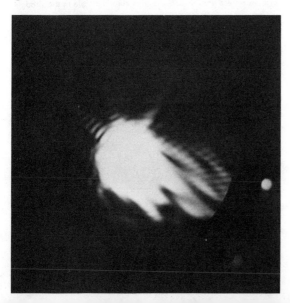

5) With careful adjustment, the blob should become a clean disc devoid of rings. If you go too far, the rings will reappear and the disc will become a blob again.

If you never get that nice clean disc, then you probably have an improper matching of objective power to pinhole size. Check the chart in the equipment section.

In actual practical use, you may never get a *perfect* circle of clean spatially filtered light, due to such things as the difficulty of exactly matching the laser beam's diameter at the point the spatial filter is located, or to the effects of double reflections from a beamsplitter placed before the spatial filter, or to a pinhole not being perfectly round, or to a dirty objective. The light should be fairly clean, though, since that's why we are going through this whole exercise; we are assuming you've just shelled out a couple hundred bucks for this thing. Make sure you get your money's worth!

Spend some time practicing with the spatial filter until it comes easily. Incidentally, you should begin to realize, due to the sensitivity of its precise alignment, that the spatial filter is always the last component to be placed on the table in a given setup. Once it's adjusted properly, you don't want to blow it out of position.

variable beam splitter

As we mentioned in the equipment section, this component is essential for obtaining proper beam ratios without having constantly to redo setups by replacing lenses, etc. Advanced setups are almost impossible to complete without it.

The variable beamsplitter (attenuated disc variety) is simple to use. Best results are obtained when the disc intercepts the beam at an angle of about 45 degrees and the beam travels through the upper half of the disc at a point clear of any obstructions caused by the disc holder or mount. The ratios are adjusted by grasping the disc by the edge and simply rotating it, taking care not to change the angle at which it intercepts the beam. The reflected beam may be thrown off if this angle changes, possibly frustrating the results of a complicated setup. Before using it in an actual setup, you may wish to practice rotating the beamsplitter until you can do it smoothly without causing the reflected beam to shift.

The variable reflectivity mirror surface is clearly visible in the shadow.

collimating mirrors

This mirror is essential for the making of rainbow and focused white light holograms. It is optional for others. We like to use it for all, however.

The use of the collimating mirror is simple. However, it is probably the most massive component in your system, so it should be mounted securely, and care should be taken in moving it around: you don't want it to fall down on the table and get sandblasted.

The action of the collimating mirror depends on its focal length relative to the placement of a lens or spatial filter. A lens placed at the focal length will spread the beam out so that it hits the mirror. Light reflecting from the mirror will no longer spread noticeably, but will maintain a constant size no matter where it is directed.

If the lens is placed closer to the mirror than its focal length, the beam will diverge (although not as much as without the mirror). The rate of divergence can be controlled by the distance selected for lens placement.

If the lens is positioned at a distance greater than the focal length of the mirror, then light reflecting from the mirror will begin to converge. This is a valuable property utilized in making white light transmission holograms, particularly those which are achromatic. As with divergence, the rate of convergence can be controlled by the lens placement distance from the collimating mirror.

You may wish to experiment with a beam directed at this mirror and placement of a lens at various distances. If you happen to have a parabolic mirror and want to use it for collimation but don't know its focal length, you can find it easily enough. Simply move the lens in the beam path until you find the point where light keeps its size as it's reflected from the mirror. This can be checked by using a white card, and measuring the diameter of the reflected light at several distances from the mirror. The lens distance at which this occurs is, of course, the mirror's focal length.

Try placing the lens at other distances, and measure the reflected light at various places. In this way, you should begin to develop some feel for how the collimating mirror operates.

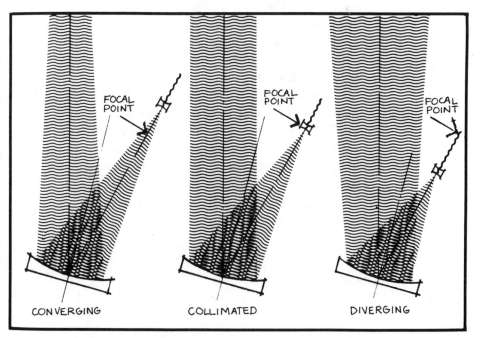

FOCAL POINT FOCAL POINT FOCAL POINT

CONVERGING COLLIMATED DIVERGING

advanced techniques

You may wish to note the following differences between these setups and those with only the simpler optics and/or smaller plates mentioned earlier.

1) The reference beam is always the beam going *through* the variable beamsplitter with the object (or master illumination in copy setups) beam reflected from the beamsplitter. With the use of the spatial filter on the reference beam, however, it is much easier to adjust ratios without disturbing the precise alignment necessary, if the position of the beam remains unchanged by going through the beamsplitter.

2) The use of an 8"x 10" plate usually requires a greater degree of care than a 4"x 5" plate. The larger plates are expensive, and the larger area is more open to the effects of stray light. For these reasons, it is recommended that all 8"x 10" shots be preceded by 4"x 5" (or smaller) test exposures to ascertain whether a good hologram is in the making. This can be done without disturbing the setup, by using a 4"x 5" holder which fits inside the 8"x 10" holder.

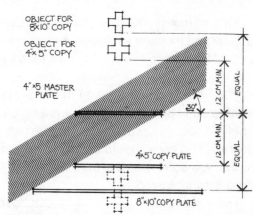

LOCATION OF OBJECT

OBJECT FOR 8X10" COPY

OBJECT FOR 4"x 5" COPY

4"x5 MASTER PLATE

4"x5" COPY PLATE

8"x10" COPY PLATE

12 CM. MIN.   EQUAL

12 CM. MIN.   EQUAL

30°

3) If you envision making a large (8"x 10") image plane copy, the object may have to be positioned farther from the plate during the shooting of the master than if only 4"x 5" copies are to be made. This is so that in the copy setup, the master illumination light will not strike the larger copy plate. This distance can be tested beforehand as described on page 215, by using two 8"x 10" holders instead of two 4"x 5" holders.

If you feel that only 4"x 5" plates will be made from the 8"x 10" master (usually preferred in the case of smaller objects), then do this test using an 8"x 10" holder with a card in the 4"x 5" holder. In this case, the holders will be closer together than in the case with two 8"x 10" ones and the final image plane hologram will have a slightly wider angle of view.

"Engine #9" refelection hologram by Dr. Stephen A. Benton, Polaroid Corp. Research Labs, 1975. Photo © Daniel Quat courtesy MOH collection MOH.

Nick Phillips, photo © MOH.

Professor Victor G. Komar with his holocinema screen, MK1, at the All-Union Cinema and Photographic Research Institute, USSR, 1978. Photo courtesy S.A. Benton and MOH.

Pulsed laser holographic portrait of Dennis Gabor by R. Rinehart, McDonnell Douglas Electronics, 1971, photo © S. Borns, 1975, collection MOH.

transmission master holograms

8"x 10" master for use in making reflection copies.

1. position laser
2. place object
3. set up plateholder
4. position reference mirror
5. position variable beamsplitter
6. measure distance
7. place object mirror
8. place spatial filter
9. position object lens
10. position 50:50 beamsplitter
11. place object mirror #2
12. place object lens #2
13. measure light ratios
14. load plate
15. calculate exposure time
16. allow to settle
17. shoot

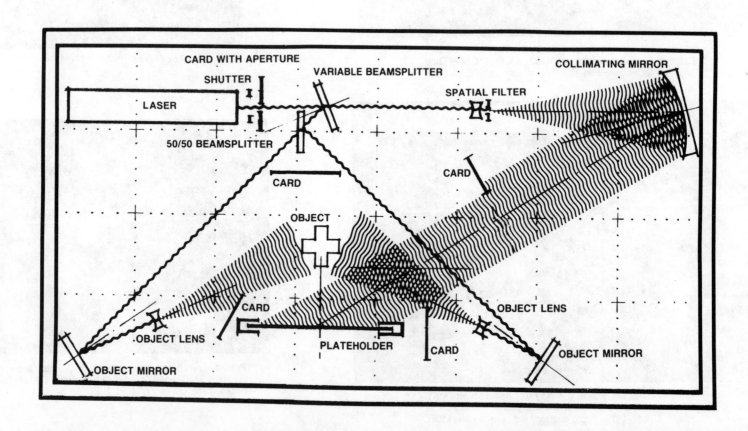

1. laser
2. plate holder

Place an 8"x 10" plateholder 30 cm in from the side of the table.

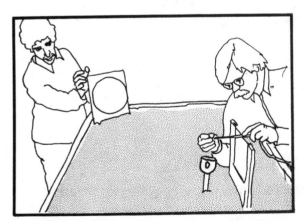

3. object

The object can then be positioned at least 12 cm from plateholder.

4. reference mirror

The ominous collimating reference mirror can now be used to transfer the beam to the plateholder at a 30 degree angle.

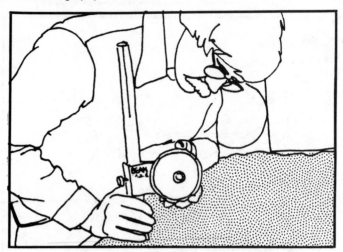

5. variable beam splitter

Intercept the beam with a variable beamsplitter so the beam travels through without obstruction.

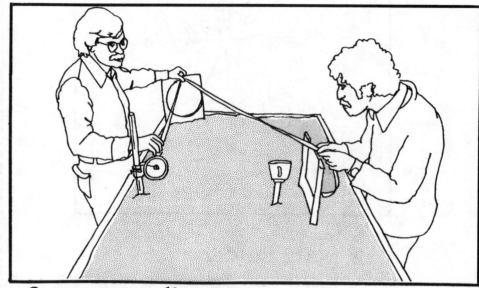

6. measure distance

This reference beam path is the distance from the beamsplitter to the collimating mirror to the plateholder.

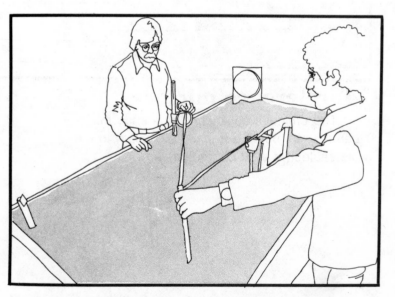

7. object mirror

The positioning of the object mirror is determined by matching the reference beam distance with that of the object beam path (beamsplitter to object mirror to object to plateholder).

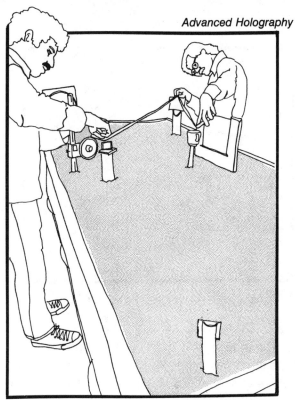

8. spatial filter

Place the spatial filter between the variable beamsplitter and the collimating mirror and adjust to get a clean spreading disc of light (Note: Do not be overly concerned with the distance between the spatial filter and collimating mirror. Place the spatial filter wherever convenient to obtain an even spread on card in plateholder.)

9. 50:50 beam splitter
10. object lens

Set up the second object beam by splitting the first object beam with a 50:50 beamsplitter positioned near the variable beamsplitter. The first object beam will continue to travel through the 50:50 beamsplitter, while the second object beam will be formed by directing light to a second object mirror and then to the object. Measure the distance from the variable beamsplitter to 50:50 beamsplitter to second object mirror to object to center of plateholder. Adjust the second object mirror, if necessary, to make this distance equal to the first object beam distance (as well as to the reference beam distance).

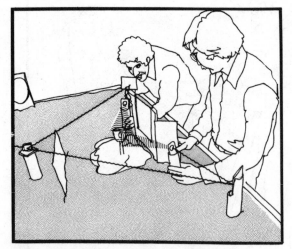

11. object mirror #2
12. object lens #2

Place lenses in the object beams to illuminate the object properly. Card off any unwanted light, especially light which may go directly from the lenses to the plate.

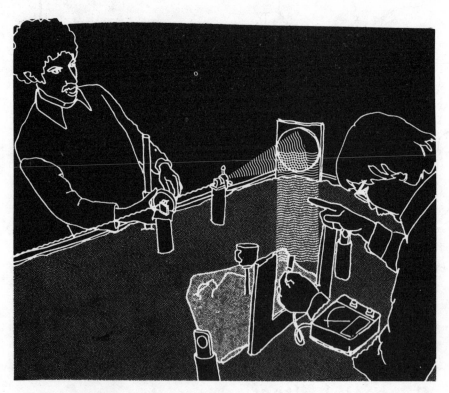

13. light ratios

Take reading of both object and reference light, rotating the variable beamsplitter to obtain the desired 4:1 ratio for transmission holograms.

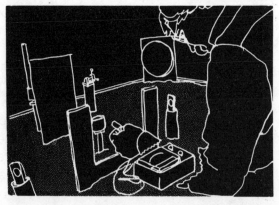

14. load plate
15. check exposure time
16. settle
17. shoot

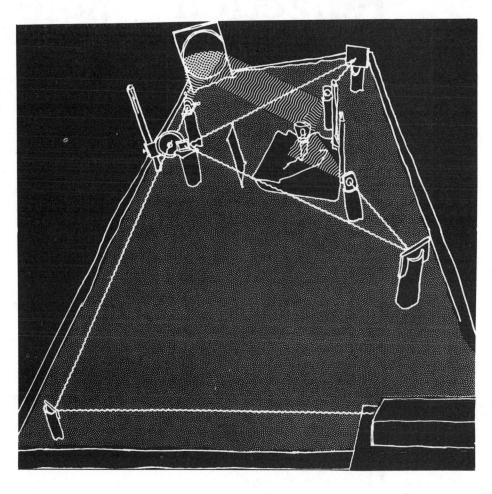

After you're certain that the ratios are correct, that no beams are left obstructed yet unwanted light is carded off, that the distances are correct, and that the components are securely settled, turn off lights and load a plate (remember: emulsion towards object).

18. process as per transmission instructions (p. 151)

reflection holograms

8 x 10 regular reflection - direct from object.

1. position laser
2. place object
3. set up 8"x 10" plateholder
4. position reference mirror
5. position variable beamsplitter
6. measure distances
7. place object mirror
8. set up 50:50 beamsplitter
9. place object mirror #2
10. place object lens #2
11. position spatial filter
12. measure light ratios
13. load plate
14. calculate exposure time
15. allow to settle
16. shoot

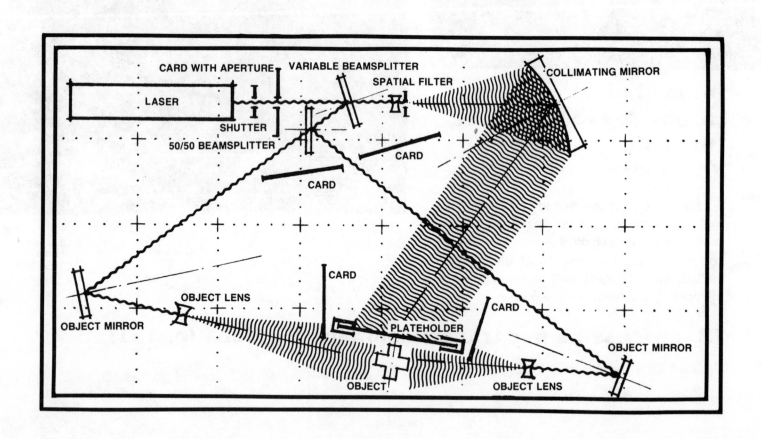

1. laser
2. 8x10 plateholder
3. reference mirror
4. variable beam splitter

Place components on table in configuration shown starting first with the plate holder, and and then the collimating mirror. Run the beam first to the collimating mirror, then on to the plateholder, so that it intercepts the plate at a 60 to 70 degree angle. Next, add a variable beamsplitter.

5. object

Place object on opposite side of plate from components, as close to plate opening as practical, while allowing some room for object light.

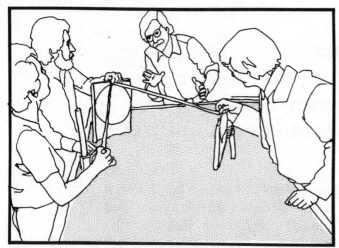

6. measure distances

Measure distance from variable beamsplitter to collimating mirror to plate (reference distance).

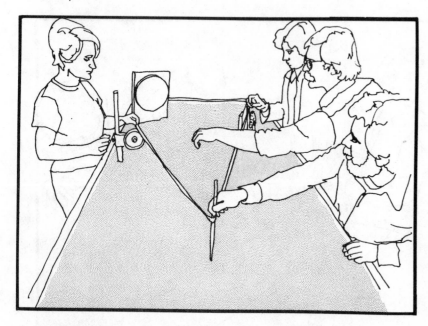

8. 50:50 beam splitter

Set up second object beam using 50:50 beamsplitter and second object mirror. Measure distance from variable beamsplitter to 50:50 beamsplitter to object mirror to object to plate and equalize with other distances.

7. object mirror

Set object mirror in same plane as plate holder, and so that the distance from variable beamsplitter to mirror to object to plate equals the reference distance.

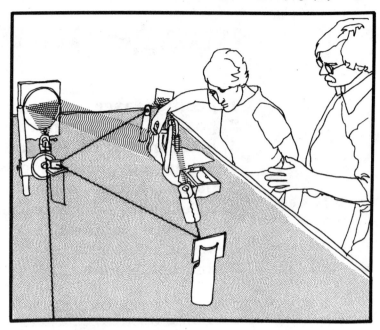

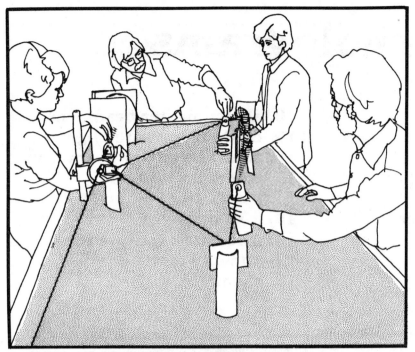

9. object mirror#2
10. object lens #2

Place lenses in object beams and adjust for best object illumination. Position and adjust spatial filter in reference beam.

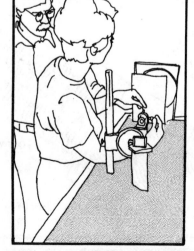

11. spatial filter

12. light ratios

After adjusting collimating mirror so reference light evenly covers card in plateholder, take light readings and adjust ratios with variable beamsplitter, if necessary, to obtain desired 2:1 ratio for reflection holograms. Card off unwanted light and recheck everything before loading plate.

13. load plate
14. check exposure time
15. settle
16. shoot
17. process as per reflection instructions (p. 171)

image plane holograms

1. position laser
2. position transfer mirror
3. place master plateholder
4. set up copy plateholder
5. place reference mirror
6. adjust variable beamsplitter
7. measure distances
8. set up master beam
9. find image plane
10. adjust master beam lens
11. card off unwanted light
12. adjust reference beam lens
13. take light ratios
14. set shutter
15. load plate
16. calculate exposure time
17. allow to settle
18. shoot

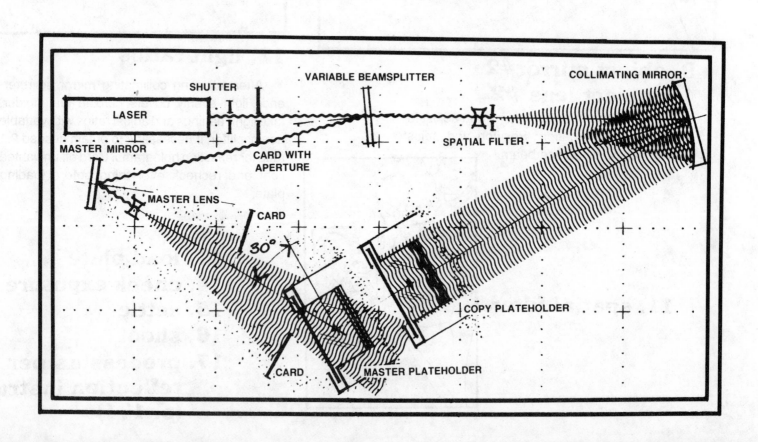

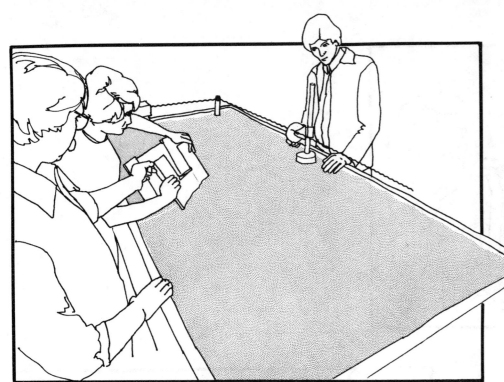

1. laser
2. transfer mirror
3. master plateholder
4. copy plateholder

First place the master and copy plateholders on the table as shown. They should be near the center of the table alongside one of the long edges. The plateholders should be tilted back and face towards a far corner of the table.

5. reference mirror
6. variable beam splitter

Place the collimating mirror on the table in the far corner and direct the beam towards the card in the copyholder. The beam should intersect at about a 60 degree verticle angle, and a 90 degree horizontal angle. Add the variable beamsplitter about midway between the transfer mirror and collimating mirror, directly across the table from the plateholders.

7. measure distances

Measure the distance from the variable beamsplitter to collimating mirror to copy plate (reference distance).

Set master illumination mirror at proper position for brightest reconstruction of the master, and so that distance of VBS to mirror to master to copy equals reference distance (see simple image plane setup directions if you forget how to do this).

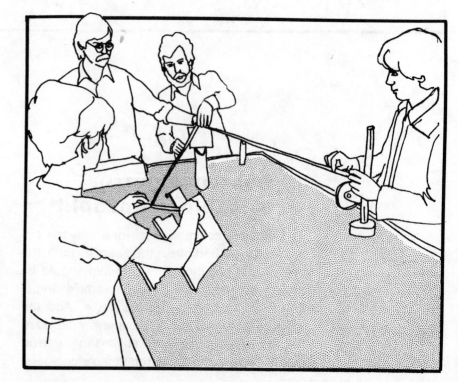

8. master beam
9. find image plane
10. master beam lens
11. carding off
12. reference beam lens

Place a lens in master illumination beam to light master plate evenly (Note: the better the spread of light on the whole master, the greater certainty there is that the copy will be viewable over the widest possible angle). Card off any master light which reaches copy plate directly. Make sure that the image reconstructs at the desired copy plane, that it is well centered, and that the two plateholders are parallel.

13. light ratios

Take light readings and adjust ratios to achieve 3:2 for reference:object (object readings should be off brightest spots from object).

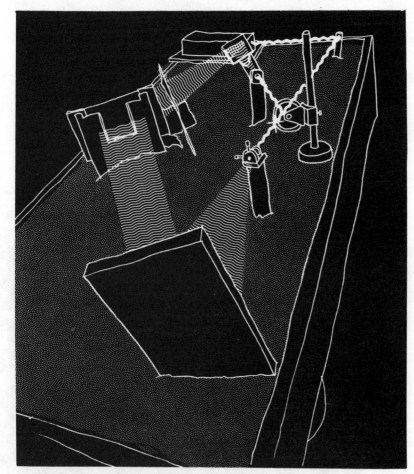

completed image plane setup

14. set shutter
15. load plate
16. check exposure time
17. settle
18. shoot

19. process as per reflection instructions (p. 171)

pseudo-color reflection holograms

by John Kaufman

the setup

Use the geometry for an image plane reflection hologram. Up to three reference beams may be set; set the angles to register the playback of the colors (red: 40 degrees from the perpendicular, yellow: 44.5 degrees, green: 51.5 degrees, blue: 44.5 degrees). For further color registration, the pinhole is set at various distances from the copy plate, as shown in the diagram.

AC is a line perpendicular from the plates through the center of view. AB=BC. CD must intersect all reference beams. Place pinhole at those intersections. Color registration will be most nearly correct along the center of view at C.

Object light on the master plate should be narrowed to only 3" or 4" vertical parallax for better brightness control.

exposures

Before each of up to 3 exposures, swell the Agfa 8E75 emulsion in the desired color-related concentrations of triethanolamine (red: 0%, *yellow*: 10%, green: 12%, blue: 18%). Carefully, wipe plate with a cellulose sponge to avoid streaking, then dry. Exposure is determined mostly by trial and error. Some variables for exposure are a) pinhole to plate distances, b) alter the sensitizing of the plate with higher concentrations of triethanolamine, and c) the desired color balance. For achromatic mixing, much less exposure is given to green and yellow for which the eye has more sensitivity.

development

Use the development procedure described on page 160. Develop dense (nearly opaque) in D-19 for 2 to 3 minutes. Stop in water wash. PBQ bleach.

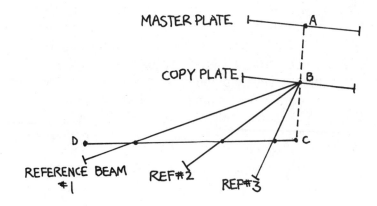

*This is a method for changing the color of reflection holograms as discussed on page 175. In this case, multiple exposures are used to produce separate images in different colors.

rainbow holograms

A regular transmission hologram when viewed with a white light yields a fuzzy multicolored smear where the image should reconstruct. In reality, all of the wavelengths in the white light are individually diffracted to slightly different positions, resulting in the blur.

In 1969, Steve Benton of Polaroid reported an exciting method which got around this problem. First, a normal transmission hologram is made of an object, utilizing a good collimated reference beam. This hologram is used as a master for a *transmission* image plane copy with the following concept: the master is illuminated with only a narrow horizontal slit of light. Since each point on a hologram can reproduce the entire image, the whole image is image-planed into the copy. However, only from the *vantage point* of the slit. Vertical parallax is eliminated by this, although horizontal parallax is preserved over the length of the slit. If the copy hologram is viewed in laser light, the eyes must be located at the real image of the slit to see the image. The hologram now behaves like a window with the shade drawn; only a narrow horizontal strip, ripped out of the middle of the shade, allows you to see through.

If you move your head up or down, the image disappears. Now, imagine viewing this hologram in two different colors of light. Each slit will be diffracted to a slightly different position - and there will be two places where you may position

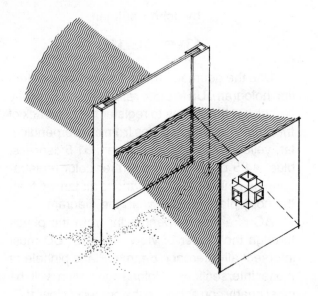

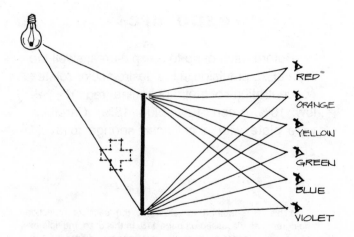

your eyes to see the image (imagine another strip ripped out of the window shade).

Now view the hologram in white light. All of the wavelengths will reconstruct their own slits, all slightly displaced vertically with respect to one another. Since you are now seeing with so many wavelengths, the slits will all run into each other; yet the dispersion is only in one dimension. The image remains sharp and clear since the eyes effectively see a given part of an image through one slit at a time. As you move your head up and down, the color of the image appers to change throughout the spectrum (thus the name rainbow hologram). Yet the image remains sharp and can exibit great depth.

In effect, the hologram is selectively filtering the white light while all that is sacrificed is vertical parallax, which our two horizontal eyes usually don't miss anyway. Rainbow images are often extremely bright, especially if the hologram is bleached. This is because all of the frequencies in white light are being used to form the image.

It is important that overhead (or underneath) reference beams be used in the making of both the master and the rainbow copy, to insure even diffraction over the dimensions of the slit. To avoid the use of cumbersome overhead devices for mounting the components, the master is made with the holder tilted in the sand to achieve an overhead reference.

In the copy setup, both master and copy are placed on their sides so that the horizontal plane of the table becomes vertical with respect to the holograms. The holograms "think" the light is coming from overhead without having it actually do so. This greatly simplifies the positioning of optics, allows much greater flexibility, and assures a more stable setup.

There is one new optical component needed. This is the device used to produce the narrow slit of light which illuminates the master in the copy setup. One of several methods may be used to accomplish this.

A slit or line of laser light will be formed when the beam strikes a cylindrical surface. This can be demonstrated easily by borrowing the metal rod holder that keeps toilet paper in place (remember to return it), and shining the beam from your laser on it. The roll holder can be used as the actual slit optics if you wish, though most do not have sufficiently smooth surfaces to generate a slit free of noise.

Good quality reflective slit-makers usually must be custom made professionally. This involves having a lab "mirrorize" a glass rod or tube with an optical coating. This can often be expensive yet some labs will do this at reasonable cost. Ours are made from glass cylinders between 1'' and 4'' in diameter.*

With coating services generally difficult to come by, however, the lens type slit maker may be easier to construct. A good quality cylindrical lens can be used. Also, we have found that excellent results can be obtained by filling an ordinary good quality clear test tube with glycerine, sealing the edge by siliconing the cork on, and then mounting the tube horizontally to a piece of masonite for a mount.

In use, a mirror will direct the beam through the horizontal tube where it will spread out into a vertical line; (we recommend that the slit be vertical in order to work with the master and copy placed on their sides). This method usually yields a good quality, low noise line, of the type normally obtained only with better quality more expensive optics.

In case none of these are available, or for some reason do not work satisfactorily, a regular collimated master illumination beam may be used, and a mask placed so that only a strip of the master is illuminated. This is how the original Benton hologram was made. In this case, a great deal of light is wasted. To obtain short exposures, the use of slit-making optics is preferred.

*See Suppliers Addresses section under "cylindrical mirrors for rainbow holograms".

CYLINDRICAL LENSES
LABORATORY QUALITY

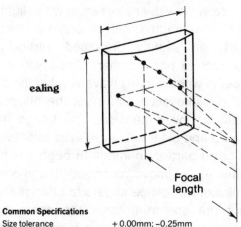

ealing

Focal length

Common Specifications

Size tolerance	+0.00mm; –0.25mm
Center thickness tolerance	±0.25mm
Edge thickness tolerance	±0.25mm
Focal length tolerance	±2%
Back focal length tolerance	±2%
Material	Crown glass
Index of refraction	1.523 (n_d)
Transmission	See curve page 464.
Surface	80/50
Coatings	Lenses are supplied uncoated. For coating capabilities see page 457.

CYLINDRICAL LENSES — LABORATORY QUALITY

Catalog Number	Focal Length mm	Length (L) mm	Width (W) mm
23-7404	40	60	15
23-7412	40	130	15
23-7420	60	60	20
23-7438	60	130	20
23-7446	80	60	30
23-7453	80	130	30
23-7461	100	60	50
23-7479	150	60	50
23-7487	200	60	50
23-7495	250	60	50
23-7503	300	60	50

Both depth and color-spread in the rainbow copy depend upon the dimensions of the slit, and there is an inverse relationship of sorts between them. The narrower the slit, the greater the depth gained, yet the less color smear will be obtained. A slit which is too narrow, however, will result in a "grainy" image. Ideally, we wish to see the whole object in a single color for a given viewing position, and have it resolve great depth. This can be achieved by experimenting with various slit widths for different types of images, and maximizing both effects with a compromise slit size. For starters, try a width of about ½ cm.

The color spread can also be enhanced by using film to make the rainbow copy, and curving it slightly along the axis of the slit. For the exposure, the film is taped to a curved piece of rigid metal or plastic, which is made to slide into a plateholder. When viewing the finished copy flat, it will "stretch out" some of the colors.

Another good rule to follow is to make certain that the edges of the slit illuminated on the master are parallel and well defined. We recommend that a "cut out" vertical line mask be positioned so as to insure a good-quality slit, even when using slit-making optics. This will prevent "fuzzy" slit edges or scattered light from striking the master.

The use of collimated light becomes important to rainbow hologram-making. Some holographers take great care to insure that the reference beam for the master and the slit used in the copy setup are both well collimated. The setups described below reflect this view. It is, however, also possible to use a slightly converging reference beam for the master. When used in the copy setup, this orientation will "assist" the slit image in diverging towards the copy plate. Although often desirable, it is not essential that the slit be collimated as well.

Rainbow masters need not be made with a full 8"x 10" plate. Since only a strip is illuminated when used in the copy setup, enterprising holographers with an eye for economy may wish to cut an 8"x 10" lengthwise to make four 2"x 10" plates. This, of course, must be done carefully in the dark so practice first on some regular 8"x 10" glass until you get the hang of it. A plateholder should be made to hold the strip-plate at the desired height.

Both the rainbow master and copy are processed like any good transmission hologram. Because such a small portion of the master is used to generate the copy, we highly recommend the use of low noise processing techniques. The use of a developer like Neofin Blue, along with very careful exposure and processing, will ultimately yield rainbow copies of less grain and greater depth resolution. A low noise master allows the use of a narrower slit to achieve these ends. As with others, the rainbow master should generally not be bleached, unless a low noise formulation such as P.B.Q. is used. The rainbow copy is usually bleached to produce an extremely bright white light viewable image. Now for the steps:

rainbow master

The laser beam's path is re-routed by **(1) the transfer mirror,** to run down the side of the table. A slight depression or hole made in the sand will cradle **(2) the object** to be holographically imaged (black cloth optional). Approximately 10 - 12 inches* in front of the depression, mound up sand and plant **(3) the plateholder** in this elevated sand pile. Tip the plateholder toward the object at an angle of 30 - 45 degrees. A collimating mirror used as **(4) the reference mirror** must intercept the beam at its exact center and, in turn, direct the light toward the plateholder. Place **(5) a spatial filter** (or lens) in the sand at the exact focal length of the collimating mirror. If your collimating mirror is 10 inches in diameter, the column it creates must measure 10 inches (no more, no less) anywhere along the reflected reference beam path. By checking this beam, you can be assured the spatial filter is properly placed, and the reference mirror is collimating properly. Add **(6) the variable beamsplitter** between the transfer mirror and spatial filter. The resultant beam can be bounced off first, **(7) the first object beam transfer mirror,** then onto **(8) the first object beam mirror.** This beam is one of two that will illumi-

nate the object. To obtain the second beam, place **(9) a 50:50 beamsplitter** between the VBS and the first object transfer mirror. The split beam will travel toward **(10) the second object beam mirror** which directs this second object beam to the object. Both **(11) object lenses** can be positioned as shown to spread the object beam paths, and to provide even illumination for the object. The beam distances must, of course, be equal.

The reference beam equals the distance from the variable beamsplitter to the reference mirror to the plate.

The first object beam distance equals the distance from the variable beamsplitter to the object transfer mirror to the first object mirror to the object to the plate.

The second object beam distance equals the distance from the variable beam splitter to the 50:50 beamsplitter to the second object mirror to the object to the plate.

The ratios once again are 4:1, and scatter cards should be positioned as needed to eliminate light scatter.

---

*It is important that the object be located at least this far from the master to obtain a rainbow hologram with a good viewing angle.

*The collimating mirror used in this setup had a focal length of 54.6 inches, which determined the placement between the spatial filter and the mirror.

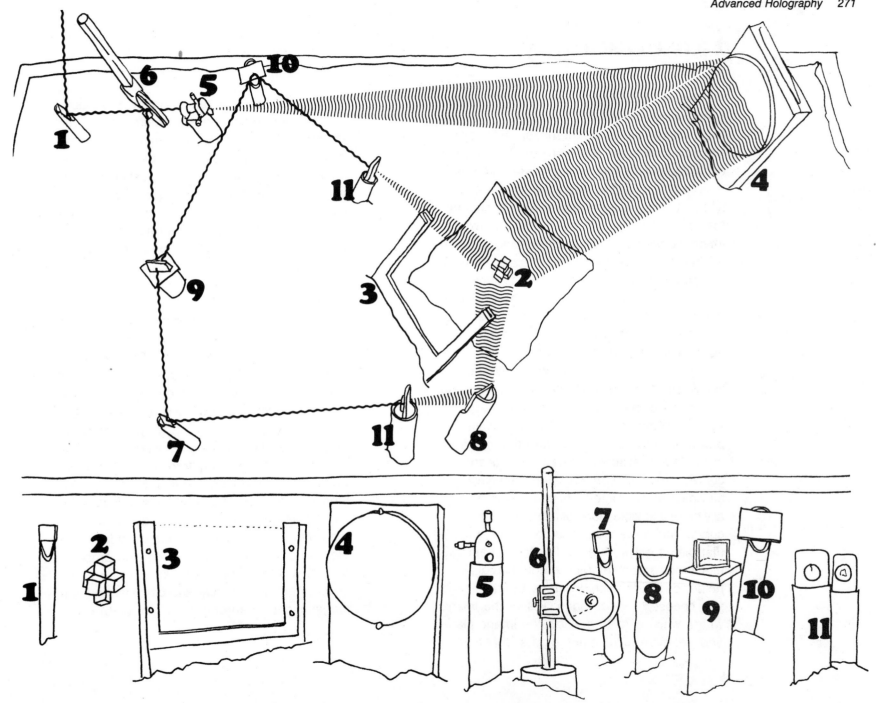

rainbow copy

The laser beam's path is re-directed by **(1) the transfer mirror** placed in the middle of the table to **(2) the reference beam mirror** on the opposite side of the table. This mirror can be either flat or collimated. **(3) The master plateholder,** preferably 8"x 10", can be placed as shown. As seen in the drawing, **(4) the variable beamsplitter** is placed relatively close to this plateholder to intercept and split the beam traveling between the transfer and reference beam mirror. Now, **(5) the collimating mirror,** can be positioned beside the reference beam mirror. At the focal point of the collimating mirror, place **(6) the slit optics.** To maintain the exact focal length distance, adjust the variable beamsplitter to direct the beam to the slit optics, and then onto the collimating mirror. This mirror can now be pivoted to direct the collimated slit to the middle of the master plateholder where it should illuminate and project the brightest part of the image toward **(7) the rainbow holder.** This horizontally curved piece of sprayed black aluminum or plexiglas wedged into a plateholder will eventually carry a flexible piece of film taped inside the curve. The curve enhances the rainbow curve (flat plates will work as well). Once again, the exact placement of this image plane plateholder depends upon which part of the image you choose to focus upon. Bisect the image with a white card to preview the image plane and estimate placement of the copyholder.

Measure the master beam distance: variable beamsplitter to slit optics to collimating mirror to center of master plate to center of rainbow copy holder. The reference beam distance should be an equal distance (adjust accordingly): variable beamsplitter to reference mirror to center of the rainbow copy plate. **(8) The spatial filter,** intercepting the reference beam, must spread out the beam enough to cover the copy plate.

Card off to prevent stray light from reaching the plates from the slit optics or variable beamsplitter. No part of the collimated slit that continues on through the master plate should reach the rainbow copy. The reference beam should not reach the master plate. Black felt placed on the sand between the two plates will also reduce scatter. The light ratios are 4:1 (reference:object beam). Most types of holographic plates or film may be used for the rainbow hologram, ranging from high resolution 8E75 to low resolution, yet fast, SO 253. If using film, it is best taped to the plateholder, but can be sandwiched between plastic. Settling time with the use of film should be at least triple that of using plates. Expose and process as you would a bleached transmission hologram. View with a point source of white light (i.e. sunlight, or an unfrosted light bulb).

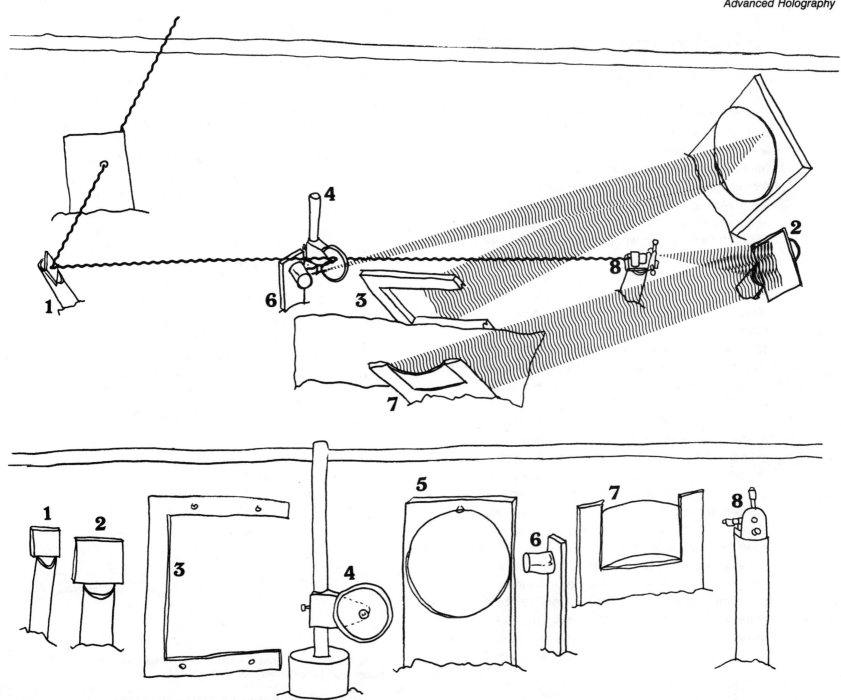

alternate rainbow techniques

One of the difficulties in obtaining a good quality rainbow hologram, as we have mentioned, is to properly compromise the width of the illuminated slit to obtain good image depth, while maintaining low noise and good color dispersion. Normally, if the slit width is too small, speckel results (the hologram is noisy). If the slit width is too great, speckle is reduced, however the image will be "fuzzier". Blurring can be controlled to a certain degree by keeping the image relatively close to the film plane. However, as distance increases, it becomes more difficult to keep the image sharp and the speckle at an acceptable level. Even the best compromise means living with a certain amount of the speckle, and blurring. Recently, Emmet Leith and Hsuan Chen*, working at the University of Michigan, devised a brilliantly simple procedure to overcome this problem. A cylindrical lens is placed over the slit in the master on the side of the master facing the copy plate. The focal point of the lens is located at the copy plate. The width of the slit illumination on the master can be relatively wide (e.g. one inch or more). Since the cylindrical lens is to be used in the vertical direction only, the image plane can still be positioned in the same way as with an ordinary rainbow hologram (remember, with a rainbow hologram, only horizontal information is retained for perception of depth). Since the slit is taken from a wider perspective, resolution is improved (less speckle). Yet, very deep images can be produced without blurring, since the cylindrical lens reduces the vertical information to a thin line to prevent dispersion.

One of the minor drawbacks to the procedure is the increase in astigmatism of the image, an inherent quality of all rainbow holograms. However, since the improvement in the image quality is so great, the astigmatism is considered to be of little significance. Although any good quality cylindrical lens can be used, Leith and Chen recommend fabricating one out of plexiglas in order to save money and also obtain a desired selection of width and focal length. Saxbe** briefly discusses the making of plastic immersion lenses, or more detailed information can be obtained by contacting Sharon McCormack, San Francisco School of Holography, 550 Shotwell St., San Francisco, Calif. (Tel. 415-824-3769).

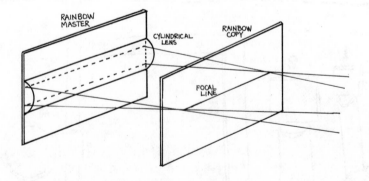

*Leith, Emmet N., Chen, Hsuan, "Deep Image Rainbow Holograms", *Optics Letters*, Vol. 2 #9 April 1978

**Saxbe, Graham, *Holograms* p. 105 (see bibliography)

pseudocolor rainbow holograms

This is a relatively new technique involving the use of multiple exposures to vary the colors of different parts or objects in the hologram with respect to one another.

We recall that the color of the image in a rainbow hologram changes depending on our viewing position. In a regular rainbow, the entire image will appear in one given color for a particular viewpoint, or at least everything in a given horizontal band will be so. It might be nice, however, to make a hologram of a scene with individual objects of different colors. We might imagine a hologram of a "staged" sunset, with a red sun setting into a blue lake with a green tree on the shoreline. This kind of "pseudocolor hologram is possible, and opens the medium to the effects of color expressionism often found in painting and creative photography.

A multicolor rainbow is made by making separate masters of the individual objects we wish to have appear in different colors. To make the rainbow copy, we first make an exposure with one master in place. Next, we remove that master and replace it with the next one. We do *not* remove the copy, but change the angle of the reference beam upon it by slightly turning the filmholder. We now make a second exposure onto the same film. Since the two images were separately referenced, they will reconstruct separately. And, since the two reference beams were from different angles, the two images will go through their color shifts "out of synch." The result is, for any given viewing position, two differently colored objects.

This process can usually be used to record up to three or sometimes four separately colored images in the same scene before the quality begins to suffer. Even more can be added by "cheating" a little and recording some on separate pieces of film. The various pieces are mounted on top of one another for a great number of variations.

Another method, first used by Poohsan Tamara and later described by Graham Saxbe, involves setting up three masters in different positions simultaneously, keeping the position of the copy fixed, and producing exposures by moving the slit illumination from master to master for separate exposures.

All masters are slit holograms (e.g. ½"x 10") and are shot with 30 degree reference beams. A decision is made as to which master will provide a red image, which will be green and which blue. Between the master exposures, objects can be removed or replaced or positions shifted, etc.

To make the rainbow copy, the masters are attached to a piece of glass, with the "red" master to the left, the "green" master in the center, and the "blue" master to the right. The angle between the red and green masters is 9 degrees as viewed from the copy, and 6 degrees between the green and blue. As an example, if the distance between master and copy is 10 inches, the centers of the red and green masters should be positioned 1½ inches apart and the green and blue about 1 inch apart. These separations are designed to obtain maximum color spread in the rainbow copy.

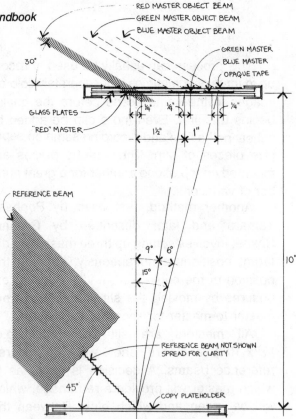

The masters are masked down to slits ½ cm. wide. One master is exposed at a time, with the reference beam remaining fixed at 45 degrees on the copy for each exposure. Since the shot is a triple exposure, each individual exposure should be for approximately ⅓ less time than light readings would indicate.

A third technique that will result in the production of new, composite, colors uses the mixture of the components of the pure spectrum. Rainbow color mixing occurs when we make a multiple exposure from the same master yet vary slightly the reference angle in the copy. In this case, the relationship between the master and copy holder must remain fixed to insure the image position does not shift between the two exposures. The reference angle is changed by adjusting the placement of the reference mirror. This is most easily done by setting up two reference paths beforehand with the beam deflected into one path with a mirror. When the mirror is removed, it will follow the second path. The other method is to cover the film, and then move the optics to set the reference beam at a new angle. The result in either case is an image that mixes two or more colors to produce a new third hue. By adjusting the angles, and relative exposures, a wide variety of hues, from the very pure to the pastel, can be produced. Even achromatic images are possible. The mixing technique can be used along with the multiple master technique to achieve an enormously varied set of effects.

Multiple exposure holography of this type is a painstaking art which involves a great deal of time and effort. The degree of color shifting will depend on the angular separation between the exposures, dimensions of the slit, the original object to plate distance when the master was made, and the curvature of the film, among other factors. A desired effect is usually achieved experimentally and not without work. As a starting point, for the multiple master technique, we recommend shifting the reference beam about 10 degrees per exposure. When making the individual masters, remember, care must be taken to properly register the various images to appear where you will want them.

Since the above techniques are so new, very little has been written about them thus far. Research is continuing into methods to accurately control the colors produced.

open aperture*

The method involves utilizing the focusing capacity of the master illumination beam. A master hologram is shot, utilizing a flat surface reference beam mirror. The beam can then be said to *diverge* as it exits from the spatial filter, and to continue to do so as it reaches the plate as the reference beam. The fringes formed in the hologram will reflect this orientation.

When the finished master is illuminated from the opposite side to form a real image in the copy setup, a converging beam which mirrors the original diverging reference beam is used. Thus, encoded in the object beam is a sort of condensing or color-corrected lensing capacity, which is image-planed into the copy. When viewed with white light, the various wavelengths diffracted near the image plane are condensed to the same positions, causing a recombination resulting in a white image. The degree of depth over which a black and white image is obtained depends on the rate of divergence/convergence; it also depends on how well the divergent reference beam (in the making of the master) mirrors the convergent master illumination beam (in the making of the copy).

Typically, however, white light diffracted over distances greater than a couple of inches will begin to smear. The major advantages of this hologram over the rainbow is parallax in all directions and relative monochromaticity. The disadvantage is loss of depth resolution.

The accompanying setup diagrams show shooting the copies on their sides, in order to avoid overhead components, as with the rainbow holograms. The most important thing to remember is to use a flat reference mirror for the master construction, and a collimating mirror for master illumination in the copy setup, while taking care to match the convergence of the second with the divergence of the first.

*We are considering sponsoring a contest to come up with a good name for this type of hologram. The only reason for the term "open aperture" is to differentiate it from "reduced aperture" of the rainbow hologram. We think a better name is needed.

open aperture master

The beam from the laser should reflect off **(1) the transfer mirror,** as usual, and follow a path parallel to the edge of the table. **(2) The object** to be holographically imaged can be placed at the approximate center of the table in a slight depression in the sand, 12 - 15 cm. from **(3) the plateholder,** elevated 15 - 18 cm. above the surface of the sand table, and tilted 40 to 45 degrees to assure that the reference beam clears the object as it travels between the plateholder and **(4) the reference mirror** (an 8"x 10" or 11"x 14" flat front surface mirror). This allows for the beam to continue to diverge (spread out) as it continues on its way (and to evenly cover a card in the plateholder). This beam is one of two that is an obvious byproduct of **(5) the variable beamsplitter.** At this point the reference beam distance can be obtained: the distance from VBS to reference mirror to plateholder. Once again, this should equal each of the object beam distances about to be set up. **(6) The first object mirror** placed as shown will reflect the other beam of light onto the object. **(7) A 50:50 beamsplitter** set into the object beam path will direct light of **(8) the second object mirror,** placed just to the right of the plateholder, collecting the light and directing it toward the object. By placing **(9) an object lens** as shown, the second object mirror will now transmit a spread-out beam of light to the object. Likewise, **(10) a second object lens** placed just to the left of the plateholder, as shown, will also spread out the beam of laser light to illuminate the object.

The last few components' placement will be determined by the necessity to equalize reference and object beam distances. Once again, each object beam should equal the reference beam distance previously measured: the distance of object beam one equals the distance from beamsplitter to object mirror one to object to plate.

Object beam two equals the distance from beamsplitter one to beamsplitter two to object mirror two to object to plate. At this point the pinhole or **(11) spatial filter** should be added to "clean up" the beam. Light readings follow a 4:1 ratio and can be taken in a similar manner to any transmission hologram. **(12) Carding off** is somewhat elaborate; both plain and pinholed cards are essential to blocking light scatter, especially the card between **(7)** and **(9).** This card allows only the needed beam to pass, and prevents any extra light that might hit the holographic plate.

Now is the time to check the beam divergence. Place a card perpendicular to the reference beam path (after it hits the reference mirror), and mark the angles at which the beam spreads out. When working on the copy setup, this same card (turned around and put in the reference beam path at the same point) should match the convergence of the beam. In other words, the copy reference beam should converge at the same angle that the master reference beam diverged. The spatial filter can be moved until the copy reference beam converges at the same rate of spread as marked on the card.

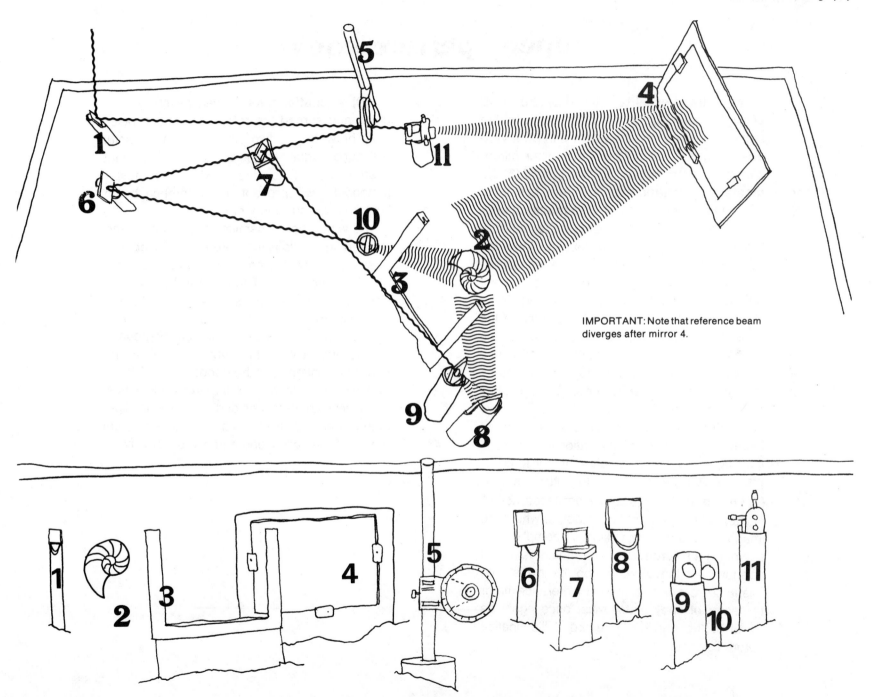

IMPORTANT: Note that reference beam diverges after mirror 4.

open aperture copy

(1) A transfer mirror placed as shown will allow the beam to run down the center of the table. By intercepting the beam with **(2) a variable beamsplitter** before the transfer mirror, part of the beam can be directed toward **(3) the master illumination mirror;** in turn, this dot of light can be spread out with **(4) a master illumination lens** as it travels toward **(5) the collimating mirror,** positioned at the opposite end of the table; (note: the placement of the lens determines whether the light will converge properly after bouncing off the collimating mirror). Place **(6) the master plateholder** in the path of the converging master illumination beam so that the hologram plays back as brightly as possible.

The light carrying the holographic image from the master plate holder to the copy plate can be considered the object beam. The now completed object illumination beam path (from beamsplitter to master illumination mirror to collimating mirror to master plate holder to copy plate) should be measured for future use.

(7) The copy plateholder can be placed 12 - 14 cm behind the master plateholder, or where the image is focused to your liking. To complete this setup, **(8) a reference mirror** should be added as shown and the reference copy beam distance (distance from transfer mirror to reference mirror to center of copy plate) must once again be equal to the previously recorded illumination beam distance.

(9) A spatial filter placed as shown will clean and spread the reference light. A black cloth between the two plates will be a helpful precaution against light scatter. An important scatter card placed just next to the master plateholder will prevent any reference beam light from hitting the master plate.

As with other transmission holograms, the desired light ratio is 4:1 (reference:object). Expose and process according to standard transmission procedures. The finished hologram is best viewed with a point source of white light in the same manner as the rainbow hologram.

Our setup shows most of the optics grouped at one end of the table, while the collimating mirror is positioned at the opposite end. This is due to the focal length of our mirror: 54.6 inches. To achieve convergence or divergence, the light being spread has to start before or after the focal length of the collimating mirror respectively.

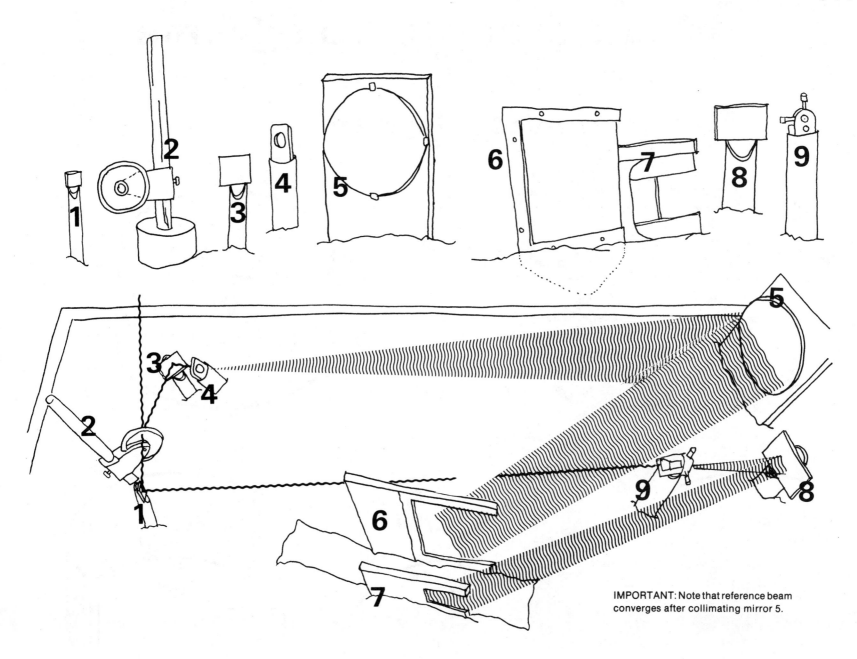

IMPORTANT: Note that reference beam converges after collimating mirror 5.

other advanced holograms

curved holograms

Curved holograms take greater advantage of the dimensional qualities of the medium than flat plates. They offer greater viewing angles, up to a full 360 degrees if desired, allowing one literally to walk all the way around the hologram and see the object from all sides. Curved holograms are a bit trickier to handle than flat plates. Besides the stability problems inherent in a thin film, one must be more aware of the existance of light scatter, and discrepancies in the angles of illumination, which change over a curved surface. There are solutions to these problems, though, as we shall see.

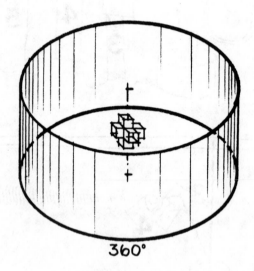

360°

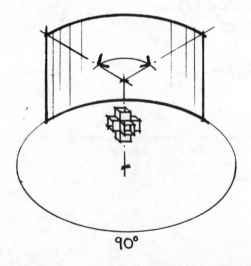

90°

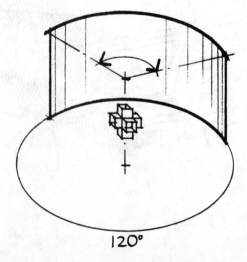

120°

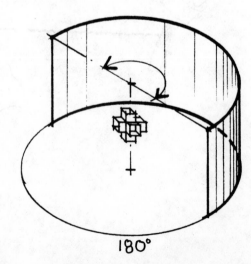

180°

360 degree holograms

The only essential new piece of equipment for 360 degree holograms is a device to hold the film. A good film holder is a plastic cylinder (available at most plastic supply houses), a few inches in diameter by several inches long. The size of cylinder selected will depend upon the dimensions of the film, which is available in widths from 35mm up to 10 to 12 inches. It is advisable to work small first to get the hang of it.

The setup can be extremely simple, employing a single beam and a lens (preferably a spatial filter). The object is centered inside the cylinder, and a single expanded beam, entering one end of the cylinder, illuminates both the object and the inside wall of the tube as a reference beam. Film is securely taped to the inside of the cylinder, allowed to settle for an hour or two, and an exposure is then made. The film is unrolled, processed as a transmission hologram, and then viewed by replacing it in the cylinder and illuminating it with the laser.

Two advantages of this particular setup: the simplicity of its optics make it suitable for very low power lasers, and, because of this, unwanted light is minimized (a constant problem with curved holograms).

The major disadvantages are the lack of control over reference-object light ratios, and the obvious fact that the object is lit only from one side.

There are a number of ways of constructing this setup. Ideally, it is convenient to place the cylinder upright on the table and to have light coming in from above, surrounding the object.

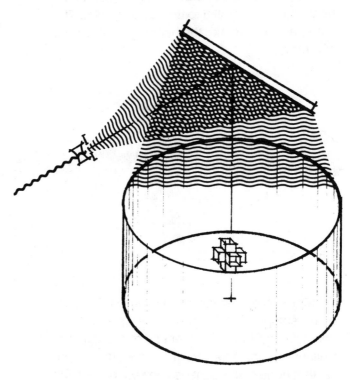

This necessitates the use of an overhead mirror to direct the light down onto the object. A mirror of this type can be constructed, or it is possible to cheat by digging a deep hole in the sand (masonite walls on four sides will hold back the sand), and placing the cylinder and object at the bottom of this hole. A mirror set just outside the hole will direct the beam down into the cylinder.

Another method is to secure the object to a base (such as a piece of glass) at one end of the cylinder. The whole cylinder construction can then be placed on its side in the sand, and the beam can run parallel to the table surface. The hologram will "think" the beam is coming from overhead.

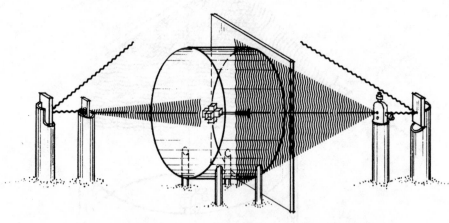

Many objects will be insufficiently lit in single beam setups. The beam can be split and one beam directed into the other end of the cylinder to illuminate the object from "below". It is essential that none of this light be allowed to strike the walls of the cylinder - it must only be used as object light.

It is difficult, but it is possible to aperture the light carefully through a proper size hole in a card, or, even better, through an old camera iris which can be adjusted.

Yet another workable single beam method to try utilizes a mirror directly "under" the object to illuminate from below.

If you happen to have access to a small parabolic mirror with a focal length of only a few inches, you might try positioning it at the end of the cylinder opposite that of the reference beam. The mirror will focus light toward the object in the center of the cylinder while helping to keep light off the inside walls. The same effect can also be achieved by substituting a convex lens for the mirror and using a split beam setup.

The point to remember with all of these setups is the same: light the side of the object away from the reference beam while keeping additional light from striking the cylinder.

Two things should be kept in mind when selecting an object for a 360 degree hologram. First, the object should be small, relative to the size of the cylinder walls. Second, the object should be relatively bright if a single beam setup is used since ratios will not be adjustable.

Holograms with curvatures of less than 360 degrees can be made, oftentimes by employing more conventional split beam techniques. Attempt to familarize yourself with plates of less curvature first (i.e., become an expert at 120 degree holograms before going on to 180 degrees).

The two outstanding characteristics to consider while constructing this type of hologram are, (1) the effects of varying surface angle to a fixed reference beam, and (2) the difficulty of object illumination without this light directly striking the curved plate which is partially wrapped around the object.

The reference angle problem can be demonstrated by diagrams #1 through #4.

In #1 the case of diverging reference beam on a flat plate, the light strikes the plate at a slightly more oblique angle at A than at B. Ordinarily, this angular difference is not great enough to affect the efficiency of the hologram significantly when reconstructed.

The angles the reconstruction light makes with the hologram will coincide with the original reference angles at these points as well, allowing proper playback of the image.

Angular control across the entire surface of the plate can be better accomplished by collimating the reference beam as in #2, insuring that reconstruction light will affect all parts of the hologram similarly.

In the case of the curved hologram in #3 however, it is clear that whereas satisfactory angles can be found at B and C, A and D don't make it. The reference beam strikes "A" at almost 90 degrees whereas "D" gets almost no surface exposure. The greater the curvature, the more this becomes a problem. Collimating the beam offers no solution, either.

One way around the problem is to position the reference beam overhead, as in #4, which is precisely what was done with the 360 degree hologram. When this is done, light strikes all points at roughly the same angle, if the curvature of the plate approximates the degree of divergence of the reflected light.

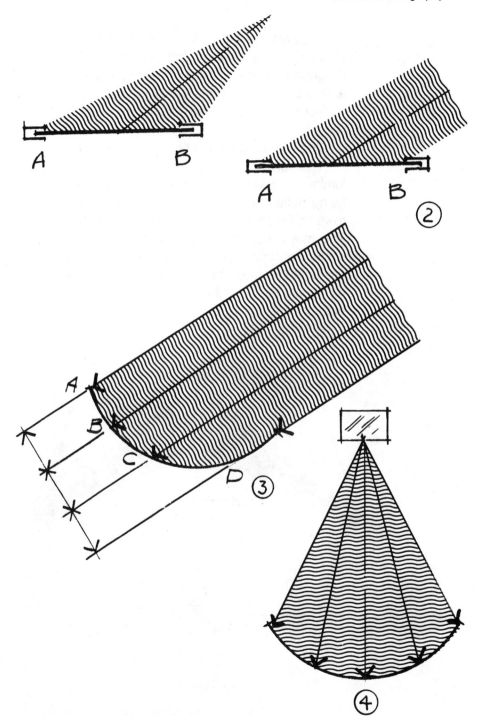

The next question is how one can manage to sneak object illumination past a curved surface. Diagram #5 demonstrates what happens when a conventional flat plate setup (solid lines) is converted into a curved plate setup (dotted lines).

The curved surface intercepts object light.

The solution is again to utilize overhead illumination. If we imagine the two object mirrors being pulled out of the page and tilted down towards the object, the light will bypass the plate. This may be simplified, if desired, by placing a single overhead object mirror at "X" behind the center of the plate. The overhead reference mirror should be located at Y - some distance behind the object.

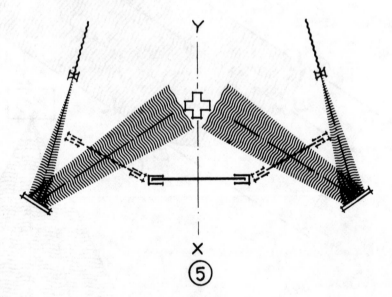

curved masters

A curved plate can be used as a master hologram to make ordinary white-light viewable copies. The major advantage of using a curved master over a flat plate is the increased viewing angle obtained in the copy. In the first diagram (A), the angle of view of the object is determined by the size of the flat plate - this will be the angular limit in the copy.

A larger plate (B), will result in a greater angle of view.

A curved plate (C), maximizes this angle by achieving the same effect as an infinitly sized plate.

Curved plate "C" sees much more of the object than either flat plate "A" or "B". The copy image from a curved master will be viewable over a much wider angle as well.

The difficulty comes in illuminating the master during the copy stage. The problem is again one of the angular relationship between light and surface. When illuminated by a point source, light will strike the master at varying angles, with only the one central axis equivalent to the original reference angle (see drawing "D"). Ideally, what is needed is the conjugate to this, or light appearing to focus at the point where the original pinhole was located, when the master was made. One would need an enormous coverging device to accomplish this (for example a very large parabolic mirror located outside the perimeter of the master).

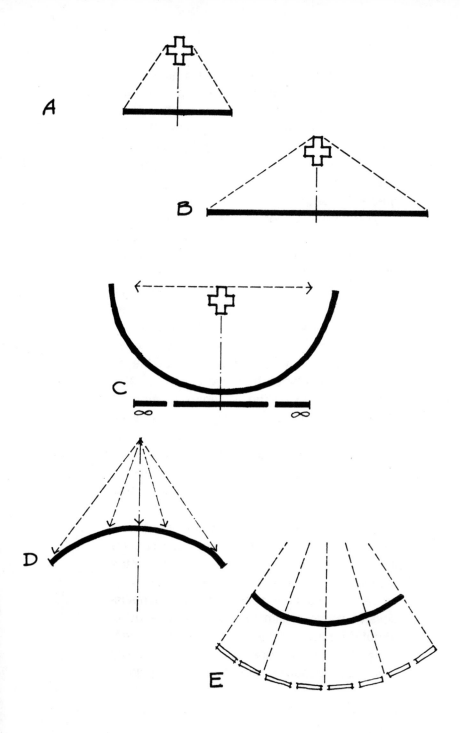

A

B

C

∞   ∞

D

E

This type of mirror can be extremely expensive and difficult to manipulate because of its size. The effect can be approximated by building an array of flat mirrors mounted in an arc similar to that of the master (see drawing "E").

As one can see, the illumination beam must be spread out quite a bit to cover all of the mirrors. For this reason, this type of hologram construction is recomended only for those with access to relatively high power lasers.

There is, perhaps, a better way to approach this problem. Suppose we make a curved master and then *flatten it out* for image planing. To make the master, we use an overhead reference beam that diverges at a rate equal to the curve of the film. Horizontally speaking, each ray is now perpendicular to the film surface. Now, if the master is flattened out, it can be illuminated with a collimated plane wave. Each part of the master will reconstruct out to the image plane, although this image will be distorted laterally. We now expose the copy hologram flat. To view the copy, it is curved in the same manner as the master, to remove the distortion. The result is a curved image plane hologram with a large view around the object. The I.P. can be made as a white light viewable hologram and the image can project some distance in front of the film plane if desired. Actually, the technique can be used on masters up to 360 degrees, by which an entire image-planed cylinder can be made.

holographic stereogram or multiplex

One of the most obvious limitations to conventional holography is the difficulty of recording live subjects or outdoor scenes. You may have noticed that we have supplied no instructions for making a hologram of your Aunt Martha and her cat. Regular holograms of this type can only be made with an enormously expensive and difficult-to-use pulsed laser. Pulsed holography is briefly discussed later in this chapter but only a priviledged few will probably ever get near a pulsed system suitable for portraits.

There is another, rather ingenious, way to make a 'people' hologram, however. The concept involves using photography first, and then storing a series of pictures holographically. This becomes a somewhat complex task, but, unlike pulsed lasers, is certainly one within reach.

What happens when we make a hologram of a photograph? We can certainly do this, however, the image will be two dimensional - just like the photo, or as flat as a piece of paper.

Suppose two photographs of a person are taken from two slightly different perspectives. If each eye sees a different picture, a 3-dimensional illusion results. This is the concept behind stereo photography. Usually special viewing systems or special glasses are needed to observe stereo photographs, to insure that each eye sees only the proper viewpoint.

Now suppose photographs are taken from many different perspectives. The pictures *together* contain all of the necessary information about the multitude of views of a subject, just as a hologram does. A hologram synthesized from all of these views is capable of yielding a 3-D stereoscopic image, which is viewable without special glasses.

This type of hologram has been referred to as a multiplex hologram, integral hologram, composite hologram, holographic stereogram and a host of other terms. Controversy over the best name for it still continues. Holograms of this type were first discussed in the literature by Robert Pole of IBM in 1967. A number of improvements were perfected over the next several years. The most notable of these was developed by Lloyd Cross in 1973 and first offered by the Multiplex Company. The Cross technique is still the most successful and most widely used for this type of hologram. Motion picture film is used to provide the multiple photographic views for the intermediate stage between live subject and finished hologram. It all works like this:

In order to easily and smoothly record all of the angles of view, a fixed motion picture camera films a subject located on a rotating table (or the subject can remain fixed and the camera can be moved at constant speed). Typically, each successive frame of film represents a change of about ⅓ degree in the viewing position. Conventional filming techniques may be used, including fades and dissolves, animation, superimpositions, and zooms. Even artifical computer-generated images can be utilized.

A hologram is made by running the film through a specially designed optical printer. Each frame of movie film is projected onto a cylindrical lens, which condenses the image down to a vertical slit only about a millimeter wide, with a length corresponding to the size of a finished hologram (usually about 9 - 12 inches). The slit is focused near the plane of a piece of holographic film mounted on a curved platen. A reference beam is brought in from overhead (or from below), and an individual condensed slit hologram is formed. Each suceeding frame of film is made into this kind of hologram and is sequentially laid down by slightly moving the platen before each new exposure.

Optical Printer to produce holographic stereograms. Photo by Sharon McCormack.

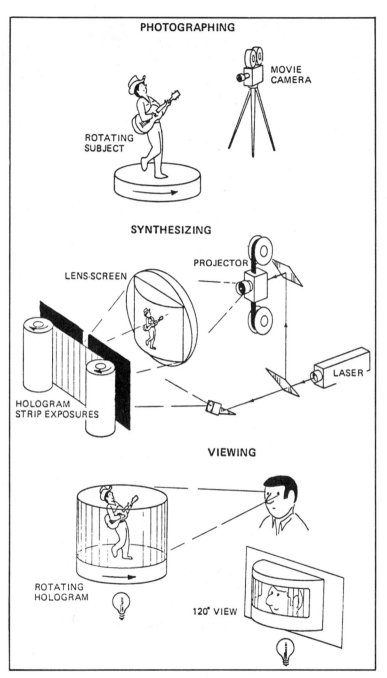

At the end of the process, as many as 1080 individual slit holograms (in the case of a 360 degree recording) have been exposed side by side (Sometimes it is necessary to print each frame twice, resulting in over 2000 holograms on one piece of film! It can take over nine hours to make the first 360 degree print!) The film is then processed as a normal hologram, and contact copies may be made by using a copy cell.

When the finished piece is illuminated, each individual slit hologram projects a virtual image which is two dimensional, just like the movie film. Each eye sees a slightly different slit, however, and we perceive three dimensions. In addition, the Cross printer's optics are designed to reduce vertical information in the same manner as with rainbow holograms, allowing the multiplex to be viewed with an unfrosted white light bulb with a vertically positioned filament. The result is a dynamic, exceedingly bright rainbow hologram of live or outdoor subjects in motion and in startling 3-dimensions.

Holographic Stereograms are among the most often displayed commercial holograms. We see them in store windows, at trade shows, museum openings, etc. For many of us the first hologram we ever saw may have been of this type. Recently, additional technical improvments have been made. Bill Molteni and Rudie Berkhout have developed achromatic stereograms which are viewed flat, rather than curved into a cylinder. Utilizing improved optics, Lloyd Cross has perfected larger images

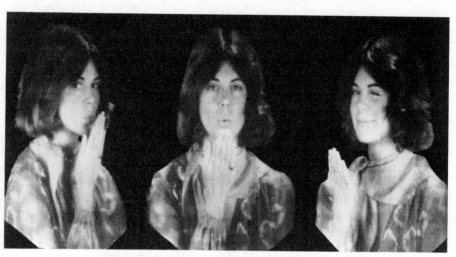

"Kiss II" holographic stereogram by Lloyd Cross and Pam Brazier, Photo © Daniel Quat, collection MOH.

which can actually be imaged at or beyond the film plane, and the lines separating the individual slit images (often an annoying distraction) are barely visible. Full color stereograms are not far behind.

Because of the startling effect of 3-D in motion, and the use of cinema in making the hologram, stereograms are often referred to as "holographic movies". In reality, they are holographic zoetropes more closely related to 19th century card-flipping devices than true motion pictures. In the world of holographic cinema, Edison has not yet arrived on the scene.

making a "Leslie" stereogram

The building or use of a multiplex printer is far beyond the scope of this book. However, duplicating one of Lloyd Cross's early experiments is not. Prior to perfecting the printer, Lloyd utilized a series of 35mm slides taken of a woman, from a variety of angles of view. Each slide was individually projected onto a diffusion screen. A mask, in the form of a vertical slit was placed over the holographic plate. A different slit hologram of each projected slide was then formed slide by slide on the one plate. The result was a crude example of what the stereogram would later supply. The hologram became known as the "Leslie type" (named after the woman, Leslie, who appeared in the first one). Leslie holograms are not too difficult to make, if you have a little patience. The following can be an interesting project:

you will need:

- Movie camera (preferably 16mm, but use what you can get hold of. Old regular eight or double-super eight cameras are best if you use the 8mm format, since the film transports in these are better than the workings of modern super-8 cartridges.

- Turntable, or some method of turning subject uniformly.

- An old projector you can cannibalize.

- An old typewriter (yes, that's right) you can cannibalize.

- A diffusion screen at least 8''x 10'' (preferably larger), for rear projection.

- Materials for constructing a special plateholder as described below.

- Regular holographic optical components.

filming:

When filming, there should be a consistent relationship between the speed of the subject's rotation, the speed of the camera, and the size of the final slit hologram that will be produced from each frame. This will have to be determined experimentally for best results. As a start, suppose our final hologram will consist of 40 slits, and we wish to record about 120 degrees of view of an object. One frame of film should then be shot for every 3 degrees of rotation. If we use a camera running at 24 frames/second, then we can compute the correct turntable speed as follows: 24 frames/sec. X 3 degrees/frame = 72 degrees/second.

72 degrees per second at 360 degrees per revolution means 360/72 or 5 seconds per revolution. This equals 12 revolutions per minute (rpm).

Since 12 rpm turntables are rare, you might look for an old phonograph with a 16 rpm setting which is close enough. The other way this can be done is by single frame animation - a turntable can be rotated 3 degrees by hand for each shot. In either case, a tripod is essential, as it is important that the camera be absolutely fixed during the filming process. Without this precaution, the images in each slit hologram may not be properly registered with respect to one another.

equipment preparation

The back of the projector needs to be open to provide access for the object beam to illuminate the film.

The typewriter will be used to move a slitted mask with respect to a fixed plateholder. If the plateholder is constructed so that it is affixed to the base of the typewriter, and a mask attached to the movable carriage, when the spacebar is hit, the mask will move precisely the same distance relative to the plate each time.

Helpful hints: find the exact distance the carriage will move per stroke and make the slit this width. Otherwise, the slit holograms will either overlap or have blank spaces between them. When fitting the mask to the plateholder, make sure that the mask is almost flush to a plate in the plateholder, to prevent shadowing of the area to be exposed.

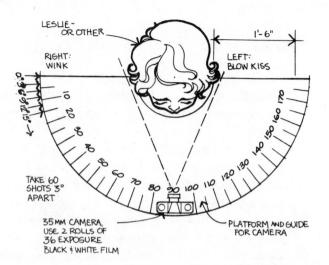

HOLOGRAPHIC STEREOGRAM MADE FROM SLIDES RATHER THAN MOVIE FILM

Build a setup by using the diffusing screen as the object, and bring the reference beam in from overhead. During the exposure, it will be necessary to advance the film and hit the typewriter space bar. The film advance may be difficult in the dark, but a stop may be built onto the flywheel from the drive motor of the projector so that it may be rotated by hand the same distance every time.

The rest is up to you. With a little luck and ingenuity it will work. It did for us.

The finished hologram will operate like a regular transmission hologram. It can be used as a master and treated normally for image planed copies.

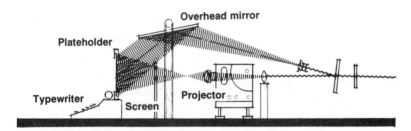

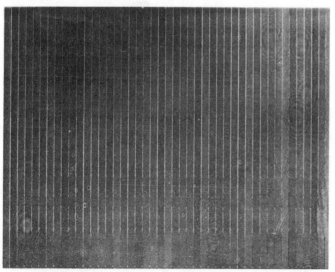

Lesliegram made with the above setup. Hologram and photo
by Bob Schlesinger.

troubleshooting

problem

Images not well registered from slit to slit even
though camera and projector didn't move during
filming or projection.

solution

Try filming and projecting with the camera and
projector on their sides. This should alleviate the
vertical registration problem which is usually due
to imperfections in the film transport mechanisms. With the devices on their sides, you
will be able to see whether the bottom and top
edges are aligned with the sides of the film.

problem

Some slits are bright, others are dim or are not
there at all.

solution

Since many holograms are being recorded on
one plate, it is easy for motion to occur during
exposure of one or more of them. Allow longer
settling times between shots (at least 2 - 5 minutes). Be sure the film is properly positioned in
the film gate so that the full image is being projected onto the diffusion screen.

pulsed holography

Although the pulsed ruby laser predates the continuous wave helium neon laser, the ruby is still much too expensive and difficult to use for the average holographer. Mention should be made of it, however, since pulsed holograms have no stability requirements. Not only is an isolation table not necessary, but holograms of live people, animals, speeding bullets, or anything else moving (just about) can be made. Unfortunately, the full laser systems needed to do this kind of work usually cost in excess of $100,000 (less if you have connections).

The stripped-down ruby laser consists of a synthetically-grown ruby crystal rod, with finely polished parallel ends, and an absolutely homogenous core. The highest purity is essential for the beam to be of good quality. Two parallel mirrors are positioned at either end in the proper position. A power supply directs energy to a flashlamp which, in turn, stimulates the atoms in the ruby rod to form a population inversion. The result is a powerful burst of light which can be made as short as a few billionths of a second with the use of a device called a "Q switch". Additional ruby rods may be used to amplify the pulse, a technique commonly used in holography.

Since it is difficult to "set up" a light of such short duration, a helium neon laser is usually used to map out the beam path. Because of the intense nature of the light, special optics must be used. The use of a spatial filter is not practical so the beam-expanding lenses must of superb quality. Another consequence of the intense nature of pulsed laser light is the potential hazard to the subject's eyes. A diffusion screen must be used to disperse the subject illumination, and care must be taken to insure that no reference light or scattered light reaches the eye.

Notwithstanding the difficulties involved, good quality holographic portraits have been made in the U.S., France, Great Britian, Sweden and the USSR.

PULSED LASERS

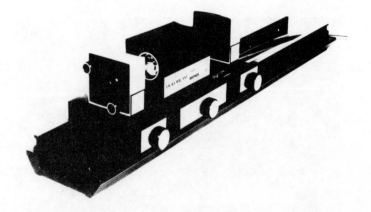

Holographic movies have been made experimentally utilizing a pulsed laser to sequentially record frames of a film. When processed, the film is played back and a three-dimensional moving holographic image is generated. One of the obvious problems with this type of film is the difficulty in viewing it; images cannot be projected onto a flat screen as can conventional films. As with other types of holograms, the images can be seen only by a relatively small number of people at a time. The size of the image is limited by the size of the film used to record it.

Experiments are now underway, however, to develop special screens which may accommodate a projected real 3-dimensional image. Although perfection is possibly years away, work is proceeding in both USA and the Soviet Union. Rumor has it that the Soviets have had greater success with this concept thus far, though their systems are overly cumbersome and expensive.

full color holograms

True color holography has been accomplished; however, it is extremely difficult to achieve. Full color can occur from the mixing of the three primary colors: red, green and blue. Color photography works in this way; different layers of the emulsion are sensitive to each of these colors. When exposed to white light or the colors that make it up, the various mixtures are recorded. Red objects look red, yellow ones yellow, etc.

We are faced with a number of problems when we attempt to do this in holography. The first is the relative difficulty of obtaining the equivalent of coherent white light, or even a laser which produces each of the primary colors. The krypton laser can do this, yet at a price usually in excess of $30,000. An argon laser teamed up with a helium-neon will produce red, green and blue, yet even the smallest argon is at least $5,000 - $6,000 now (we're not even certain since the prices seem to increase every day).

To make holograms with red, green and blue light we need films or plates sensitive to all three colors. Most holographic materials are sensitive to only part of the spectrum. For example, 8E75 will work with red light; however, 8E56 must be used for green. The one readily available emulsion which is suitable for holography and is more or less equally sensitive over the full range is the Kodak 649F. This emulsion is extremely slow, however, usually requiring 10 or more times the exposure of 8E75.

Full color is most difficult to achieve with transmission holograms. We recall what happens when we view a transmission hologram in a street lamp - several images are produced of various colors, all displaced with respect to one another. A full color hologram would need three separate exposures, one for each primary color. Upon viewing, each color would produce a separate image from each of the three interference patterns for a total of nine. Rather than full color, we would have a lot of separate unmixed color images all slightly "out of synch".

The problem is called "cross talk" or cross modulation. Cross talk can be minimized or avoided by choosing separate reference angles for the various colors, while still illuminating the object with all colors. This can become a difficult task with three independent beams as each must be sufficiently displaced with respect to one another, yet be allowed to intercept the plate at reasonable angles without getting in the way of optics, object, etc. The setup can be simplified somewhat by using two of the three primary colors. The result will give a close approximation to full color upon viewing. The accompanying diagram demonstrates one method of producing this type of "two-color" hologram. In addition to taking care to independently match the red object and reference beam distances as well as those of the blue, allowance must be made for the different sensitivity of the film to the different colors. Light readings and ratios should be taken for each color separately. Upon viewing, each reconstruction beam must be of the proper color and intercept the plate at the original reference angles, resulting in a "double illuminated" transmission hologram. Another method is to independently expose each of the colors. This makes it easier to compensate for the differing sensitivity of the film. Needless to

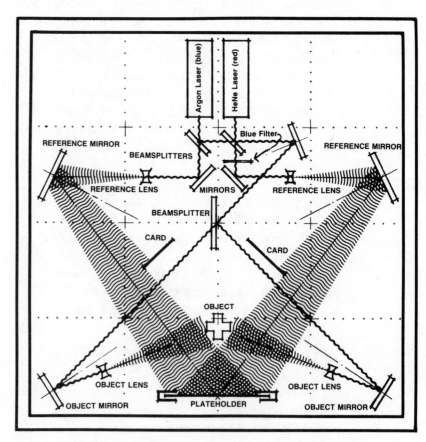

TWO-COLOR TRANSMISSION HOLOGRAM

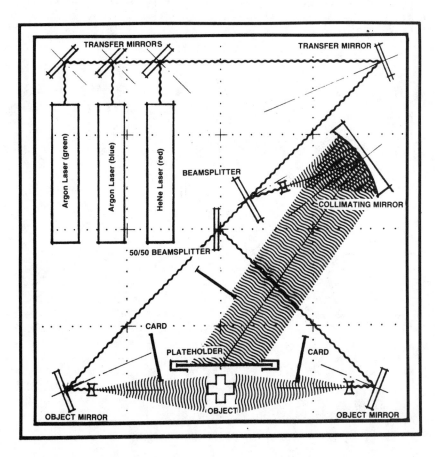

THREE·COLOR REFLECTION HOLOGRAM

say, one might easily conclude that this type of hologram is really not worth the bother - we'll leave it to you to decide, providing you happen to have all the necessary equipment. Several other techniques for producing color transmission holograms have been investigated, yet they don't appear to be any easier. Most require the use of special diffraction gratings, diffusion screens, or lenticular devices. As previously mentioned, rainbow hologram color-mixing can be used to obtain true color. Precise matching of change in angle with chromatic shift is difficult, however, and the viewers eyes still must be located at precisely the right position for true color registration. Reflection holograms seem to show more promise for the further development of color techniques. Since, by nature, reflection holograms are wavelength selective, cross talk is not much of a problem. One simply needs to make the three exposures, one with each primary color, and then process the plate so as to reclaim the original hues. PBQ processing appears ideal for this, or the gelatin may be expanded using triethanolomine (review reflection processing). Again, care must be taken to account for the varying sensitivity of the film to different colors during exposure. For additonal reading (about color holography), see "Optical Holography", chapter 17 (see bibliography).

theory

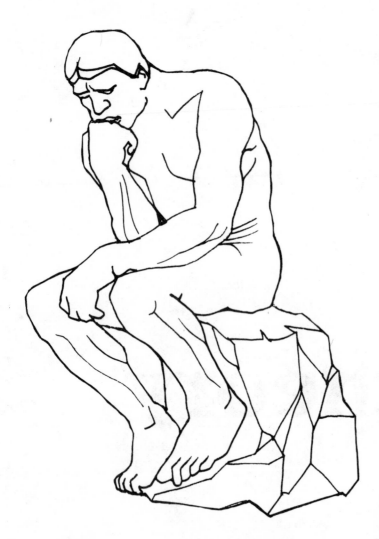

Every holography book worth its salt (and even those that are not) has a theory chapter. We are all intellegent readers and naturally curious as to how and why it does what it does. Yet, most of us are turned off by leafing through volumes of mathematical derivations of wave equations, Fresnel - Kirckhoff integrals and convolution operations (say what?). The question is usually: "How much do I need to know to do holography?" If you've followed the progression of this book, it should be apparent that really all that's needed is the ability and desire to read, and then go make holograms.

The question actually becomes: "How much do I wish to know?". Each reader will have different needs in this department and different expectations as to what we will supply to fill them. Obviously, the chapter cannot serve as the be-all and end-all to everyone. Besides, a full treatment of the subject might add another several hundred pages to the book making it too expensive to buy and too heavy to lift, right?

Even so, we believe it is important to explain the basic goings-on in such a way as to remove any lingering misconceptions that holography can only be attempted by technical people.

With this in mind, we have decided to focus on those questions most often asked by students without science or math backgrounds. Readers desiring more in-depth treatment of specific areas, from either technical or conceptual viewpoints, should consult the bibliography at the end of the book.

There *is* a method to our madness. We have found that holography can best be explained by discussing the properties and interrelationships of light source, object, and observer in the ordinary viewing process, and then applying these properties to the making and viewing of a holographic image. First, we need to know something about light; what kinds exist, how it is produced, and what we observe it doing in various situations. This includes a comparison of the qualities of laser light with those of ordinary sources. Next, we observe how light affects us and explain what really takes place during the process of seeing. This becomes extremely important since, as we shall see, the reason we see a 3-dimensional holographic image is essentially the same as the reason we see dimensions in reality. We take ordinary 3-D for granted without understanding why we see that way, yet are suprised when something like a hologram provides the same effect. The mystery needs defusing.

With this under our belts, we tackle the heavy stuff. We discuss the effects of light encountering different surfaces or materials. We wish to know about the properties of reflection and refraction since holograms often cause light to act in similar ways. Next, we explain how an ordinary photograph is made so we will be able to appreciate how it differs from a hologram.

At this point we have accumulated enough background to approach the actual making of a hologram by action of the property of interference of light. The characteristics of laser light which allow interference to occur are covered as are discussions of stability and film resolution.

In order to explain how the image is reconstructed after the hologram is made, the property of diffraction is investigated. We now begin to understand how certain types of holograms work; why some need laser illumination for viewing while others operate with ordinary white light, why certain effects occurred during the making and viewing of our holograms, and we also discuss some of the limitations of the process. Mention of more advanced techniques including color and fantasies of the future follow.

what is light?

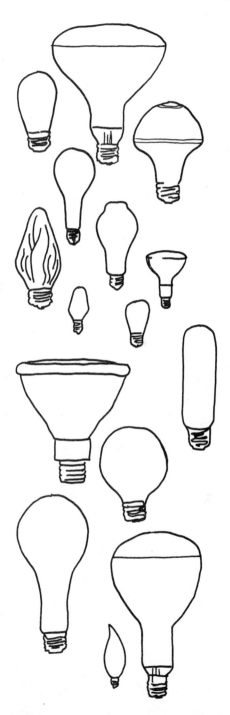

The truth is, no one really knows. The question is a basic one along with "What is matter?" or "What is energy?". The problem is our difficulty in closely observing light; the funny thing is, we can't actually see it (you might laugh, what are we trying to pull here?). Yet think about it a second. As you read this page, light is reflected from it towards your eyes. You see the page, yet where is the light between you and the book? Try pointing it out. The truth is, light itself cannot be seen. As with matter and energy (of which light is but a special case), we only observe the effects indirectly.

One reason for this failing on our part is the speed at which it travels. At 186,282 miles per second (3×10^8 meters) in a vacuum (or a few miles per second less in air), it holds the all-time observable record. Nothing seems to travel faster, and all light travels at the same constant speed in space. Imagine traveling to the moon in less than two seconds; or from the book to your eyes in a *billionth* of a second! No wonder we can't witness light traveling through the air, zipping through windows, bouncing off things and generally going about its "light" business.

Throughout history, theorists have tried to describe the nature and behavior of light. Plato thought that light was emitted by the eyes like "streamers" seeking out

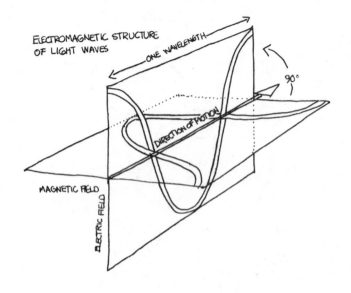

ELECTROMAGNETIC STRUCTURE OF LIGHT WAVES

ONE WAVELENGTH

90°

DIRECTION OF MOTION

MAGNETIC FIELD

ELECTRIC FIELD

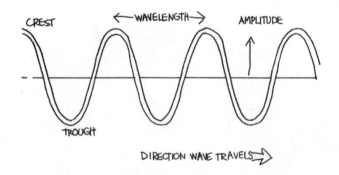

CREST

WAVELENGTH

AMPLITUDE

TROUGH

DIRECTION WAVE TRAVELS

and locating things. Euclid also agreed, explaining: "How else can we explain that we can't see a needle that has fallen on the floor until we seek it out and our eyes have fallen on it?"

Modern concepts of light have been shaped by such scientific biggies as Isaac Newton, Christian Huygens, Thomas Young, Robert Hooke, James Maxwell, Max Planck, Neils Bohr and, of course, Albert Einstein.

Earlier researchers were faced with the dilemma of explaining light as either wave or particle. Sometimes light seemed to act like a shower of tiny bits moving at high speed, and, at other times it seemed to behave as a wave, much like the ocean variety.

In the mid 1800's, wave theory appeared to have the edge with the development of Maxwell's equations. His theory demonstrated how energy could be propagated through space in the form of vibrating electric and magnetic fields. Light was just one portion of this energy, representing a very small section of an "electromagnetic spectrum." All of the various types of energy were waves traveling at the incredible speed of 186,000 miles per second.

Even so, the wave-particle duality remained an enigma until Max Plank in 1900 proposed that electromagnetic waves radiated in discrete small bundles or "quanta". This Quantum Theory of Energy, which is used to give the currently accepted view of light, enables energy to act as both wave and particle, depending on the situation and how it is looked at. Basically, the theory says that, although light can act as a continuous wave, the energy is still some whole number multiple of a single lowest value. This unit, or quantum, is called a photon.

If we make a chart showing the variation of intensity of the energy of a photon as it moves through space over time, it comes out looking like a sine wave. The sine wave, (from the Greek word for circle) can best be explained by using the circle as a study of its anatomy. Picture a wave as two parts of one circle which has been cut through its horizontal diameter and unfolded to form one crest and one trough. The circle starts as a zero point and contains the entire cycle from off (zero) to on to off to on and back to off again. One wave length embodies a complete cyclic change; a gradual progression of ascendance and descendence.

The particle-like photon exibits wave-like behavior. At any given time the light is in a stage of going on or going off as it goes through its cycle. Our visual world, however, doesn't seem to flutter like an old silent movie or pulse like a strobe light. . .it doesn't seem to but it does; it just doesn't reach our brains that way. The explanation goes back to light and its speed. So rapid (in the range of a billion billion cycles per second) is the on and off process, that our eye "sees" (meaning our brain perceives) it as continuous light; unable to make the distinctions between light and dark as quickly as it happens, we see an average of the two, or a blending. In reality, we live in a world that is totally blacked out 50% of the time. Imagine that!

You have probably heard of other types of energy waves that are carried through the universe and also consist of electromagnetic energy. Light, cosmic rays, gamma rays, television signals, radio waves, infrared rays and x-rays are included in this spectrum. How do we differentiate between the various types of radiation?

For each, the length of the wave (that is, from start to finish of a complete cycle) is different. Radio waves can have lengths of hundreds of meters, whereas the wavelengths of cosmic rays may be a billionth of the size of tiny atomic particles. Usable light falls into a narrow median section of the spectrum with wavelengths of about 400 to 700 nanometers (billionths of a meter). The various waves also differ in the *rate* at which they cycle, or their frequency.

Electromagnetic waves show a continuous range of frequencies (frequency being described as the number of wavecrests passing a point in one second). The frequencies of light run from about 400 to 800 trillion (10^{14}) waves per second. Photons are not lazy and are *never* found at rest (that's a fact). Don't forget we said the speed of light is constant, so it follows that the shorter the wavelength, the higher the frequency.

If you spent a half hour pondering over the last sentence in the last paragraph you are a good, honest, thinking person and you will do well because of your qualities. You people that breezed over it are either scholars (we welcome you) or

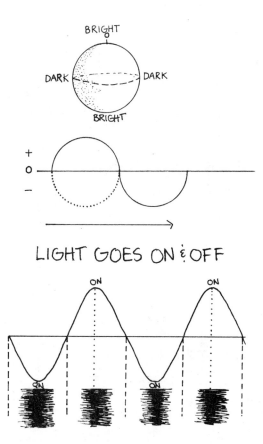

LIGHT GOES ON & OFF

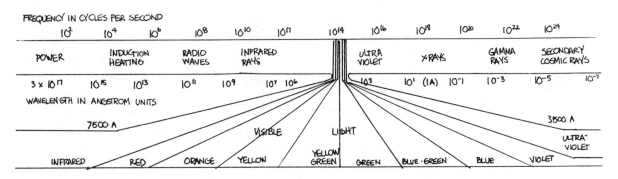

C IS	λ nm.	μ CYCLES/SEC.	COLOR
ALWAYS	650	4.6×10^{14}	RED
186,000	600	5×10^{14}	ORANGE
MILES	570	5.3×10^{14}	YELLOW
PER	520	5.8×10^{14}	GREEN
SECOND	470	6.4×10^{14}	BLUE
	430	7×10^{14}	VIOLET

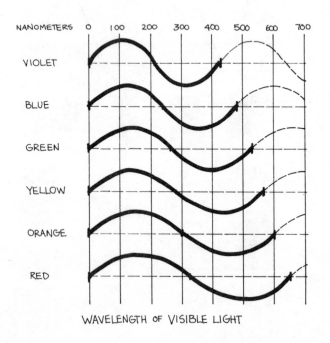

WAVELENGTH OF VISIBLE LIGHT

liars (repent now! Since you have paid for this book, we will consider you a born-again scientist and forgive you). The relationship of the speed of light to its wavelength and frequency obviously needs to be explained in simpler terms. Here is the best way we can do this:

$$C = \lambda\nu$$

Before you attack this page with a blast from a high power laser (you must be furious to see some math after we promised to do without), let us explain. This one's easy: "C" is the speed of light, which is constant - it doesn't change. The "λ", called "lambda", is the wavelength. The "ν", called "nu" is the frequency. Thus, the speed of light equals the wavelength and the frequency multiplied together. Since the speed of light is always fixed, the relationship between the frequency and the wavelength must also be. Let's look at it another way:

We will have to resort to fantasy: suppose you are sub-atomic microman (or microperson if you are offended). It's a nice day and microman decides to set up an observation point and watch some photons go by. We find, especially down here at microman's level, that light waves are very much like ocean waves. Microman selects his observation point so that a photon is at its full brightness (i.e., on the chart it is at its highest point or top of the wave crest) as it passes his observation point. What if it doesn't peak at his observation point? - It has to! Remember, photons down here are not particles, and this is more like picking a spot and watching the crests of waves on the ocean pass through a observation point. Microman starts his incredibly fast stopwatch and times the interval between peak "brightnesses" of the photons as they pass. He notices two different kinds of photons. They are both traveling the same speed (of course), yet one takes almost twice as long to go by as the other. He then measures it and finds that it has almost twice the length as the other. No wonder! Since he has nothing better to do, microman decides to watch a lot of other photons. Each one has a slightly different wavelength and, as a result, different numbers of each pass per second. They all travel at 186,000 miles per second so the more frequent ones are shorter and the less frequent ones are longer. Sounds true and it is true. As wavelengths get longer, frequency gets lower. As frequency gets higher, the wavelengths get shorter. Thank you microman.

In the light spectrum, the shorter wavelengths/higher frequencies are perceived as blue or violet. Red is much slower, and the wavelengths are almost twice as long. All the other colors help compose a continuum of frequencies from one side of the spectrum to the other. One way to look at it is that light is all the same energy, and color is just the rate at which it goes on and off.

Frequency and wavelength are independent of *amplitude*, the next characteristic we wish to consider. This is the brightness of the light, or graphically it is seen as the height of each crest or depth of each trough, from an imaginary horizontal line around which it vibrates. The amplitude of a desk lamp's light is much greater than that of a match and the sun's amplitude is bigger than both of them.

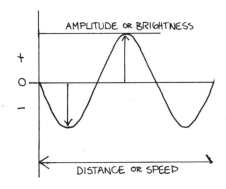

The amplitude of light changes over distance. Light is brighter closer to the source, and dimmer as you get further away. This is because the light spreads out as it travels. The energy in the whole wave front stays the same* yet, since it covers a larger area, the intensity at a given point must be less. For instance, if we are located twice as far away from the light, the intensity will be only ¼ as much. This is known as the inverse square law, which states that the intensity of light falls off as the square of the distance. Three times the distance will be $1/3^2$ or $1/9$ the intensity.

In order to visualize just what the light is doing in this case, imagine a wave front as the surface of an expanding balloon. As the surface moves farther from its starting point, the skin "spreads out" and the intensity (thickness) is rapidly diluted. Another consequence is that a section of the surface appears flatter relative to a similar piece at a point of less expansion. The same phenomenon occurs with light; the wave fronts begin curved, yet flatten out over distance. Yet the speed, frequency, and wavelength are unchanged due to distance or spread. As we shall see, both the amplitude and the phase (which has to do with what part of the wave is passing at a given point) are important in holography.

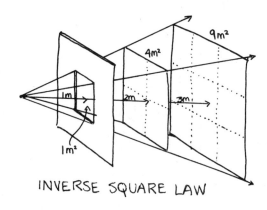

INVERSE SQUARE LAW

*Except that lost to the environment.

how is light produced?

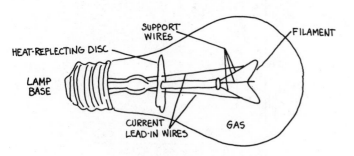

INCANDESCENT LAMP

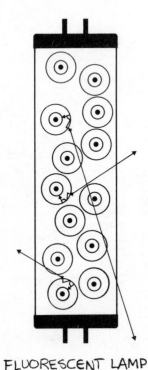

FLUORESCENT LAMP

We are well aware that light can be produced from a variety of sources including the sun and other stars, fire, incandescent lamps, fluorescent lights, phosphors, and, of course, the laser. The mechanism for these may vary somewhat, yet the principle behind all of them is essentially the same.

Let's go back and visit microman, who has now gone to look at a typical atom in a light bulb filament. He sees a myriad of electrons, frantically zipping around, yet each appears to keep its distance from the center of the atom. Someone now comes along and turns on the electricity. Before he gets zapped, microman sees the energy knocks some of the electrons into higher orbits. Since electrons in higher orbits have greater energy, he realizes that the electrons must have absorbed some of the energy from the electricity. The atom is said to be in an excited state. Yet, all good things must come to an end. So suddenly, without warning, the unstable atom emits a photon. The electron returns to its normal position and the atom resumes its ordinary "relaxed" state. These photons make up the light we see emitted from the bulb. Light is caused by the movement of electrons from higher to lower energy levels. Microman is pleased.

Each element has its own number of electrons and energy levels. Since there are many paths the electron may take to jump from level to level, a given atom may emit photons of many different frequencies. The overall characteristic color of a given light is caused by the combination of all of the individual frequencies. White incadescent light is formed when frequencies comprising most of the visible spectrum are combined fairly equally. Light from a mercury-vapor street lamp has a bluish-white tinge, which means that, although it emits light of many colors, it has a preponderence of blue and violet frequencies.

The sun, which contains some of all the elements, is essentially a white light source. Overall color characteristics of given stars vary somewhat, depending upon their overall composition and density (more closely packed atoms will influence their neighbor's energy levels to a greater degree, leading to more frequency combinations).

A neon sign, as we know, emits a characteristically red color. This is because the average difference in neon energy levels is proportional to the frequency of red light. (This actually becomes an important factor in understanding how the laser works.)

All of the types of light we have discussed thus far have one thing in common (besides the fact they are visible) - they are *incoherent*. What do we mean by incoherent light? We recall that when microman observed the emission of light from an excited atom in a light bulb filament, the photon was produced *spontaneously*. We could not be sure which electron would go from which to what energy level and exactly when it would take place. If we now look at the whole filament with its countless trillions of atoms, we observe an incredible confusion. Atoms are releasing different frequencies at slightly different times, causing them to be out of phase (two waves are out of phase if their crests and troughs don't line up in the same place) and traveling every which way.

Incoherent light can be compared to handfuls of gravel thrown into a pond. The resultant cacophony of waves interacting with one another represents the nature of most of our light sources. Unfortunately, incoherent light is not usually of much use for forming holograms.

There is a second method of emitting energy from an atom. This involves stimulating it with an electromagnetic field of the proper frequency. The emission induced by this process will have the same frequency and phase as the energy used to stimulate it. If this stimulated emission now induces emission from more atoms, both the new and old waves will be in phase and of the same frequency. The result is an increase in amplitude of the light (amplification). This is coherent radiation.

Prior to 1960, there were no known sources of true coherent light. The development of the laser (Light Amplification by Stimulated Emission of Radiation) offered, for the first time, a source of light of exactly one single frequency. Although originally touted as a solution looking for problems, the laser now has applications too numerous to mention. It provides the "purest" kind of light known.

Coherent light can perhaps best be explained by illustration. If you could arrange for a million people to drive identical 1971 red Volkswagens through a tunnel at a constant speed keeping an arbitrary but standard distance between car bumpers, you would have a fine analogy to coherent light. If, on the other

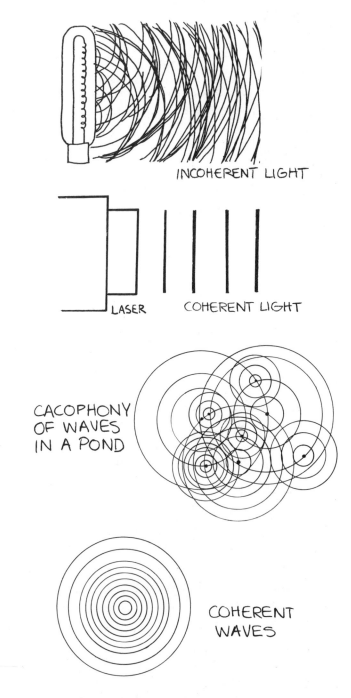

INCOHERENT LIGHT

LASER   COHERENT LIGHT

CACOPHONY OF WAVES IN A POND

COHERENT WAVES

hand, you gave a million people a million different vehicles: DeSotos, Winnebagos, Chevy Vans, Mopeds, Studebakers, etc. and then let them loose on the Bonneville Salt Flats with the instructions to choose their own direction and go, the resulting chaos would remind you of incoherent light.

It's amazing we can do anything with incoherent light. Imagine trying to draw a picture with 3,000 differently colored pencils held in your hand at once. The pencils, or multicolored and displaced frequencies of light, are unmanageable and conflict with one another. A much clearer picture is to be had if we throw away 2,999 of the pencils. We will use this concept later to demonstrate why a laser must be used to make the hologram. But first, back to more about the laser itself:

what is a laser?

There are a great many different kinds of lasers in use these days and more are popping up all the time. The most popular is the helium-neon laser, first invented in 1961, which is also used most often for holography. We shall use it as a model to explain how lasers work.

A laser of this type consists of a glass tube filled with a precise mixture of the two gases, helium and neon. The tube is sealed, yet remains transparent at both ends. At one end, a 100% reflective high quality mirror is placed. At the other end is placed another mirror which is only about 98% reflective or just a wee bit less. The placement of the mirrors is critical - they must be perfectly parallel and spaced a distance apart which is an *exact multiple* of the wavelength the laser will produce. Remember, the wavelengths vary by only a few billionths of a meter, so if the mirrors are only slightly out of alignment, it won't lase. A very precise art, this laser building.

If we wish to produce coherent light, the proper energy must be radiated into the tube to cause electrons to reach higher energy levels. This the power supply does, by first stimulating the helium atoms. The neon atoms are then elevated to a higher energy state by collisions with the helium. As the excited neon atoms return to their relaxed or *ground state*, red photons similar to those produced in a neon sign are released. However, these photons in turn stimulate neighboring atoms. The process continues until a majority of the atoms are in the excited state. This is an unnatural condition and is known as a *population inversion*.

Most of the light produced escapes from the sides of the tube; however, a very small amount begins to cycle back and forth between the two mirrors. Since the mirrors are positioned an exact multiple of wavelengths apart, each adds on top of the next, greatly increasing the amplitude. Only the wavelengths that are exactly

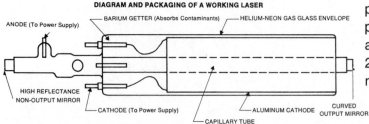

DIAGRAM AND PACKAGING OF A WORKING LASER

ANODE (To Power Supply)
BARIUM GETTER (Absorbs Contaminants)
HELIUM-NEON GAS GLASS ENVELOPE
HIGH REFLECTANCE NON-OUTPUT MIRROR
CATHODE (To Power Supply)
CAPILLARY TUBE
ALUMINUM CATHODE
CURVED OUTPUT MIRROR

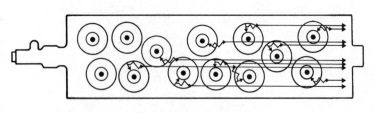

LASER

the right length, in this case 633 nanometers (6328 angstroms for those who are more precise), are amplified between the two mirrors. Finally, the light becomes bright enough to exit through the mirror which is not completely reflective, and we see a beam of coherent, single frequency light.

There are, of course, other kinds of lasers, although many are not suitable for holography, are too expensive for most people, or are difficult for most to use properly.

The first laser, built in 1960, did not use a gas-filled tube, but rather a solid-state ruby rod. Thie artifical ruby (unlike the natural gemstone) is carefully machined and polished with the ends exactly parallel. This laser emits a short pulse of light rather than a continuous beam, since the atoms go from their excited state to ground state all at once. This property is often valuable (see pulsed holography).

A laser produces a relatively small amount of light, however, since it is concentrated into a narrow beam, it is extremely intense, and does not obey the inverse square law. This is why the unspread beam can be harmful to the eye; our lens will focus this intense light on a tiny spot, causing lesions. Most HeNe lasers put out only a few milliwatts or thousandths of a watt while, by comparison, a dim Christmas tree bulb produces about one watt. A laser which produces a beam of one watt or more is capable of burning things. Incredible, huh?

There are, of course, many types of lasers other than the helium neon gas laser and ruby pulsed. A few of the more popular appear below. The range of different wavelengths they emit is also noted in the spectrum found on the back cover of this book.

RUBY CRYSTAL ROD
PHOTO FLASH TUBE
COHERENT LASER LIGHT
PULSED RUBY LASER

	TYPE OF LASER	PRINCIPAL OPERATING WAVE LENGTHS (μm)		DISTINGUISHING CHARACTERISTICS	PRESENT AND FUTURE AREAS OF APPLICATION	
Gas	Helium Neon (He-Ne)	0.6328 3.39	1.15	Highly monochromatic. Visible output. Rugged. Simple. Long life. Inexpensive. Low power.	Interferometry Information processing Military systems	Optical alignment Holography
	Helium Cadmium (He-Cd)	0.4416	0.3250	Highly monochromatic. Visible and UV output.	Photochemistry Spectroscopy Display	Biochemistry Data recording
	Carbon Dioxide (CO_2)	10.6		High power. High efficiency. IR output. Detector limited.	Materials working Communications	Military systems Medicine
	Argon (A)	0.5145 (strong) 0.4965 0.4765 0.4658 (weak) 0.3638	0.5017 0.4880 (strong) 0.4727 (weak) 0.4579 0.3511	High power (CW). Visible and UV output. Selection of many wavelengths. Highly monochromatic. Inefficient.	Display Spectroscopy Medicine Pump for parametric devices	Holography Information processing Military systems
	Krypton (Kr)	0.6741 (strong) 0.5208 0.4762 0.3564 0.6471	0.5682 0.4825 (weak) 0.4680 0.3507	Similar to argon but with the range of wavelength extended into the red, thus providing white light but with reduced efficiency.	Display Spectroscopy	Holography Medicine
	Nitrogen (N_2)	0.3371		UV source. Inefficient. Very short pulses.	Photochemistry Spectroscopy	Biochemistry Data recording
Crystal and Glass	Ruby	0.694		High peak power. Moderate spatial coherence.	Materials working Medicine	Military systems
	YAG: Nd	1.064		High average power. Highest efficiency of optically pumped lasers. Compact laser head.	Military systems	Materials working
	YAG: Nd doubled	0.421	0.532	Efficient. Compact laser head.	Military systems (especially under water) and pump for parametric devices	
	Glass: Nd	1.06		Glass can be formed in any shape and size desired. Low cost of glass compared with crystals. Excellent optical quality. Highest peak power. Low average power. Relatively broad spectral line.	Materials working Plasma research	Military systems
Semiconductor	GaAs/GaAlAs — Single diode Arrays of diodes	— 0.85—0.91		Very small. Incoherent arrays practical. Inexpensive. Efficient. Poor coherence. Frequency temperature sensitive. Low power.	Military systems Security systems (intrusion alarm, night vision aids, etc.)	Short range communication
Liquid Dye	Pumped with an N_2 laser Pumped with a flash lamp	0.36—0.65 0.32—1.0		Tunable over broad range. Liquid medium facilitates cooling.	Remote sensing (i.e., air pollution) Photochemistry Medicine	Photobiology
Chemical		1—30		High energy and power (energy provided by chemical reaction).	Military applications	Materials working

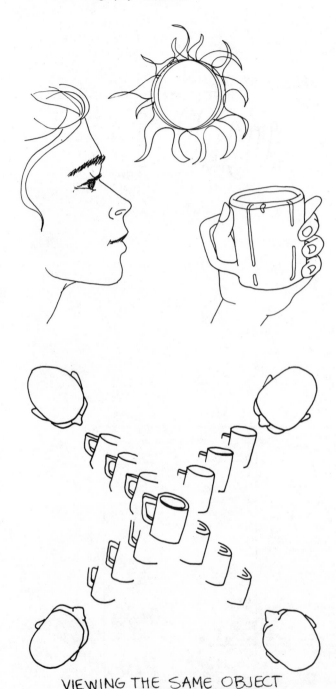

VIEWING THE SAME OBJECT
FROM DIFFERENT POSITIONS

perception of light

Now that we know a little about the mechanism and production of light, let's consider how we actually see it (or rather, its effects). First of all, there are three things required to see a given object. Can you name them? (All readers who successfully answered this question without peeking ahead to the next sentence are immediately nominated for placement in the Wowie - Zowie Holography Hall of Fame). The answer is light (of course), an object (also obvious) and...an observer (not so obvious). We hope you don't think we are insulting your intelligence - the point is an important one.

At the risk of sounding somewhat facetious, we would like to ask that you perform the following exercise: find a room that can be made completely dark (i.e., one where you can turn off all the lights with a flick of the switch so that you can't see a thing). Next, in the middle of the room place an object, a coffee cup for example. With the light on, look at the cup. *Really* look at it; its surface, angles, outline, every bit of visual information you can take in. Now turn off the light. Now what do you see? Repeat this several times. Now, do we really see the cup itself or *just the light reflected from the cup?*

We relate to light in the same manner a fish relates to water: it surrounds us yet we take it totally for granted. The visual process is wholly dependent on our ability to see light reflected from objects, not the objects themselves. Light is the medium by which *information* regarding characteristics of the object is carried to us. Another experiment: again look at the cup in the middle of the room. Where is it? In the middle of the room, right? Now walk to some other place in the room and look at the cup. In fact, walk around and observe it from numerous positions. Can you still see it? Why? Because the light reflected from the cup can travel to any part of the room (unless there is an obstruction). Now, can't we say our sense of "cupness" is located everywhere in the room? At the same time, some other object, be it a table or doorknob, also fills the room with its reflected light. So here we are, walking around a room filled with cacophony of jumbled reflected wave fronts, all passing through each other like so many individual waves on a pond sprinkled with rain. Yet each individual wave maintains its integrity and individuality as it shares the same space with all the others. It is really quite incredible that our eyes and brains are able to make some sense out of all this and can actually perceive things.

The eye is a remarkable organ. The iris works like an automatic diaphragm regulating the amount of light passing through the pupil to the lens. The soft adjustable lens has the ability to focus light instantly to the retina, usually in the area of the fovea.

At the retina, photosensors convert the energy into a weak electric current. This current travels via the optic nerve to our brain where a complex of billions of nerve cells compares these signals to those of memory and prior experience. This is where we experience the vision of reality, in our minds. The eye is indeed the portal to the world around us.

One of the very powerful realizations we are seized with is the extent to which our perception of an object differs from the reality of it.

We start with the object - information via light is reflected from it in all directions. Our tiny eyes pick off a minuscule portion of this information and cause it to be changed once more by focusing it to a spot on the retina. This in turn undergoes another change to electrical impulses. And still another follows when it is received and interpreted by the brain. It is incredible how much we miss in what we see. No wonder we can fool ourselves by looking at a holographic image!

The fact that we have two eyes results in a rather remarkable phenomenon, the perception of depth, or dimension. Each eye picks out a slightly different perspective on a scene or object and the two are combined in the brain to yield an *illusion* of depth. Each eye really sees only two dimensions. Don't believe it? Try sitting in a chair with your head as still as possible and look across the room. Now close one eye, and imagine for a moment that you are looking at a picture of the room rather than the room itself. Now slowly open the other eye. You may be almost startled by the depth seeming to pop out of nowhere.

Our eyes are separated by approximately 2½ inches. This causes our visual field to be made up of two overlapping two-dimensional images. The right eye sees around the right side of an object, while the left eye fills in information not seen by the right eye. The integration of these two views occurs in the brain, and results in *parallax*, or the sense of 3-D.

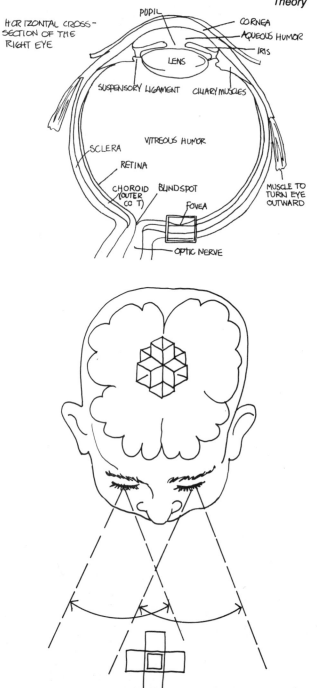

HORIZONTAL CROSS-SECTION OF THE RIGHT EYE

PUPIL
CORNEA
AQUEOUS HUMOR
IRIS
LENS
SUSPENSORY LIGAMENT
CILIARY MUSCLES
VITREOUS HUMOR
SCLERA
RETINA
CHOROID (OUTER COAT)
BLINDSPOT
FOVEA
MUSCLE TO TURN EYE OUTWARD
OPTIC NERVE

FORMATION OF 3-D IMAGE IN THE BRAIN FROM PARALLAX

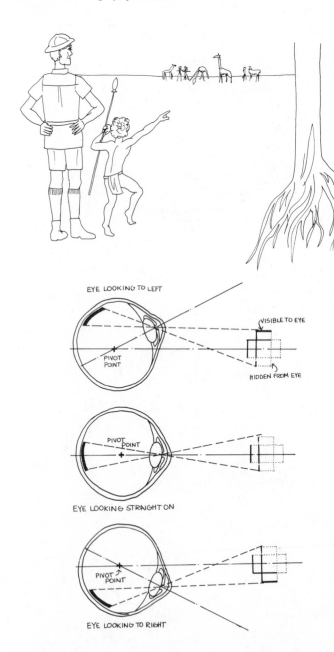

EYE LOOKING TO LEFT

VISIBLE TO EYE

PIVOT POINT

HIDDEN FROM EYE

PIVOT POINT

EYE LOOKING STRAIGHT ON

PIVOT POINT

EYE LOOKING TO RIGHT

PIVOTING SINGLE EYE CAN SEE THREE-DIMENSIONAL VIEWS

Another revealing exercise: draw a vertical line on a cup. Hold it out at arms length and look at it with the left eye (cover your right). Now slowly move the cup to the right till the line is just out of vision, then open the right eye and take notice of the different angle of view you gain of it. Switch eyes, and you'll soon see how two eyes working together create true three-dimensionality. Three-dimensional viewing may also result from the effect of one or more secondary visual cues, most of which are acquired by learning. When one object "overlaps" another and partially blocks it, it becomes apparent that the former is closer than the latter.

Distance from the viewer will establish size and scale. A six-foot man looks smaller as he moves further away. His known size helps to establish the scale of the objects in his vicinity, even though the distance "shrinks" him. As the man approaches and fills more of our visual field, he appears larger, relative to the other objects. We know from experience that his size doesn't actually change; we attribute the change to a shift in position in this dimension.

A case in point is the discovery of an African tribe of pygmies living in the densely foliated heart of the jungle. Because of their environment, they never had the opportunity to see things at a distance. A group of anthropologists took them out into a clearing at which time they became very excited by what they saw. When asked why, they pointed to the horizon and said, "Look, men and animals smaller than us!". They lacked familarity with long-distance vision and scale.

Another cue is the effect of the atmosphere on distant objects. Mountains, clouds, and cityscapes often take on a bluish tinge since the air generally reflects blue light. We learn to associate these shades with dimension, adding great depth to our perception of a scene.

A single eye can provide dimensional viewing through motion. This is why you may still have a sense of 3-D when closing one eye; the other eye may not be absolutely still. The moving eye sends a multitude of different views to the brain where they are integrated in a manner similar to that of two eyes.

In all cases, our minds continually receive and assimilate these ongoing sets of differing images and visual cues to provide us with the sense of shape, depth and substance of reality. True realistic three-dimensional viewing is thus an integration of these various factors. And where is the 3-D actually? It's in our minds! We project 3-dimensions out into the world around us - they are produced entirely in our heads. And, with the miniscule amount of information our eyes suply, only seeing certain frequencies and from very limited viewpoints, what we see and what's really out there may be quite different.

reflection

Thus far we have discussed how light is produced and how it is interpreted by the eye and brain. Time now to get down to the technical stuff (we'll keep it simple). Our concern first is what happens to light in transit (when it interacts with various types of materials).

Usually, uninterrupted light will travel in a straight line. There are cases, however, when light is made to bend. One of these involves using a polished reflector such as a mirror.

The law of reflection states that the angle of incidence is equal to the angle of reflection, or , in other words, a ray striking a mirror will leave it traveling at an equal angle, but in an opposite direction.

The characteristics of the light: phase, amplitude, frequency, etc. are generally unchanged by a mirror. A change in direction is the only significant alteration. Any information the light is carrying is not changed; this is why we can see people, things, etc., while looking into mirrors.

Diffuse reflection is another story. Most objects are diffuse reflectors, as their surfaces are not smooth enough to prevent significant alteration of the light. Light striking a diffuse object will be reflected in many directions, while some of the light will be absorbed. Objects have characteristic colors depending upon which frequencies are being absorbed and which are being reflected for us to see.

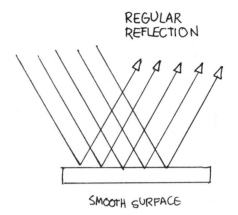

REGULAR REFLECTION

SMOOTH SURFACE

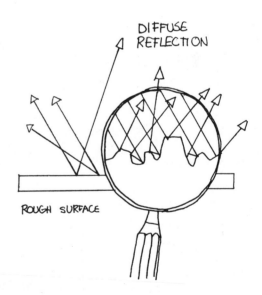

DIFFUSE REFLECTION

ROUGH SURFACE

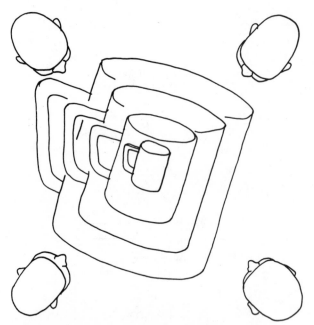

EXPANDING OBJECT WAVEFRONT
REACHES ALL VIEWERS

When we observed the coffee cup, we were looking at a diffuse object reflecting light back into the room in all directions. When we make a hologram, we also wish to use a diffuse object, so that light will be reflected to all areas of the plate. Since we are using a red light, we select objects which reflect this color well.

Another interesting characteristic of a diffuse reflector should be noted. When light strikes an object, each infinitesimal point on the surface reflects its own wave fronts. Just as we saw that the cup reflected light to all points in the room, so too we might say that each and every point on the cup also sent light off in all directions. Try this: make a small dot on the cup with a magic marker. Now place it on the other side of a window and look through the glass at the cup. If you look through the upper right-hand corner of the window, you can see the spot. So, too, with the lower left corner. *Light reflected from that spot must be reaching all parts of the window.* This is extremely significant. Would you be surprised if we now asked you to break the window, or board it up except for a small piece (you don't actually have to do this, least we be criticized for promoting destructive behavior). Looking through the piece, you can still see the whole object. For exactly the same reason, you can break a hologram and see the whole image by looking through a small piece. A hologram operates exactly like a window.

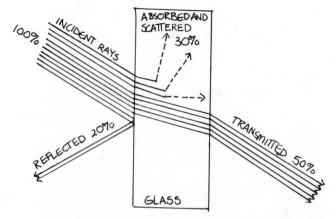

REFRACTION OF LIGHT BY GLASS

refraction

Light is also bent while traveling obliquely through transparent material. This is due to a change in the speed of the light as it interacts with the new material. Light travels more slowly through a dense substance, and thus bends somewhat to compensate. After it leaves the material, it resumes its normal speed and the direction again changes to accommodate this. Refraction is responsible for many illusions. Try inserting the edge of a straight ruler into a bowl of water, and observe the ruler appearing to bend. Refraction is also responsible for underwater objects appearing to be closer to the surface than they really are, and for other phenomena, including mirages. Perhaps the most notable effect is that of a rainbow , or a prism. The various colors or frequencies are separated since each frequency interacts with the prism or raindrop to a different degree, and changes to a different speed. When the surfaces are arranged so that the light bends twice in the same direction, a fairly large *dispersion* results[1].

Refraction is used to advantage in the manipulation of light through a lens. There are two general types of lenses. When a beam of parallel light travels through a lens with convex surfaces, the light converges to a point. This is called a *converging* or *positive* lens. When the light travels through a lens with concave surfaces, the beam spreads out. This is called a *diverging or negative* lens.

All lenses have a *focal point.* The focal point of a positive lens occurs where the light actually converges to a spot. Since a lens has two surfaces, there are two focal points, one on either side. The *focal length* is the distance from the center of the lens to the focal point. The focal lengths of various lenses vary, depending on the degree of curvature of the surfaces, and the relationship between the two surfaces themselves of a given lens.

A negative lens also has a focal point, even though light is not actually focused to one. Since light passing through a negative lens diverges, it will *appear* to come from a point behind the lens. This imaginary focal point is located a "negative" focal length from the center of the lens. Most of the lenses we use in holography to spread the beam are of this type. Positive lenses may also be used, since the light will spread *after* it converges to a focal point. However, a more rapid spread is usually obtained with a negative lens. The rate of spread is greater with a short focal length lens, and is less with one of longer focal length. Thus, for example, if we are using a lens with a -28mm focal length and the light does not cover a great enough area, we may try a -13mm or similar replacement.

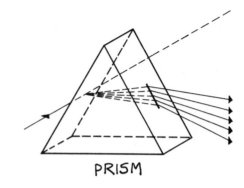

PRISM

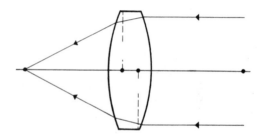

POSITIVE LENS

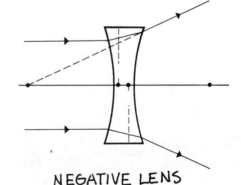

NEGATIVE LENS

---

[1]An excellent discussion of the formation of a rainbow, as well as most other characteristics and qualities of light, will be found in *Conceptual Physics* by Paul Hewitt p. 452 (see bibliography for more information).

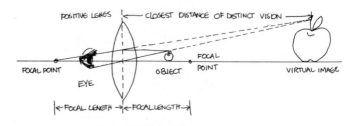

VIEWING WITH A LENS

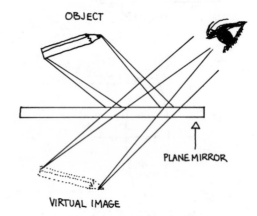

VIEWING WITH A MIRROR

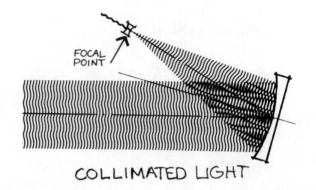

COLLIMATED LIGHT

Positive lenses may form either *real* or *virtual* images. A virtual image is formed if the distance between the lens and an object is less than the focal length. The image is called virtual since it is not physically located anywhere, it only "appears" to be, and cannot be projected onto a screen or other surface. A magnifying glass works in this way; the object appears to be enlarged but in reality is not. A real image is formed if the object is located beyond the focal point. The image is real in that it can actually be focused onto a surface. The lens of the eye as well as the camera lens both operate in this way. Unlike the virtual image, the real image is inverted with respect to the original object. Compensating optics are used in devices such as binoculars to re-invert images. Our brains provide this service to our own optical systems. We actually see the world upside down!

Negative lenses differ from positive ones in that *only* virtual images may be formed. These lenses can never actually focus light to a point so that they can never image onto a surface.

Mirrors can also form real or virtual images. Ordinary flat mirrors form virtual images, since the reflected scene only appears to be located behind the surface. Concave mirrors, however, can actually focus light in a manner similar to positive lenses, and form real images. Actually, for certain applications, convex positive lenses and concave mirrors can be used almost interchangeably. A telescope, for example, can utilize either a lens or a mirror as its major light-gathering objective.

Collimators, which are really telescopes running in reverse, may also use either lenses or mirrors. Basically, a telescope gathers light by causing parallel rays to converge. A collimator, on the other hand, takes light spreading out from a point and converts it into parallel rays.

Some of our holographic setups utilized collimated light provided by a large concave parabolic mirror. A large lens could have also have been used for the purpose, but this would have been extremely expensive. The cost of a concave mirror can be less than one tenth that of a comparable lens.

Lenses or curved mirrors can also have an effect on the *shape* of the wave front of light. Earlier we discussed the analogy of light spreading from a point and taking on the appearance of an expanding balloon. The surface is curved and is thus considered to be a *spherical* wave front. Lenses will greatly affect the degree of curvature of the front. For example, light from the laser can be considered parallel for the most part. The wave fronts are flat or are *plane waves*. When the beam is passed through a lens, the light not only diverges (negative lens) or converges (positive lens), but also becomes curved in a direction opposite that of the lens surface. Conversely, a spreading curved wave front can be converted into unspreading plane waves, which is what occurs with the use of the collimator. Why do we care about curved or plane waves? For one thing, curved wave fronts, if severe enough, will cause image distortion. This occurs noticeably in photography, when a fisheye lens is used. In holography, a curved wave front may affect different areas of the holographic plate differently. With some, such as the rainbow hologram, this can become a real problem, and the use of a collimator to correct it is valuable. With many types of holograms, however, it is usually not critical. The curvature will be greater the closer one gets to the lens, so it is usually best not to position the reference lens or spatial filter too close to the plate.

Imperfections or distortions in image formation by optics are called *aberrations*. *Spherical aberration* occurs because light passing through the edge of a lens does not focus at precisely the same point as light passing through the center. Similarly, light from a point will only form an approximate plane wave front when passed through a collimator. Eventually the light will slowly spread out.

Chromatic aberration also occurs in lenses, since different wavelengths of light will refract differently. This is not a problem for us when using a lens with the laser since we are using only one wavelength.

At this point, we should have a conceptual understanding of how light is produced, observed, affected by objects, and manipulated by lenses and mirrors. We now wish to consider how a photograph is made.

When we "take a picture", we have three things: the camera, a subject, and light. The camera consists of a positive lens array (several are usually used to reduce chromatic and spherical aberration) and a film plane which is located beyond the focal point. A real image can thus be projected onto the film. Photographic film is simply a gelatin-coated piece of plastic. Suspended in the gelatin or emulsion are silver bromide crystals which have the unusual property of decomposing when exposed to light (the silver is reduced to its nascent state for those of you into chemistry).

Suppose we take a picture of a person wearing a white shirt with black stripes. Light reflecting from the shirt bounces off in all directions. The white areas of the

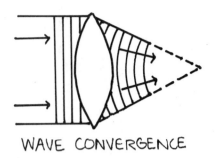

WAVE CONVERGENCE

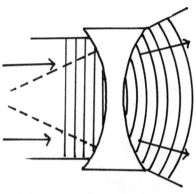

WAVE DIVERGENCE

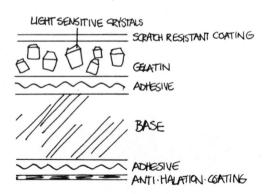

PHOTOGRAPHIC EMULSION

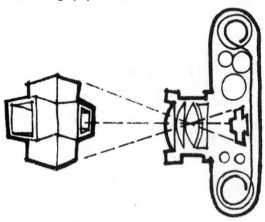

HOW CAMERA SEES AN OBJECT

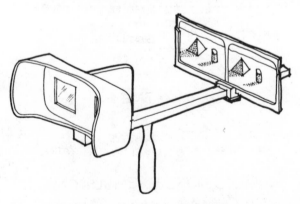

STEREOGRAPH VIEWER

shirt reflect more light (greater amplitude) than the black stripes which absorb light. Now we place the camera at a given point to attempt to "capture" the effect of a small portion of the reflected light. The light passes through the lens, is focused to a point, and then spreads out to form an image on the film. This image contains information about the relative brightness/darkness of areas of the subject. The brighter areas will cause more of the silver bromide to decompose than the darker stripes. When the negative is developed, the areas of greater exposure turn darker than the less bright areas that were very dark as clear spots on the film. We can say then that the negative is an *amplitude recording* of the subject, although in reversal (white becomes black and vice versa). When a print is made from the negative, the blacks and whites are again reversed and the proper shades are back to normal. (In color photography, this is done by using the reversal primary colors, their complementary hues, and combinations thereof to achieve other colors.)

What else is recorded in the photograph? Why, the recognizable shape and form of the subject, of course. All of the black and white particles when looked at "en masse" actually *look like* the person we took the picture of. Yet something is still missing. Yes, you guessed it. There's no actual *dimension* to the photograph. Why is this?

We recall that the sense of dimension or parallax occurs in the mind when views from two different perspectives are combined (from two eyes, or one eye in motion assimilating different views). Yet the photograph is made from *only one single perspective*. Remember, the lens, in order to make an image, has to focus the light through a single point. *This then limits the perspective of the entire photograph to that of a single point* Each point on the original subject is then focused to a single point on the film. When we observe the finished print, even though we use two eyes, information from all other viewpoints has been lost. Both eyes see as if located in the same point and the picture appears flat.

Stereophotography attempts to overcome the problem by adding one more point of view. When a stereo-pair is made, two photographs spaced apart a distance equal to that between our two eyes are taken. When the left and right eye each sees its respective viewpoint only, the mind is "fooled" into seeing dimension. To work properly, a special viewer or glasses must be worn to make certain the proper picture goes to the correct eye.

Besides the inconvenience of using special viewers, stereophotography, in utilizing only two viewpoints does not give us the same *kind* of dimension we find in reality. The two viewpoints are fixed; we can't move our heads and see around objects. The dynamics of how much of our field of view an object occupies is gone; the motion of a closer object relative to one further away is lost. We would have to add a great many more viewpoints to achieve these phenomena. Can this be done? You betcha.

Remember, it's the light which acts as the carrier of visual information. The photograph stores part of the information - the amplitude and even the frequency/wavelength, in the case of a color photograph. Yet an important physical quantity has been omitted. This is the *phase* of the light, the relative positions of the wave fronts. This is really what's happening out there in the world before the lens in our eye distills it down to a concrete image. The hologram records both phase and amplitude. The word hologram, meaning "whole message" or "whole recording" is justly named. In order to understand how and why the hologram does this we need to investigate two more properties of light:

interference & hologram formation

Earlier we spoke of the way in which light takes on the character of waves. Suppose we have two waves which "run into" each other. What happens?

Waves of any kind superimposed on one another produce a new wave that is different than either original wave alone. The crests and troughs of both will combine either to form greater crests and troughs (constructive interference) or smaller ones (destructive interference). This can be illustrated by sitting in a bathtub and slapping the surface of the water with both hands to start up two sets of waves. Watch what happens when two crests meet; they form a new crest equal to both added together. Similarily, if a crest meets a trough they cancel each other out.

Light waves work in a similar fashion. Ordinary multifrequency incoherent light is already a very complex wave form. When this light meets another complex

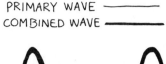
PRIMARY WAVE ————
COMBINED WAVE ————

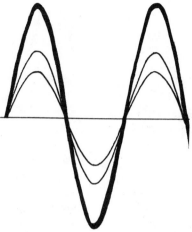
CONSTRUCTIVE INTERFERENCE

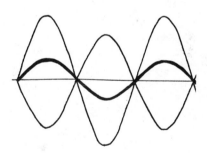
DESTRUCTIVE INTERFERENCE

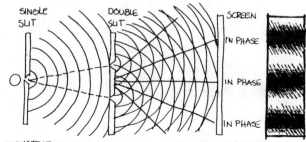
DEMONSTRATION OF INTERFERENCE CAUSED BY LIGHT FROM SINGLE SOURCE PASSED THROUGH TWO APERTURE

INTERFERENCE OF LIGHT
DIFFRACTING THROUGH
TWO APERTURES

form, interference takes place. Yet, the effects are only recognizable as changes in hue. This incredible jumble of frequencies provides almost no clue as to the relative positions of its component parts. This makes recording a hologram impossible in most cases. It really is like drawing a picture with a few thousand pencils.

However, when light of the same frequency combines, very dramatic effects occur, as visible as those in the bathtub. If the two waves meet "in phase," which means the crests and troughs line up in the same position, the new wave formed will have an amplitude equal to both originals together...if, however, they are 180 degrees out of phase - each will cancel the other out and the result is a much weakened wave.

In actuality, when two given waves meet, there will be areas of light and darkness. The size and shape of the fringe is determined by the rate at which the two waves go in and out of phase. This rate can sometimes be rather slow, resulting in large visible fringes of the type we observed after setting up the interferometer (p. 48 - 49).Fringes will also be formed if a coherent wave is allowed to pass through two or more fine, closely spaced slits (see diagram).

If we place a light-sensitive plate or film in some area where the wave fronts are interfering, we may record an interference pattern. The interference pattern, then, gives us a picture of the relative phase relationship between the two waves. By storing the areas of light and darkness formed by the combining of the two wave fronts, we capture all of the characteristics of the individual waves which we could not do independently.

Why does this work? Remember, light travels so quickly that we cannot hope to notice the individual positions of photons, yet we can observe an aggregate effect. Since coherent waves are all in step, exactly alike, each succeeding wavelet takes the place of the preceding cycle, for as long as the light is generated. We thus have what amounts to a standing wave pattern - energy moves through the system yet the waves "appear" immobilized. Now, if this wave front interferes with another, an apparently stationary pattern is formed storing the characteristics of each. Either can be recalled by stimulating the pattern with the other, or in effect "unforming" the interference pattern. What an amazing concept!

A hologram is nothing more than a stored interference pattern. This pattern is formed as a consequence of reflected light from an object interfering with a reference beam. We recall that any spot on a diffuse object reflects light in all directions. Therefore, each and every spot on the object will establish its own interference pattern with the reference beam.

Let's return to the example of the coffee cup for a moment. We imagine the small dot we made with a marker, sending light off into the room. Now, if we illuminate the cup with coherent light, that spot will generate a single frequency wave front of itself. We now place a holographic plate nearby, so that the light from the spot will be visible from all parts of the plate. We then add a pure beam of light, the reference beam, to cause an interference pattern. A set of fringes is generated so unique that *it can only be caused by reflected light from that particular spot being located where it is.* And, the plate continues to see the spot from any place on its surface. Now we also realize that any other spot must also form its own interference pattern, and an aggregate pattern of all of the individual spots must exist. If we add up all of the spots, we discover that reflected light from the whole coffee cup is stored all over on the plate, by virtue of an enormously complex interference pattern.

We have found the above to be an extremely difficult concept to grasp thoroughly at first, so don't feel embarassed if you find that a puzzled look has crossed your face and you need to read it over a few times.

Another way of thinking about it is to again consider the hologram a window. The interference pattern is simply a method of "freezing", for later playback, all of the qualities and relationships between the various waves of light.

We can see, then, that photography and holography differ fundamentally. In photography there is a point-to-point registration of image to original object. In holography there is none: each object point can be seen from a multitude of positions. In photography, a lens is used to form an image of an object. Holography uses no lens for this purpose and no image is actually formed at the plate - only a strange pattern which *has the capacity to play back an image* under the proper conditions.

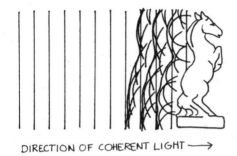

DIRECTION OF COHERENT LIGHT ⟶

COHERENT LIGHT REFLECTED FROM AN OBJECT

WAVEFRONTS REFLECTED FROM ALL POINTS ON OBJECT

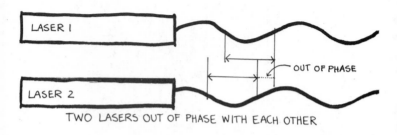

TWO LASERS OUT OF PHASE WITH EACH OTHER

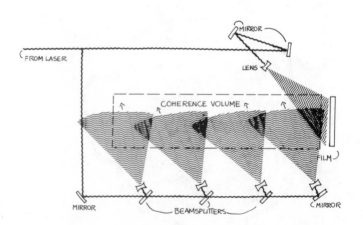

MULTIPLE COHERENCE VOLUMES

Photography stores information about the amplitude of the light. Holography stores both amplitude and phase information in order to store *all* of the visual characteristics of the object. Both photography and holograms are *developed* similarly, as the chemistry is essentially the same. However, whereas the individual light and dark spots on a photograph are of an image, the very fine light and dark areas on a hologram are microscopic interference fringes. These fringes by themselves bear no recognizable relationship to the image of the object.

It is essential that both the reflected light from the object and the reference beam be mutually coherent to form a proper interference pattern. Not only must a laser be used to make the recording, but the reference and object light must come from the same beam. If we attempt to use two different lasers to produce the reference and object illumination, the wave fronts might be out of phase, and interference would not occur.

Keeping light in phase is not an easy task. The laser is the only light capable of this, yet, even so, laser light can easily get out of phase with itself. How so? Well, we should be aware that the term "coherent" is a relative one. Even laser light is not perfectly coherent, only nearly so. If two parts of the wave front are separated by a sufficient distance, the waves will no longer be perfectly in step with one another. This is the reason we equalize the beam paths in the various setups. The distance these paths can vary before they are no longer in phase is called the *coherence length*. The coherence length can become a factor, not only for individual beam path distances, but for reference and object distances in single beam holograms as well. If we review the single beam setups described in this book, we will see that they are all arranged so that both the object and reference light travel *approximately* the same distance to the plate. We can say then that the difference between these two distances is within the coherence length of the laser. The area in which this occurs is the *coherence volume*.

The difference in length traveled by light illuminating the front and back of a sufficiently large object can also pose a problem. If the object is larger than the coherence volume, then parts of the object may appear very dim or may not show up at all, due simply to unequal path lengths. This may be remedied, as in the accompaying illustration of multiple coherence volumes, by using a number of object beams. The distance the light travels from the first beamsplitter to the plate is equal for each separate path, yet is arranged so that different areas of a large volume are illuminated.

Considerable light is lost when splitting the beam so many times, in addition to spreading it out over such a large area, so this type of setup is really only practical with fairly high-power lasers.

The other method of increasing the coherence length of the laser is by use of an *etalon*. The etalon is a specially designed transparent substrate which is positioned inside the laser cavity, usually between the rear mirror and end of the tube. It can increase coherence length from a few centimeters to a number of meters. The etalon, however, can reduce the laser's output by 30-50%. Only certain lasers are designed to accept an etalon, and these are usually the more expensive, high power, devices.

The type of coherence we have just been describing is an example of *temporal coherence*. Another type, *spatial coherence*, is necessary for the making of holograms. This refers to coherence across the thickness of the beam, or the transverse mode. We mentioned in the laser-purchasing section the necessity of obtaining a laser operating in the TEM_{00} mode. This refers to the lowest transverse mode, which means the light is in phase across the dimensions of the beam. If we spread the beam through a lens, and observe it projected on a wall or card, we should notice a fairly uniform spot of light. If, however, the spread beam has a dark spot in the center, the laser is said to be donut-moding or in the TEM_{01} mode. The dark spot is caused by destructive interference occurring in the beam itself. It is difficult to make holograms with this type of beam or any of the higher modes. *The only interference we want is that which we cause to be formed between the pure reference light and reflected light from the object.*

Understanding the mechanics of interference of light also allows us to appreciate the need for rigid stability while making a hologram. If the relationship between the two wave fronts changes during the exposure, an interference pattern will not be recorded. Why not? Remember, the film must "see" the well-defined pattern over the course of the full exposure time. A slight shift in the position of one wave front with respect to the other will effectively wipe out the areas of constructive and destructive interference. This shift need only be a small fraction of the wavelength of light for the fringes to be obliterated. Any motion of the object, holographic film, or relative positions of mirrors, etc. will cause this to

TEM_{00}

TEM_{01}

TEM_{MULTI}

SET·UP

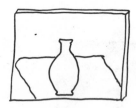

FINISHED HOLOGRAM

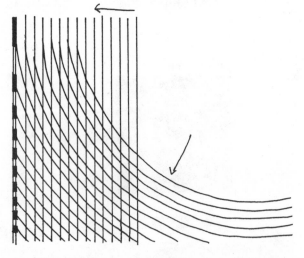

INTERFERENCE BY TWO ANGULARLY
DISPLACED COHERENT WAVEFRONTS
(TRANSMISSION HOLOGRAM FORMATION)

happen. And remember, we are talking about only a few billionths of a meter! It is no wonder then, that we must make holograms on a special vibration isolation table with all components and objects rigidly affixed. It explains why glass is more reliable than flexible film as a recording medium. It also explains why we cannot make ordinary holograms of live subjects. It's not simply a matter of not being able to sit still enough, either. Even the motion of blood pulsing through the veins is enough to prevent an interference pattern from remaining over the course of a normal exposure time.

In photography, if motion occurs, we get a blurred image. In holography, we get nothing, only a dark spot or "hole" where the moving object was positioned. Blurred images in holography have nothing to do with motion. They are usually caused by either scatter or noise produced by the emulsion, or by a slight chromatic aberration in the case of white light viewable holograms.

The reason we use special films with high resolution capability can now also be explained. Since the fringes are so tiny, the film must be able to distinguish clearly between the microscopic areas of light and dark. The resolving capability needed depends upon the recording geometry used to make the hologram and the wavelength of light. Fringe size and spacing are a result of the angular separation between the two wave fronts. When we construct an interferometer, we align two beams almost on top of each other to produce very large visible fringes. There is really no angular separation between them. We can "take a picture" of this fringe pattern with ordinary photographic film if we wish, since it need not resolve to any great degree. However, as the angle between the two beams increases, the size of the fringes decreases. This occurs quite rapidly, as a great many fringes will be formed with separations of only a few degrees. At this point, the fringes are too small to be observed with the eye. As the angle increases to that normally found between an object and reference beam in a given setup, the fringes are microscopic, though there are an enormous number of them. They are capable of storing the tremendous amount of information contained in a 3-dimensional scene. Several thousand individual fringes may be found per millimeter and the film used must be capable of resolving them. This is done by insuring that the silver particles suspended in the emulsion are very tiny, much smaller than those in normal photographic films. This is a difficult type of film to manufacture, but Agfa HD emulsions typically contain particles as small as 30 nanometers. The Soviets have produced some emulsions with particles as small as 10 nanometers. In all, the smaller the size of the particle, the better the film will resolve the fringes, and the better quality hologram will be produced. Larger particles do not resolve as well; if they are too large, they may not pick up the change from one fringe to the next. Larger particles are also more prone to the effects of "scatter" or "noise". This overall graininess or haze is caused by irregularities in the surface of the particle which tend to send some of the light bouncing around where it shouldn't. The definition of the fringes is affected, as is the clarity of the image upon viewing.

Those of you who *are* math fiends will be pleased to see we have included the formula for determining fringe spacing:

$$d = \lambda/(2\sin(\theta/2)) \qquad \text{or} \qquad 2d\,\sin(\theta/2) = \lambda$$

where "d" is the fringe spacing and θ is the angle between the two wave fronts.

Obviously, one need not memorize the above formulas to be able to make good holograms; however, they do provide some hard-core theoretical basis for the "why" of the hologram working the way it does. The formula is known as the *Bragg Condition* for the constructive interference of diffracted light. (So named, we like to think, for those who like to Bragg about how much theory they know.)*

The Bragg condition sets the requirements for the film resolution and stability needed during the recording process. The only factor which can be varied to affect these criteria is the reference/object angle. Also, because this angle is a function of the point of intersection of the reference wave front with every object point wave front, the ring spacing becomes the written record of the light amplitude and phase from the whole object.

*Actually named for W.L.Bragg who derived it.

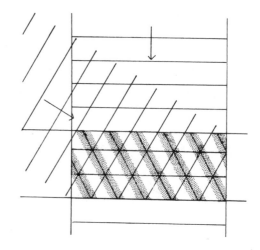

CONSTRUCTIVE INTERFERENCE BY BRAGG DIFFRACTION

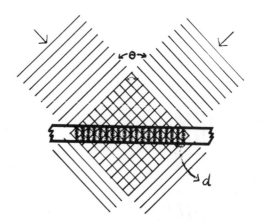

We should also be aware that the fringe spacing will vary across the surface of a given hologram. Because of the geometry of the wave fronts, they will generally be smaller on the side closest to the origin of the reference beam, and will gradually increase in size towards the other end of the plate. The film used, of course, must be of sufficient resolution to record all areas of the hologram.

One more note about fringes: the swirly gray patterns you may observe on your hologram are *not* the fringes recording the object - they are caused by imperfections in the optics, dust, etc. The real fringes, remember, are microscopic, you can't actually see them. The swirly stuff can be of use while attempting to *visualize* the real ones; they might not look too much different. Neither, though, resembles the object in any recognizable way.

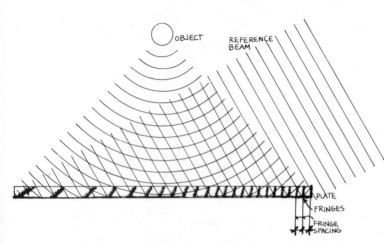

FRINGE SPACING VARIATIONS

diffraction and hologram reconstruction:

Earlier we mentioned that refraction was responsible for bending light as it traveled through a medium, as with a lens or prism. There is another method by which light can be made to bend.

Normally we think of light traveling in straight lines. If the light encounters an object, a shadow is formed. The straighter the light, the more distinct a shadow. If we look closely at the edge of the shadow, however, we will notice that it is blurred. This is because light will bend very slightly around the edge of an object.

This bending of light around corners is known as *diffraction*.

When light passes through a hole, it will be diffracted around the edges. If we make this hole very small, on the order of the size of the wavelength of light, an interesting phenomenon occurs.

We would expect that, if light passed through a small opening, the waves would continue on through as shown in the diagram at left.

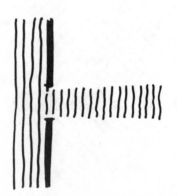

DIFFRACTION NOT TAKING PLACE

This, however, is not the case. What actually happens is as follows:

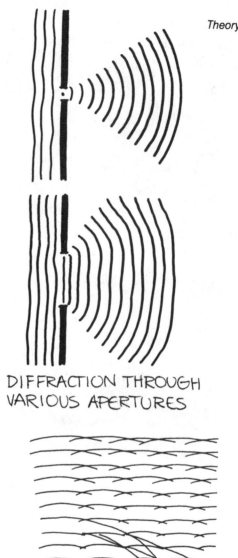

The small hole acts as a point source for a new set of waves which are bent toward the edges because of diffraction. The size of the opening will determine the amount of diffraction; the smaller the hole, the greater the bending of light. Larger holes will bend it less.

If we now add more holes to the barrier, each individual hole will act as its own source for a new wave front. If we consider a medium that has a great number of these holes, all of different size and configuration, it's easy to see that an extremely complex aggregate wave form can be generated out of the effects of all of them together.

The hologram, of course, provides this kind of medium. We recall that the hologram is composed of millions of very tiny light and dark spots, representing the fringes of the original interference pattern. When we shine light onto the hologram, the dark spots act as barriers, yet light passes through the tiny clear holes, diffracting as it does. The interference pattern and the new wave fronts duplicate those originally present when the hologram was formed. Now when we see the hologram, we see the very same complex wave front that was originally reflected from the object - only now it is being produced by a lot of little bitty holes. As far as we're concerned, it's hard to tell the difference. The new waves spread out to fill the room, just as they did when the object was there. Each of our eyes makes a separate image out of two perspectives of this wave form and 3-dimensions are synthesized in the brain *just as they were for the real object.*

We recall that in the case of a lens, light bent by refraction can make either a real or virtual image. Light bent by diffraction can also form these kinds of images. In fact, it is possible to make holographic lenses or other holographic elements (HOE's) and this is rapidly becoming a popular practical use of the technology. In the case of making holograms of 3-dimensional objects, we can consider their images to be real or virtual. When we view a normal transmission hologram, the image of the object appears behind the plate. The light used to form the image passes through the plate to our eyes. The light itself doesn't actually form an

DIFFRACTION THROUGH VARIOUS APERTURES

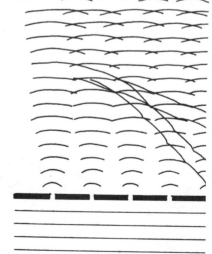

DIFFRACTION THROUGH A HOLOGRAM

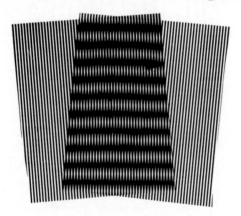

CHANGE IN FRINGE SIZE BY CHANGING
ANGLE OF INTERSECTION OF 2 BEAMS

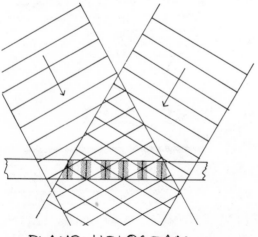

PLANE HOLOGRAM

image (try sneaking a look around to the area behind the plate - nothing is there). We can say that a virtual image is behind the plate. The hologram is really an extremely sophisticated lens bending light in just the right configuration to cause this to happen. We couldn't hope to accomplish this with any ordinary type of lens, of course.

When the hologram is flipped around so that light passes through the other side, it actually focuses and a real image is formed. The image can be seen on a screen or, in the case of the image plane hologram, used in place of an object for a second hologram. The image is *pseudoscopic*, or inside out. The virtual image is *orthoscopic*, or everything is in its proper place (we urge you to start using these words every day to impress your friends).

Actually, in both cases, the position of the image doesn't change; that part of the virtual image that appears closest to the plate remains so when we view the real image. However, we "view" the real image backward so it appears inside out (it is really *we* who are pseudoscopic).

The same kinds of real and virtual images are formed with reflection holograms, only now the plate is acting like a mirror. Yet, as in the case of the transmission hologram, we can obtain real or virtual images, not by curving the surface, but because of the characteristics of the given interference (diffraction) pattern.

Image plane holograms may have part of the image displayed in front of the plate and part behind. In this case, both real and virtual images are contained on each side of the plate (although one is orthoscopic and one is pseudoscopic). The real image is the part out in front, the virtual image appearing behind the plate. This is true no matter which side you observe, although the real part will become virtual, and vice versa as the plate is viewed from the opposite side.

It is possible to produce either *plane* or *volume* holograms. A plane hologram is formed when the size of the fringes is large compared to the dimensions of the emulsion. We recall that this occurs when there is very little angular separation between the reference and the object beams. Gabor's original holograms were of this type. They were also "in line" holograms, because of the geometry used to provide as little separation as possible.* (This small separation also allowed him to use the less coherent sources available before the advent of the laser and the lesser film resolving capabilities of the time.)

*We did not discuss the making of an in line hologram in this book as it is not practical for 3-dimensional objects.

Plane holograms of any type, whether in line or separated, are characterized by fringes which vary across the plane of the film surface. One characteristic of plane holograms is the arbitrary choice of angle of illumination light upon it to cause maximum diffraction.

The holograms of the type we describe making are volume holograms. This type is formed when the fringe spacing is smaller than the dimensions of the emulsion, so that some of the fringes become "stacked up" through the thickness. The object and reference separation in our setups were large enough to form the small fringes of this type. All volume holograms act according to Bragg's law, which, as we already mentioned, provides the basis for constructive interference of a diffracted wave front. A consequence of fulfilling the Bragg condition is that the angle of illumination of a hologram must closely approximate the original reference angle. We know this is true by experience; to view the holograms we make, each has to be properly positioned with respect to the light. Why is this so?

Basically, a given ray passing through a plane hologram is diffracted by only one fringe. In a volume hologram, the same ray will encounter a number of fringes. Because the spacing of the fringes is set when the hologram is made, each will diffract light in phase through to the next only if the light travels the same distance between them as before. If we change the angle of illumination, a wave diffracting out of one fringe may meet the next one out of phase. In this case, it will *not* add constructively and will simply be absorbed by the emulsion.

Although it is sometimes frustrating to have always to make certain the light and hologram are properly angled with respect to one another, the Bragg phenomenon can be used to an advantage. Since the hologram will only reconstruct with the light at one particular angle, it becomes possible to make another hologram *on the same plate* which will reconstruct at another angle. In fact, a number of completely different scenes can be recorded on the same plate and each played back independently by changing the position of the viewing light. These multi-channel techniques also provide the basis for rainbow color mixing. In this case the scenes are close enough to overlap, yet vary enough to cause relative shifts of the rainbow spreads. When using these techniques, one must be aware of the angle over which a given image will reconstruct before it fades out and is replaced by the next, or overlaps to the desired degree. This angle will vary as the fringe spacing, emulsion thickness, and hologram-forming geometry changes.

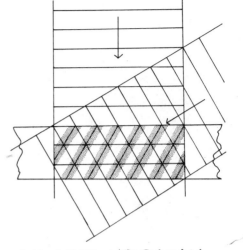

VOLUME HOLOGRAM

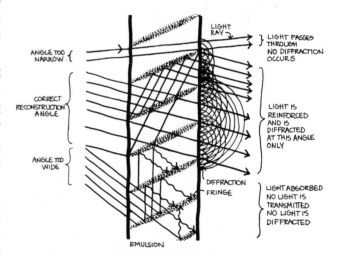

BRAGG DIFFRACTION TRANSMISSION HOLOGRAM ACTS AS A LENS

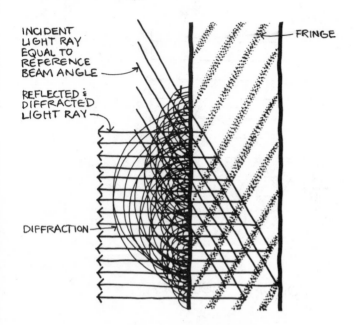

INCIDENT
LIGHT RAY
EQUAL TO
REFERENCE
BEAM ANGLE

REFLECTED &
DIFFRACTED
LIGHT RAY

FRINGE

DIFFRACTION

BRAGG DIFFRACTION
REFLECTION HOLOGRAM
ACTS AS A MIRROR

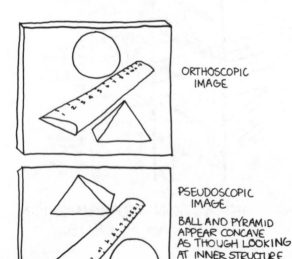

ORTHOSCOPIC
IMAGE

PSEUDOSCOPIC
IMAGE

BALL AND PYRAMID
APPEAR CONCAVE
AS THOUGH LOOKING
AT INNER STRUCTURE

Most transmission holograms actually exhibit both plane and volume characteristics. A given hologram will be labeled one or the other, depending upon which features predominate.

A special case of the volume hologram occurs with the reflection hologram. This leads us to one of the most often asked questions: "Why can you view a reflection hologram in ordinary light, yet you need a laser to see regular transmission holograms?"

To make a reflection hologram, we allow the reference and object beams to approach at opposite sides of the plate. This hologram, then, is almost purely a volume hologram; all of the fringes are stacked on top of one another, formed by standing waves through the emulsion thickness. Also, because of the great angular separation (up to 180 degrees) the fringes are extremely small, or approximately half the size of the wavelength of light! Fifty to one hundred or more may be stacked up through an emulsion only a few microns thick. (This explains why very stable systems and films with very high resolutions are needed for this kind of hologram.) After the hologram is made, the fringes act like a stack of closely spaced planes which both diffract and reflect light back in the direction from which it comes. Reflection holograms are both angle and wavelength specific. This is because the planes are so close and there are so many of them. The Bragg condition is only met for the one wavelength which matches up to the precise spacing between them. All others are absorbed. The reflection hologram then acts as a resonant reflective filter, selectively reinforcing one color at the same time as it plays back an image. The illumination light can be multicolored or white, it doesn't matter. A clear image of one color is still obtained.

Since the planes are similar to those formed in Lippman color photographic emulsions, (but for different reasons) these holograms are often called Bragg-Lippman holograms. We've even heard "Bragg's angle - Denisyuk - Lippman - Volume - Phase - Amplitude - Standing Wave Hologram," but we prefer, "Reflection Hologram".

Reflection holograms are more prone to the effects of processing than transmission holograms. When a reflection hologram is developed and then fixed, the unexposed silver particles are removed (clear spots). This makes the emulsion thinner, after it is dried, than it was originally. The spacing between the fringes is smaller. This causes the hologram to play back with a color of shorter wavelength. A reflection hologram made with the longer red will play back with the shorter green color. Triethanolamine is used to swell the emulsion back to its original thickness, which restores the original distance between the Bragg planes.

When the processing sequence does not call for the use of a fixer, as in reversal bleaching, this shrinkage is only minor. We thus get a red or orange image reinforced. The other colors, including the green in this case, are absorbed.

Transmission holograms, on the other hand, do not ordinarily* filter the light. Many wavelengths are capable of constructively diffracting through the same areas. When we view a regular transmission hologram with white light, each wavelength is diffracted, although to a slightly different position. Green frequencies diffract faster than red, and blue even more than green. In other words, each color is attempting to produce its own image, and each is displaced slightly with respect to the other. The overall result is a multicolored blur as all images run into one another. The only way to obtain a clear image is to use only one wavelength for reconstruction, hence the term "laser viewable transmission hologram". Actually, highly filtered non-laser light can be used to view transmission holograms, although the image will not be quite as sharp as with the single frequency laser. Oftentimes, a mercury arc lamp with a narrow-pass interference filter is used to display transmission holograms, since these lamps provide a lot of light concentrated into some fairly narrow ranges of the spectrum.

A good way to observe the effect of different frequencies on the hologram's image-making capabilities is to view a transmission hologram with a mercury street lamp at night. You will be able to see how several discrete images of different colors are generated, all displaced with respect to one another. The colors of the various images correspond to the various spectral lines of the lamp. The blue-purple image will appear as the farthest from the plate since these higher frequencies diffract faster than the yellow, for instance, which will appear closer.

Holograms are usually categorized as being made in either *absorption* or *phase* media. Silver halide films normally produce absorption holograms. This means that some of the light is absorbed during reconstruction, in order to produce the diffraction which corresponds to the original interference pattern.

*See rainbow and open aperture sections.

LIGHT SOURCES AND THEIR RECONSTRUCTED IMAGES

LIGHT SOURCE   RECONSTRUCTED IMAGE

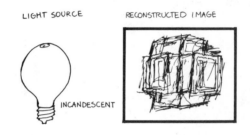

INCANDESCENT

MERCURY VAPOR

BLUE
GREEN
RED

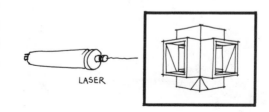

LASER

Remember, the hologram is made up of lots of little holes - the dark ones absorb light and the clear ones let it through. Both transmission and reflection holograms work this way; however, since "reflections" are also absorbing light while acting as wavelength filters, they are characteristically very inefficient. Only a small percentage of light is left over to be diffracted out to form an image.

Phase materials, on the other hand, do not operate with little black and white holes, but rather the fringe variations correspond to changes in the index of refraction of the emulsion from one point to another. The phase media is transparent, so much less light is lost, and we obtain a brighter image. Silver halide absorption holograms can be converted to phase holograms by bleaching, and so that's why we had you do that. Bleaching converts the silver particles into transparent silver salts which both diffract and refract the light. It's easy to understand why bleaching is especially important for the very inefficient reflection hologram.

Some materials, such as dichromated gelatin, are phase media to begin with. This is why these holograms can be exceedingly bright, as very little light is lost. Pure phase media can theoretically diffract 100% of the incoming light; however, in actuality some light is usually lost. Holograms with 70-80% efficiency are not uncommon, however, and with a bright viewing light they can be dazzling. Dichromate reflection holograms can also be made to "broadband" or to reinforce selectively several individual colors at once. More of the viewing light is channeled into the image that way, making a brighter image, and it allows the mixing of wavelengths to produce a wide variety of hues. Although not "true color" since the colors are not necessarily related to the object's original hue, it is possible to obtain golds, silvers, whites, and a variety of pastel shades. These are all determined by the relative mix of all of the individual, separately produced wavelengths, independently reinforced by their own set of Bragg planes. This capability is much more difficult to achieve in silver halide reflection holograms, yet is currently under investigation by a number of holographer-chemists. Full color holograms have, of course, been made and they are explained in the advanced holography section.

the future

It is easy to list the fantasies we wish holography would fill. We imagine eerie full-color images floating in front of us in the middle of the room, or projecting out of our TV sets. We imagine beaming images of ourselves to our friends over telephone lines, or taking holographic shapshots of the family that develop before our eyes. Holographic roadsigns or billboards could dot the highways. Complaints about them as eyesores could be delt with quickly "tear them down" by simply turning off the light. One *could* go on.

It's hard to say what will happen. Certainly a few major technical break-throughs would help. The biggie would be a new efficient, and cheap lasing medium. All right you laser companies, get on the ball! We need something that will lase many colors, put out several watts of light, be the size of a flashbulb and cost about as much! (we shouldn't hold our breath.)

We need improvements in chemistry - how about a grainless film like dichro-mated gelatin which is sensitive to all colors, is about as fast a Kodak plus-X, and experiences no deterioration from moisture, heat, pressure or anything else.

We need someone to figure out how to transmit a hologram over television more quickly. It would now take two hours to send the information contained in one frame that now passes the screen in 1/30 of a second.

Well, it may be quite a while before any of the above become realities. But don't take our word for it. O.K. sports fans - get out there and prove us wrong!

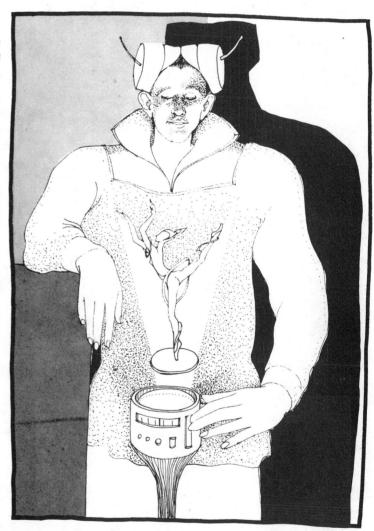

art & holography

"Moire Screens" by John Kaufman. Reflection hologram, 1981.

"Egg and Jewell" by Randy James. Rainbow hologram, 1978.

"Holography may be the greatest advance in imaging since the eye." - Asimov, I.

"Our newfound artistry in controlling light's detailed structure - the waves themselves - may vastly change our view of the world." - Cochran, G. and Buzzard, R.

"The most beautiful thing we can experience is the mysterious. It is the source of all true art and science." - Einstein, A.

introduction

The topic of Art and Holography has, as might be expected, been somewhat controversial in nature. Art, in any form, has always involved a particularly personal experience, both on the part of the artist and those persons who come into contact with his or her work. None the less, our culture still insists on diffferentiating between art which is "valid" and that which is not, especially when it involves new techniques, technology or a new medium. We would like to avoid too heavy of a soporific discourse on "what is art?", a question on par with "what is reality, man?", yet some discussion of how holographic art has become embroiled in this question is necessary (actually, if there were universal acceptance of holographic art as such, this chapter would probably be unnecessary, or at the very least be titled differently).

In this chapter, then, we will need to investigate the nature of holographic art and the creative process, what the controversy seems to be about, some conparisons and contrasts with other media from both current and historical perspectives, the implications of it all, and then suggestions of a few of the techniques which may be used to enhance the nature of ideas that you, as the artist, may wish to express holographically. This last section will include such methods as double exposures, multiple channel holography, matte screening, collage, as well as the manipulation of pure light by producing diffraction grating, or using highly reflective materials.

Is creative holography possible?

This has been a question asked for some time. It underlines a rather apparent confusion as to the capabilities of the process as a result of historically how it has been used. We recall that display holograms were born in the research lab and first used as a method of visually duplicating all of the dimensional characteristics of an already existing object. Most display holograms have been produced with this in mind. In fact, the first hologram made by most people is usually of some rather mundane little trinket. Yet, because of this, some critics with a lack of imagination and some who should know better have been apt to make rather negative generalizations about a process by viewing one or two poor examples.

The question usually is this: if a hologram simply produces a dimensional image - is it any different than the original object? Why not simply make the object? It is true that a number of exhibitions of holograms have been characterized by this approach - relying on the ability of the process to wow the viewer with wizardry rather than appeal to an aesthetic sense. "The medium of holography too often has seduced its audience with technical sophistication, relying on impact of the process rather than impact of the content." This has led to statements like the following: "Holography, by its very nature, is doomed to duplicating existing forms, not creating new ones" (- S. Duffshirt, professional art critic.).

After a line like that, we might all be inclined to trade our lasers in for a nice block of marble. However, we tend to be wary of statements including "by its very nature", especialy when the "it" is so new as to make "its nature" barely definable. The rational also assumes that since some holograms merely duplicate, then all must do so. How narrow minded! Try telling Ansel Adams that his photographs should be compared to snapshots of Aunt Martha (if you have an Aunt Martha, or even happen to be Aunt Martha, no offense meant - only wish to make a point.)

Another more basic question is sometimes asked "Is holography art?" This is a rather perplexing quip. It's like saying "Is clay art?" or "Is a paintbrush art?" The hologram is a tool, not an end product in itself. A tool may or may not be used creatively, depending on the intent of the user and the nature

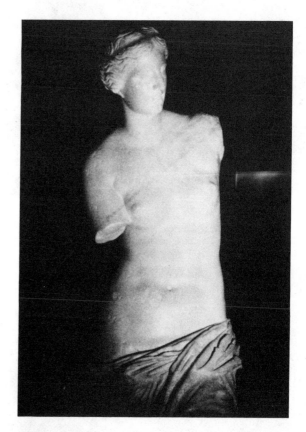

"Venus de Milo", 1 meter x 1.5 meter laser transmission hologram by Jean-Marc Fournier and Gillbert Tribillon, LOBE Laboratories, Besancon, France, 1976. Photo © LOBE Laboratories, courtesy MOH

"The Hybernation of Consciousness - Homage to Magritte" by James Feroe, Rainbow holograms (2), 1980

of the experience of those who encounter his work. The real questions are, "Which holograms are art?" and "What does it take to produce a creative hologram?"

To approach these questions, some understanding of the creative process is necessary. The artist is both creator and communicator, who works through interaction with and manipulation of his environment. As a sensitive person, the artist first observes the world around him. In his mind he makes conjectures, conceptualizes images or forms, and then translates his vision into a cognitive, positive thought. This image is then transferred to some medium. The medium may be paint, glass, clay, stone, peanut butter, or whatever. It may involve using new tools to develop a view of the world that transcends the normal or mundane. It may, in the case of a hologram or other form of light sculpture, be the medium of light. Remember, light is the medium, not the hologram, the tool used to sculpt it. Or, imagine the hologram as a "paintbrush" with the laser supplying the "paint". We can see then, that a particular hologram may qualify as art, not because it is a hologram, but because of what the artist wishes to achieve with it. Holography is simply the language chosen, as opposed to some other method. This kind of hologram must not be confused with that of the "holographic snapshot". Doing so does the artist a great injustice.

To effectively translate ideas, one must develop an understanding of the capabilities and limitations of the medium in which one works, and develop skill with the use of one's tools. Sometimes, those who are new to holography lose sight of this fact and become frustrated when the fantasy of what they think they should be able to accomplish is not realized. This was, along with a relative scarity of good information and instruction, keeping many artists out of the field. Fortunately, this state of affairs has been turned around. As more people learn what holography can do, the more the knowledge will contribute to the evolution of both skills and ideas. One of our major intents is to facilitate the development of both, in order to help transform our normal concept of visual reality.

We believe that, hidden beneath the controversy is the fact that holography is a good deal more than just a novel technique crying for acceptance. Rosemary Jackson writes:*"Holography is perhaps the most revolutionary

*from "Through The Looking Glass". Travelling show sponsored by the Museum of Holography.

visual recording medium since the prehistoric cave paintings at Lascaux. For the first time in the history of literate man, we can communicate through a medium which has the same dimensional properties and characteristics as the world in which we live.''

The implications of this unique design of space with light, in a way which can merge the dimensional qualities of sculpture with the special perspective, luminous, or collage effects of the flat media, are enormous. For example, in conventional sculpture, a piece will be presented in a given environment. One may sit the piece on a pedestal, in a box, suspended from the ceiling, in a special room with unusual lighting, or what-have-you. The space around the sculpture is important, however, it becomes a function of the locale. Space is a variable separate from the piece, which must be reckoned with. In holography, the space as well as the object can be designed simultaneously. This completely controlled volume becomes an integral part of the piece - a space, for a time, superimposed over or replacing the volume of our normal, perceived reality.

The significance of this, with regard also to two-dimensional media, cannot be underestimated. In her article, ''Cultural Implications of Holographic Space'',[*] Kalina Haran notes the following: ''Holographic space is the first major breakthrough in the experience and expansion of pictorial space since the penetration of the picture plane by Renaissance artists following Brunelleschi's discovery of linear perspective in 1425.'' Haran goes on to explain the significance of the use of perspective in the 15th and later centuries: ''Perspective became the visual equivalent to new movements in art that were changing man's conception of himself and his relationship to his world. Linear perspective encouraged man to perceive and analyze accurately his external environment from his unique, single point of view.''

Haran continues to explain the importance of this event, noting that, prior to the use of perspective, artists really didn't understand the nature and use of space, which transcends and unites objects. The development and use of perspective both influenced and mirrored perceptions of the world which, at that time, were being shaken by revelations of Newtonian concepts of space. Artists and scientists of the day worked to establish a gridwork of left-to-right, up-and-down reality within which logical, linear, sequential thinking would function.

''Stargate'', white light transmission hologram by Dan Schweitzer, 1980, photo © Nancy Safford, courtesy Museum of Holography.

''Rough Cut'' Shadowgram by Rick Silberman, 1979, photo © Nancy Safford courtesy MOH.

*see *Holosphere* Sept. 1980, vol. 9, #9, p.7

It's true these concepts were attacked or simply ignored by more recent artists working in impressionism, cubism, minimalism, or several other "-isms", yet, limitations as to what could be represented in two dimensions were becoming apparent. Haran remarks: "(We) search for pictorial conventions that reflect our changing sense of reality without (finding) such a unifying concept, a new space large enough, complex enough, to image our Einsteinian, four dimensional, cultural paradigm."

Holography can be used to express this nature of space-time since it can depict things which are not tied to particular viewpints. Clearly, the relative positions of objects or forms dominate, with the concept of a fixed view only valid from rather limited or localized perspectives.

Creating one's own space (apologies to EST) is exciting to an artist. It becomes possible, for the first time, to either radically alter our normal concept of space, or create entirely new ones. For example, double exposures allow two objects to occupy the same space at the same time. This is impossible in our normal world, yet is a concept often attempted in the flat media. Now it can be done dimensionally. It also becomes possible to carve the space up into overlapping "sub-universes", through the use of multiple exposures, matte screening, and multiple channel readout. We shall discuss how these techniques are accomplished a little later.

Ed Bush writes: "For a holographer, the imagination can become a domain of time and space in which the relationships of objective reality become plastic. Shapes can be twisted at will. Objects can loose their solidity, drifting through each other with unnatural ease. Sudden appearances and disappearances are commonplace and perfectly acceptable, in contrast to the rational world where such inconsistancies are often dismissed defensively by the perceiver. With the tools of holography, the artist has an opportunity to deal with space and time in a unified manner, instead of choosing one over the other as other media often require. So they can enter this new territory and return with pieces of it packed away in holograms."[*]

Dan Schweitzer notes: "With these techniques it is possible to create a fabric of reality containing patches of another reality...there are spaces in the fabric of our normal perception which are crammed with other universes that we do not normally deal with."[**] Anait Stevens adds, "My main interest is not

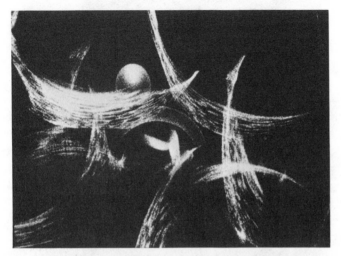

"Planet Claire" whitelight transmission hologram by Rudie Berkhout. Photo © MOH.

"Crystal Beginning", 1977 white light transmission hologram designed by Stephen Benton, produced by S.A. Benton, H.S. Mingace, and W.R. Walter at Polaroid Research Laboratories, 1977. Photo © MOH, collection MOH.

[*]from *Holosphere*. June 1978, vol. 7, #6, p.4

[**]*Holosphere* Jan. 1981, vol. 10, #1, p.3

encoding reality behind the plate - it's playing with new places and spaces and volumes having mass without weight, which is exciting for a sculptor.''

Holography is not limited to the act of creating or dividing of space. It also represents a new way to approach the treatment and control of light. It is actually rather remarkable to note that, in such a short time, holography has been used by artists for such disparate purposes - from the purely representational to the purely abstract, and combinations thereof.

Andy Pepper sees holography representing an important step in the history of light sculpture. He notes that light sculpture can trace its origin to Louis Bertrand Castel, a Jesuit priest who invented a device called the ''clavessin oculaire'', in 1734. This device produced changing abstract color images by passing sunlight through a series of prisms in a darkened room. Others continued to build devices over time to manipulate light. One example was the ''clavilux'', which consisted of various spot and floodlights controlled by rheostats and used prisims, filters, and projection screens. Its inventor, Thomas Wilfred, considered the clavilux an instrument to be ''played'', and he gave a number of performances with it in the 1920's. His work inspired others to use light in a kinetic sense, by incorporating elements of both sculpture and performance. Some worked with effects produced by the new electronic technology of the 1960's, using both light and sound synthesizers. Others, including Pepper himself, produced dimensional forms by passing light through semi-transparent cloud chambers or steam.

Pepper notes that it is interesting to observe how the interaction of the viewer with light sculpture has changed over time. Castel maintained a degree of mechanical control over his creation. Wilfred, present at each performance of his device, offered his personal control. More recent kinetic artists used an interface of control between viewer and equipment - the viewer could actually influence the results of the performance.

Holographers go a step further, giving the viewer direct control, within the parameters of the holographic process. Movement of forms or images occur with the change in position of the observer. Stepping out of the viewing zone causes the image to vanish. Different viewers can observe the work simultaneously yet, since they do so from different positions, a very personal interaction with the work occurs for each. One can observe it from a stationary

*Holosphere Dec. 1980, vol. 9, #12, p.6

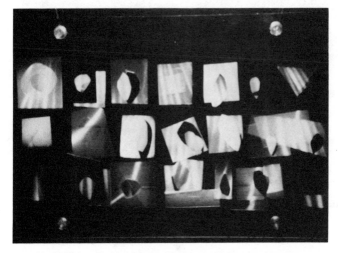

Untitled by Randy James. Rainbow Hologram Montage, 1979

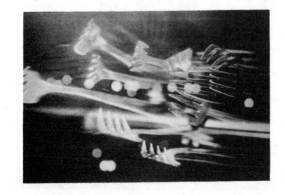

''Equivocal Forks II'', white light transmission hologram by Harriet Casdin-Silver, 1977, photo © Nancy Safford courtesy MOH.

"Round and True" by Randy James. Reflection hologram, 1981. Photo by Bob Schlesinger.

"Vacuum Hoses" by John Kaufman. Reflection hologram, 1978.

position while another can move and derive a kinetic experience at the same time. In this way the viewer becomes the choreographer - he directs the performance.

This effect is most pronounced in work done with pure light, or those which use the effects of abstract diffracton gratings rather than concrete objects. Artists such as Bruce Goldberg, Vince DiBiase, Fred Unterseher, Ruben Nunez and Harriet Casdin-Silver work in this area, Others, including Dan Schweitzer, Sam Morree, Randy James and Lon Moore incorporate aspects of both in their work. Rudie Berkhout, working with multiple color rainbow techniques, has produced some rather remarkable effects.

Holograms have also been used creatively in other ways. Some, like David Hylinsky and Michael Sowdon, combine two and three dimensional effects by painting over the emulsion. Others, like Margaret Benyon, John Franske, Ed Lowe, and Setsuko Ishii, have incorporatred holograms into larger pieces of work. Benyon, Rick Silberman, and others have experimented with the concept of "negative space" where shadows take on a 3-D sensation. These "shadowgrams", formed by the deliberate moving of a three dimensional form during exposure (or by use of an unstable object) results in a three dimensional "black hole" where the object was located. Some have used this method along with multiple exposures to position new forms in the midst of these black holes. Others, including Dan Schweitzer, deliberately modify their optical components to introduce distortions into the images. Silberman and others have utilized the unique pseudoscopic properties of holography, which may be used to turn space inside out. Jim Feroe demonstrates that it is possible to develop a composition with regular identifyable objects in such a way as to cause the space in which they are located to take on new meaning, with objects of incompatable size positioned in impossible situations. His work also takes one of the limitations of holography - that objects are not normally enlargeable or reduceable, and uses it to an advantage. John Kaufman works with ultra-realism in reflection holograms. Bill Molteni, Anait Stevens, Amy Greenfield, and many others have used holographic stereograms creatively, incorporating some of the special effects used in video or filmmaking. Many other holographic artists are currently experimenting with a wide variety of methods in an effort to both create pieces which individually express a desired effect and collectively begin to define the scope of holographic art. Most artists will probably agree that, at this point, however, the scope appears to be so vast as to defy limitation.

It seems likely that holographic art may ultimately be seen as playing as important a role in human development as perspective did in the 1400's. Perhaps the concept of unfixed, relative viewpoints, first espoused by Einstein and embraced by scientists and now demonstrated by artists holographically, will do as much to radically alter our view of the universe. As Haran suggests, "The wider implication of holography is its potential to re-educate our visual perceptions into a richer experience of our (four) dimensional world. Holographic art, therefore, offers the possibility of the esthetic experience attaining a new didacticism, promoting a new visual literacy that may influence man's perception and manipulation of his environment, possibly even encouraging a more holistic philosophy."

Techniques to Try:

Now that we've gotten you excited about the role of holography in the evolution of art, you're probably wondering how to do some of the things of which we speak. The first recommendation we have is to become better acquainted with work current artists are producing - visit the Museum of Holography and other galleries listed in the appendix. Join one or more of the societies listed and begin to get involved - you may be surprised at how much is going on and how communication will help foster your development, both technically and esthetically.

It is impossible for us to discuss all of the techniques used to produce holographic art, they are much too numerous. Instead, we shall limit ourselves to a number of general catagories which may be investigated in order to help get you on the right track.

Multiple exposures offer one of the most exciting approaches. With it, two or more objects may be made to occupy the same space at the same time, or their images made to overlap. One may also use the concept to allow orthoscopic and pseudoscopic images to appear simultaneously, or to see the

Sharon McCormack, holographic artist and director of the San Francisco School of Holography with life-size holographic stereogram self portrait printed with special large format oil-filled lens system she developed, 1979, Photo by Carol Bernson.

Facade of Museum of Holography, 11 Mercer Street, NYC, photo © Sing Sy Schwartz, courtesy MOH.

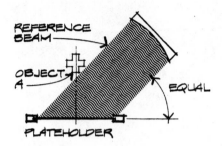

EXPOSURE NO. 1

REFERENCE
BEAM
OBJECT
A
EQUAL
PLATEHOLDER

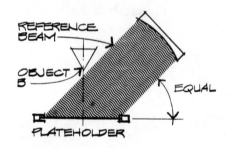

EXPOSURE NO. 2

REFERENCE
BEAM
OBJECT
B
EQUAL
PLATEHOLDER

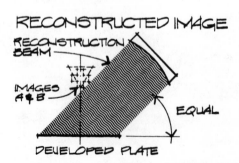

RECONSTRUCTED IMAGE

RECONSTRUCTION
BEAM
IMAGES
A & B
EQUAL
DEVELOPED PLATE

front and back of an object or scene at the same time. A multiple exposure hologram of this type is accomplished by first making an exposure in the usual way, then, changing the object while keeping the rest of the setup intact, especially the reference beam. A second exposure is then made. The process may be repeated a maximum of three or four times before the quality of the hologram will begin to suffer. Exposure times for each should be reduced slightly (approx. 20%) to compensate for the overall increase. Usually, the plate is removed between shots, with care taken to remember its proper orientation. The second object is then positioned, illuminated properly, and the plate replaced and exposed for a second time. The same effect may be achieved in image plane setups by substituting other masters for the multiple exposures.

A variation of the above is accomplished by rotating the plate between exposures. If the plate is rotated in the same plane as the reference beam, i.e. top to bottom if an overhead reference is used, a double exposure will result in both orthoscopic and pseudoscopic images appearing simultaneously. If the

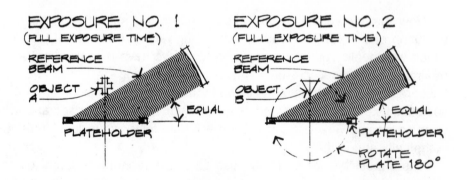

EXPOSURE NO. 1
(FULL EXPOSURE TIME)

REFERENCE
BEAM
OBJECT
A
EQUAL
PLATEHOLDER

EXPOSURE NO. 2
(FULL EXPOSURE TIME)

REFERENCE
BEAM
OBJECT
B
EQUAL
PLATEHOLDER
ROTATE
PLATE 180°

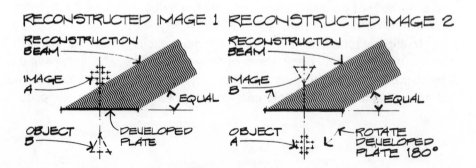

RECONSTRUCTED IMAGE 1

RECONSTRUCTION
BEAM
IMAGE
A
EQUAL
OBJECT
B
DEVELOPED
PLATE

RECONSTRUCTED IMAGE 2

RECONSTRUCTION
BEAM
IMAGE
B
EQUAL
OBJECT
A
ROTATE
DEVELOPED
PLATE 180°

plate is rotated perpendicular to the plane of the reference beam, i.e. left to right for an overhead beam, then two different images of the same type (either ortho or pseudo) can appear on the same plate, one on each side. By using the same object in both exposures yet rotating it 180 degrees for the second shot, both sides of the object will be recorded. If you view the plate on one side, you will see the front of the object. If you turn it around, the back will appear. This gives the illusion of almost 360 degrees of view in a single, flat plate.

EXPOSURE NO. 1

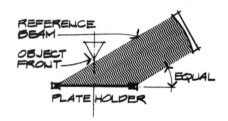

EXPOSURE NO. 2

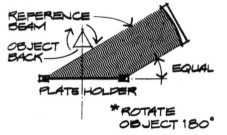

RECONSTRUCTED IMAGE 1

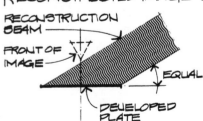

RECONSTRUCTED IMAGE 2

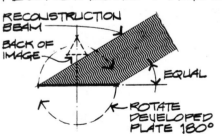

Multiple channel techniques may also be used to store more than one image on a plate. With this method, both the object and the reference beam angle is changed for each exposure. This is also known as "coding" the reference beam. The first exposure is made in the usual way. The object is replaced with another and the reference beam adjusted to intercept the plate at a different angle. By experimenting with various angular changes, either of two effects will occur when the hologram is viewed: As the plate is rotated, object #1 will disappear and #2 will appear. If the angular change is small, object #1 can be made to fade into #2 as the plate is moved. This angle will be dependent on the type of hologram made, type of light used to view it and distance of the original object(s) from the plate. This is basically the same technique used to color-mix in rainbow holograms. By careful adjustment, objects can be made to both appear and disappear and change color independently.

EXPOSURE NO. 1

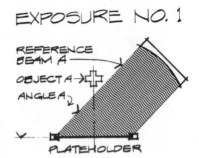

EXPOSURE NO. 2

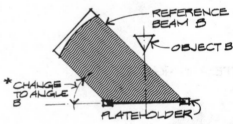

RECONSTRUCTED IMAGE

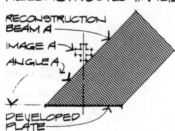

RECONSTRUCTED IMAGE

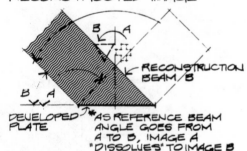

Matte screening or spacial collaging is another method often utilized. This is a special double exposure technique whereby part of a three dimensional scene is blocked out, and another scene made to appear in the missing area of the first. Although similar in concept to matte screening in 2-D media, in holography it produces an effect not obtainable by any other means. It becomes possible, by using two scenes, to look through the first into what should be the space of the second, but isn't. One of the spaces appears where it shouldn't - remaining completely distinct from the other - like a separate universe without overlapping images, yet still in the midst of the other's space.

As an example, cut a hole out of the center of a piece of cardboard or masonite. Save the "hole" and glue it to the center of a piece of glass. You should now have a positive and negative hole of the same size. Now, two exposures are taken, one through each matte. One exposure will "see" only the view through the hole, the other will "see" everything but that view. If the two exposures are made of different scenes - the resultant effect will be of two different spaces each viewed through certain parts of the plate. By changing your viewing position you may look through one space into what should be behind the other, yet isn't (& vice versa). In this sense, there is a spacial overlapping of viewing angle. The same effect may be accomplished by making the separate holograms on film and then cutting them up and pasting them together into a collage. Combinations with other types of multiple exposures can lead to an enormous number of effects. By using multiple mattes or small cutouts it is possible to place a large number of independent scenes on one plate. By using a moveable slit mask, animation can be produced in a manner similar to that of holographic stereograms, only using real 3-D objects.

CARDBOARD MATS

MAT A: NEGATIVE.
CUT OUT
CARDBOARD MAT

MAT B: POSITIVE.
GLUE CARDBOARD
CUTOUT ONTO
CLEAR GLASS

EXPOSURE NO. 1

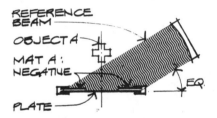

EXPOSURE NO. 2

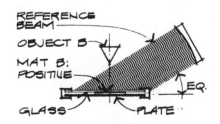

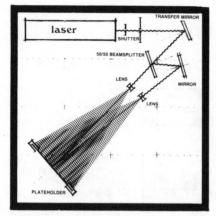

ON·AXIS DIFFRACTION GRATING

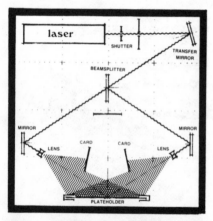

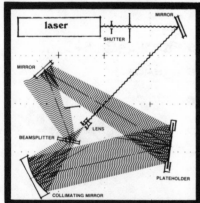

2 OFF·AXIS DIFFRACTION GRATINGS

One of the most exciting effects for many artists to work with is that of the abstract; a pure spectrum produced by diffraction grating. Gratings are made by the interference of two (or more) beams of pure undiffused laser light. This contrasts with the making of a hologram of a normal diffuse object - the subject of all of our efforts up to this point.

Diffraction gratings are among the most simple holograms to construct, and they may be designed to produce unusual bursts of very pure color and form. The most basic type of grating can be produced by using one of the diagrams illustrated.

The on axis type will yield a large number of diffraction "orders" resulting in a large ray of repeating spectrum eminating from a light source when viewed with a white light. This type can be modified by using multiple exposures of different lens arrays to achieve a variety of color bursts or wheels. The off axis type will usually produce a grating with a single order (due to the Bragg effect - see theory) resulting in a single spectrum viewed when the plate is positioned at the proper reconstruction angle (corresponding to the original angle of one of the beams). This type also may be modified by passing the beam(s) through different kinds of lenses, pieces of plastic, etc., allowing the beam to reflect off of mirror - like surfaces including crystal, aluminum foil, ball bearings, etc.

As with other holographic setups - care must be taken to equalize all beam-paths, adjust ratios (1:1 to 4:1 is O.K.) and expose based on light readings (either beam may be treated as "reference"). Diffraction gratings may be processed as either transmission or reflection holograms, although bleached transmission types are preferred due to their efficiency.

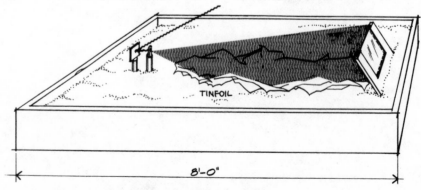

SINGLE BEAM DIFFRACTION GRATING

A synthesized wide viewing angle can be produced by joining several flat holograms together. This can often be an easier method of producing a 360 degree hologram than using curved film. For example, a series of four holograms are made. The same object, which is kept securely in place yet rotated 90 degrees for each shot,is used. An overhead reference beam is set.The finished holograms are glued together into a box and illuminated by a single light from above - as the box is rotated (or the viewer moves around it), all sides of the object will be viewable. The technique can be added to some of the others mentioned above to achieve 360 degrees of "impossible spaces".

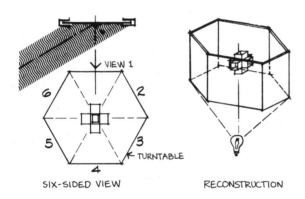

SIX-SIDED VIEW   RECONSTRUCTION

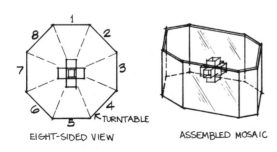

EIGHT-SIDED VIEW   ASSEMBLED MOSAIC

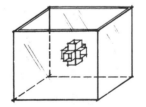

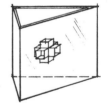

FOUR-SIDED MOSAIC   THREE-SIDED MOSAIC

360° WIDE ANGLE MOSAICS

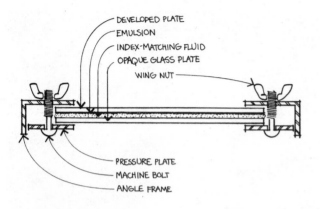

CONTACT COPYING PLATE HOLDER

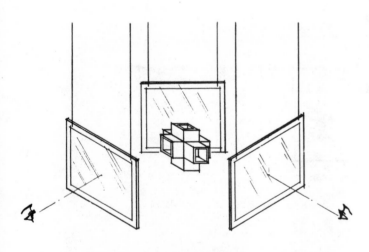

SCULPTURE WINDOW

Copying Holograms:

If you are producing an edition of holograms and are working with one of the transmission formats, be it regular transmission, rainbow, open aperture, or transmission gratings, you may wish to contact copy your hologram rather than make each one using a complex setup. This method also conserves laser light, as it can be done with a single beam. In this way, it becomes possible to produce a series of "prints" of the same hologram.

The copy is made by using an index - matching holder of the type described on page 87. The original and copy plate are indexed together and placed in the holder with the original towards the beam. The beam angle should also match that of the reference beam of the original. After the exposure, the two plates are separated and the copy processed (remember to process the correct plate!)

Hologram Display:

Displaying the final hologram is almost as important as producing it. Perhaps more than any other process, holography is completely dependent upon light for its effect. If the light does not approach at the proper angle, is of the improper type for the kind of hologram shown, or is overpowered by ambient light, a poor or nonexistant image is about all you'll have to view of even a super hologram. A little care in display makes an enormous difference, both in image quality and the overall effect of the entire piece.

Designing a display can be as simple as placing the hologram in a frame, or as difficult as building a whole sculpture around it. The type of display should be based on consideration of a number of factors, including:

1) Hologram type (reflection, transmission, rainbow, etc.)
2) Image type (real, virtual, projected, realistic, abstract,etc.)
3) Reference angle (should be above or below - side reference holograms are poor for galleries).
4) Emulsion characteristics (clear, milky, etc.)
5) Film format (flat, curved, pieces, sizes, etc.)
6) Efficiency (brightness of viewing light - effects of ambient light).
7) Viewing area (to determine positioning height).

The actual display can be fabricated and arranged in a variety of ways. Some have been mounted in clear plexiglass frames and suspended from the ceiling by wires. Others have been placed in more conventional frames. The mount should take advantage of the medium, i.e. it's often nice to leave accessible space behind the hologram to demonstrate that there is "nothing actually there". Reflection holograms, since they are illuminated from the front, can be mounted against a wall and track lighting from above or below can be used. In this case, the image appears to extend into the wall. Transmission holograms can also be displayed with frontal lighting by backing them with a mirror.

Some holograms are displayed on bases or tripods placed out in the middle of a room. In these cases, special attention should be paid to the effects of ambient light. The display should also be of durable construction, as many people, while marvelling at the fantastic effects you've created, will probably be placing their hands all over it while trying to figure it out.

Some holograms, especially those for which positioning of the illuminating light is extremely critical, are supplied with their own self-contained light. Holographic stereograms are displayed in this way. You may wish to design your own display with a light source for other holograms. This will make it easier for you to demonstrate them as you move from place to place. One of the more frustrating exercises is to walk into a room filled with florescent lights to talk to a gallery owner interested in seeing your work. Be prepared!

We hope the above comments at least help you begin on your way. Art and holography is a subject in its infancy, still, we would like to see someone devote an entire book to this subject. Perhaps one of you out there will help to realize our dream . . .

"Transparent Threats" six rainbow holograms by Ed Lowe. 360 degree installation surrounding viewer. Denver, 1979. Photo Ed Lowe

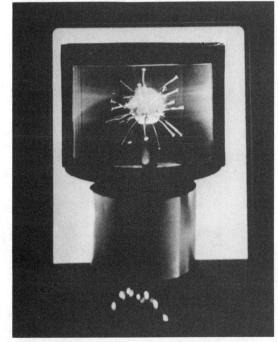

"Voodoo Dolly" by A. Kurzen. Holographic stereogram.

philosophy

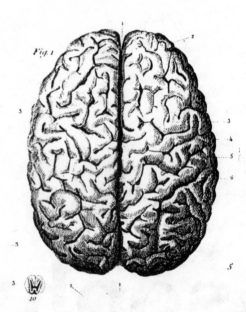

In the *Holography Handbook,* we have attempted a reasonable discussion of workable techniques for obtaining good quality display holograms with a minimum of effort and expense. We would be remiss, however, if we did not note that display holography represents only one application of a much broader holographic spectrum (sorry). Holography is used industrially in interferometry, in non-destructive testing of all sorts of materials and devices, in flow visualization, contour measurement, stress evaluation, etc. We have holographic microscopy and holographic data storage, processing and retrieval. Holograms can be made with a variety of types of radiation, including sound (acoustical holograms). Although beyond the scope of this book, excellent information is available on these applications.[1]

There are two non-optical holography topics, both gaining in popularity, which we do wish to consider briefly. One utilizes a holographic model to describe how the brain works, the other uses holographic theory to narrow the gap in understanding the interrelationships of matter and energy in our concept of the universe. Both are of extreme significance, as they may explain the workings of fundamental systems, which we ordinarily take for granted, in a dynamic new way. What is even more exciting, is that we may be able to use what we have learned from optical holography to help understand these processes.

the holographic brain

One of the great mysteries we have faced as a species is the attempt to understand the mechanics of our own brains. How do we process information, learn, receive new stimuli, reason, and become aware of our condition? Thousands of people in the fields of neurophysiology, philosophy, psychology, education, sociology, religion, etc., have attempted to address thees questions professionally. We have all probably grappled with them individually. Even with the enormous bodies of data which have been accumulated, there are still fundamental omissions in descriptions of how we acheive these basic functions.

One of the biggest puzzles is the way in which the brain stores information. No one-to-one relationship has been detected between a given brain cell or group of cells and a particular thought or memory. If this were so, it would be possible to verify, by removing selected areas of the brain and observing the loss of a particular learned trait. Yet "one of the best established yet puzzling facts about brain mechanisms and memory is that large destructions within a neural system do not seriously impair its function."[2] Lashly and others[3] first discovered this experimentally by removing 80-99% of neural structures, such as the visual cortex, in various animals. They observed, incredibly, that no effect on recognition of a previously learned visual pattern resulted. Somehow, the information was stored elsewhere.

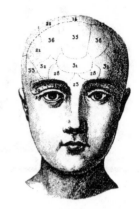

Other tests were conducted where areas of the whole brain were either removed or reshuffled in attempts to destroy a learned trait. One researcher describes in vain his attempts to make a salamander forget how to eat. "In more than 700 operations I rotated, reversed, added, subtracted, and scrambled brain parts. I shuffled, I reshuffled. I sliced, lengthened, deviated, shortened, opposed, transposed, juxtaposed, and flipped. I spliced front to back with lengths of spinal cord or medulla with other pieces of brain turned inside out. But nothing short of dispatching the brain to the slop bucket - nothing - nothing expunged feeding!"[4]

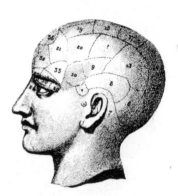

Lashley found that "while *intensity* of recall was in proportion to the mass of the brain, nothing short of removal of the entire cerebrum could interrupt recall altogether."[5] This led him to postulate that "intensity of recall is dependent upon the total mass of the brain, but memory is recorded ubiquitously throughout the cerebrum."[6]

Pribram[7] noted the astounding similarities between this concept and conventional holographic theory: "We can thus distinguish two features of holography which make it unique as an information storage device: the first is that any one of its parts is equal to the sum of its parts, because the message is reduplicated ubiquitously throughout every part of the hologram.... the second feature is that the hologram records the essence of an object, and thus, repeated superimposition of essences supplies the details, the particularities, of the object when the total hologram is illuminated."[8]

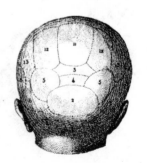

As we recall, when a hologram is made, information about the object is stored everywhere on the plate. If the hologram is broken, a tiny piece will still contain a perspective on the whole. The only way to eliminate the image completely is to throw away the entire hologram, pieces and all. Sound familiar? As a matter of fact, Rodieck[9] demonstrates "that the mathematical equations describing the holographic process match exactly what the brain does with information."

Is this more than just a coincidence? If so, then what acts as the storage mechanism? Where is the interference pattern and from what kind of light is it formed?

We recall that holograms do not necessarily need to be formed with visible light as do our holographic plates (for example, acoustical holograms or even ripples on the surface of a pond). They may be formed in the presence of any wave action. And, "It is not the presence of physical waves as such that is needed for making a hologram, but rather an interference pattern, *a ratio of harmonic relationships*"[10]

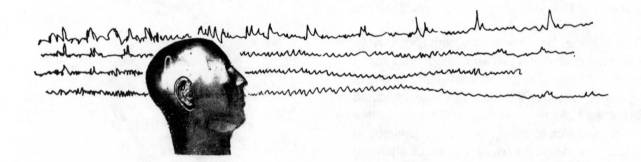

[10]The significance of the italicized statement will be evident in the discussion on holocosmology to follow.

Thus, all we need to look for is a mechanism which creates interference patterns in the brain and stores them.

Let's consider the following model: the brain is a hologram. The mind is the holographic image. The individual neurons are analogous to the grains of silver in a holographic plate. Like silver grains, single neurons carry an extremely limited perspective and have little real significance. As an aggregrate though, an enormous information storage capability is achieved.

The system would operate as follows: new sensory information is received by the brain. This new information cannot be stored by *itself*, yet it interacts and interferes with all past memory and experience of the organism. "Past experience" acts as a reference frame for the new stimuli, or object frame. Because of this, storage as an interference pattern can be accomplished. Almost immediately this new knowledge becomes a part of the reference background forming a new "reference beam". Now new information is received, and it interferes with this new reference. Thus, an ongoing cumulative learning experience is described whereby new is constantly compared to old, assimilated, and then used to evaluate new stimuli. The interference pattern that results can then be stored ubiquitously throughout the brain as would any other interference pattern. "The neural hologram (the brain) is continually exposed and re-exposed to the changing environment thus encoding a constantly shifting set of interference patterns which are 'read out' as a temporally unfolding hologram, i.e., the mind, with its constantly shifting 'model' of reality and associated thoughts, memories, images and reflections."[11]

The astute reader may ask, if the information is distributed throughout the brain, then why do certain areas appear to specialize in particular functions? One can influence vision, hearing, taste, and other inputs by stimulating the proper area of the brain. This apparent paradox can be resolved by considering that it, by analogy, a conventional holographic plate is investigated, greater fringe densities are to be found in some areas, less in others. This is why the image may appear brighter when certain areas of the plate are looked through, and dimmer where perhaps less exposure or beam ratios are present. We might imagine a similar phenomenon occuring in the brain, acting as a sort of multi-channel hologram,

with varying densities for different traits located in different specific areas. Since areas of greater density will tend to act as stronger reference sources, new inputs of a related nature will find most efficient storage at these places. These areas thus become stronger in their specialized functions by the redundant action of an ever-increasing reference frame. Now, if a section of the brain is removed, the information is still stored in the remaining areas, only the resolving capability is reduced. The deficiency can often be made up by relearning through repetition the particular trait or building a new strong reference frame. In actuality, this is what occurs in rehabilitation following a stroke.

To help visualize the holographic memory storage system in action, we can compare the cognitive process of an adult to that of a newborn child.

When an adult sees an apple, almost instant recognition occurs. The adult, having seen, tasted, or heard others describe apples innumerable times needs little new sensory input for rapid and efficient identification. The adult's strong ''apple'' frame of reference might be compared to viewing a hologram with a strong illumination beam to yield a bright image.

The baby, on the other hand, has had no prior experience with an apple to influence his first encounter with it. True, there are genetically derived cognitive processes at work which afford some degree of object perception, but recognition of the apple as *apple* occurs only through repeated exposure to it. The baby begins with a weak reference frame, but in each successive moment, cognitive interference takes place (the previous moment's experience is added to the next moment's memory, or reference frame). New information now interferes with this new product. Eventually, this ongoing process results in the production of a reference frame of sufficient strength to require very little new sensory stimulation for recognition.

One can be aware of this process at work. Both moment-to moment and moment-to-sum-of-the-past-experience interference takes place in this system. For example, the adult can easily experience mirages, or illusions, by processing visual cues with very powerful reference frameworks. It's possible, with great concentration, to see through the illusion, however, by substituting the moment-to-moment frame with the usual adult frame employing the body of greater experience. Gradually, a new reference will take hold, shattering the old illusion.

Seeing through mirages can often be a difficult exercise. What may happen, however, is that cognitive interference between the two reference frames *themselves* may occur. This should yield an almost steroscopic, or new dimensional perspective on the event, allowing greater control over evaluation of the situation. In actuality, this new interference may explain the phenomenon of consciousness, or the concept of "that little person inside the person" we all experience.

Pribram[12] discusses how "recognition holography" may work according to conventional holographic theory. Suppose when making a hologram we use a reference beam from a parabolic mirror or converging lens to make a spot. When the spot is illuminated, it will recreate the object. Also, if the object is illuminated, it will recreate the spot! A detector placed at this spot can then "identify" the object. This idea can be taken a step further by making a hologram of two objects. The hologram can be reconstructed, of course, with an illumination beam taking the place of the original reference beam. Reflected light from one of the objects can also be utilized to act as a reference beam for the other object. If the reference beam is blocked off, and the object and hologram have not changed position, even though only object light is being used, one object will image the other. We have just described an associative relationship, which, in the case of the neural hologram may explain why one thought may conjure up another.

In summary, Pribram notes that "holographic memories show large capacities, parallel processing and content addressability for rapid recognition, associative storage for perceptual completion and for associative recall. The holographic hypothesis serves therefore not only as a guide to neurophysiological experiment, but also as a possible explanatory tool in understanding the mechanisms involved in behaviorally derived problems in the study of memory and perception."[13] And, as Ferguson notes, "Pribram's theory has gained increasing support and has not been seriously challenged."[14]

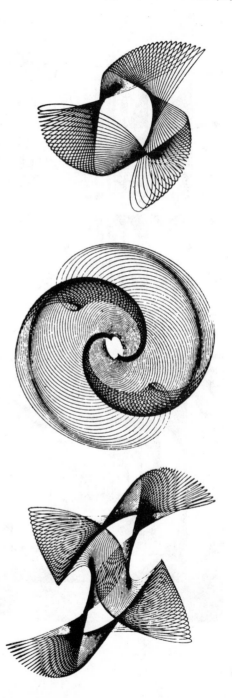

holocosmology

Perhaps the central theme in the science of physics is to attempt to describe accurately, in consistent terms, the mechanics of our perceived universe. One of the biggest difficulties in a science such as this, is that as more and more data is accumulated towards verifying a hypothesis, new uncertainties begin to arise, resulting in more unanswered questions and insufficient explanations.

Such has been the case in discussing how matter and energy shape our universe, and, more fundamentally, what this "stuff" that makes up everything really is (Just when we think we have it all figured out, along comes a black hole to crush the life out of all the old theories).

Our greatest problem might be that when we go looking for answers to how everything works, we discover we are handicapped by our own limited senses. Leibniz realized this in the 18th century, when he maintained that space-time, matter and energy were all intellectual constructs. In modern physics now we are told that "physics is simple only when analyzed locally"[15] or relative to our everyday frame of space and time.

One way we might grasp this conceptually is to take an imaginary trip down into the workings of the atom. Let's say we are standing on a good solid hardwood floor in a good solid building with a good solid foundation in the good solid earth, ok? Now we begin to shrink. The floor looms upon us, spots of dirt and dust appear large. The floor suddenly gives way to large wood fibers separated by big spaces. As we continue to shrink, we observe individual chains of molecules which appear to be endlessly linked together, surrounded by a sea of differently shaped air molecules rapidly bouncing around. We shrink some more, and the molecules give way to individual atoms. We are especially struck by the enormous spaces that exist between the atoms, yet we focus on one in order to see what it is made of. The outer shell inexplicably dissolves, and we see a tremendous empty space. After what seems like an eternity, we arrive at a tiny nucleus in the center, yet, as we continue to shrink, this apparent solid also dissolves into nothing. We are surrounded by a void.

Now wait a minute, where did the matter go?

Perhaps we can look towards Quantum Mechanics for an answer. As we recall from holographic theory, light can be described as both wave and particle. So too with all matter.

"Under the quantum theory, each quantum of matter is both wave and particle, and pervades the universe; there is no solid matter as such, but only probability densities in the continuum."[16]

According to the theory, when one observes a particle, such as an electron, the rule is to point to its exact location only as a "cloud of probability." We can't know exactly where it is at a given time, but can say it will *probably* be around such and such a place. This is very upsetting, since we are used to pinpointing the location of things. But remember that physics is no longer familiar at this micro-level. Still, we are disturbed that the fundamentals which make up everything in our "consistent solid" world *can't be pinpointed in time or space!*

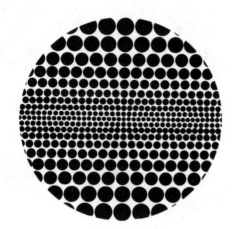

David Bohm, who worked with Einstein, remarks "what appears to be a stable, tangible, visible, audible world is an illusion. It is dynamic and kaleidoscopic - not really there."[17] Bentov carries this a step further by stating, "it seems that the real reality - the micro reality - that which underlies all our solid good common sense really is made up, as we have just witnessed, of a vast empty space filled with oscillating fields! Many different kinds of fields all interacting with each other."[18] In other words, matter is simply a special kind of energy (from relativity theory). It's clear then that the "particle" doesn't really exist, it's only a sensual manifestation; we can only perceive it as a particle when it happens to impinge upon our sensory apparatus as such.

But, but...everything we know of is made up of these particles!

If matter can be reduced to a series of oscillating fields with no determined position, then our description of the particle's location needs in no way be limited to where we "think" it should be. It's actually possible to say that for some given instant in time it may exist anywhere, or even everywhere, since there is, by nature, no limiting condition.

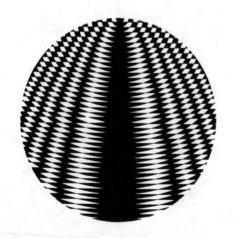

Bentov[19] as well as Tiller[20] investigates this phenomenon further. First, both make the clear distinction between our *perceived* universe, and an actual universe which may lie beyond our normal sensory apparatus. They then discuss a new perspective on one of the most basic yet least understood concepts in modern physics. Einstein postulated that the speed of light was an absolute limit, and experimental results testing relativity theory appear to bear this out. Bentov and Tiller maintain, however, that matter and energy can move at velocities

greater than light, with the speed of light acting as a boundary of the *lowest* obtainable velocity. This concept is still consistant with relativity, with experimental verification being difficult, since we can't *see* things moving faster than light. Nevertheless, experimeters are searching for these particles, called tachyons.[21]

Bentov describes the action of a pendulum, or of any other body for that matter, when its individual component particles move over distances less than 10^{-33} cm., or Plank's distance. Quantum theory shows that it is possible for a particle to move from point A to point B over these small distances without necessarily taking any time to do so. If it does indeed traverse this distance in zero time, then it must move at infinite velocity to do so! Bentov then investigates the action of a pendulum as it moves more and more slowly in its upswing until it changes direction. At one instant, its momentum is zero, which means its position is really indeterminable. This says that, for an infinitesimal period of time, the pendulum could be traveling at infinite velocity and be located anywhere or everywhere in the universe! Immediately after this instant, it appears to return to its former position.

When the particle seems instantaneously to expand to fill the universe, what happens to all of the other particles which may be doing the same thing? Do they just get pushed out of the way? Not if particles are really made up of waves. If we look at two particles now, which happen to expand out at the same time, we have a condition similar to the old model of two stones tossed into a still pond. Two wavefronts are generated which pass *through* each other, yet interact at specific points, causing constructive and destructive interference. Now, if one imagines many particles, it is possible to see how an almost unlimited number (or at least a whole bunch) of interference nodes and antinodes may be generated.

At this point, we have identified the following:
1) When examined closely, matter is found to be made up of energy.
2) We have a limited sensory experience of matter and energy, since we are used only to viewing enormous aggregates of these phenomena. We may not see what they are truly made of.
3) Matter cannot be precisely located in space and can be shown possibly to exist anywhere, and even everywhere, if it travels at very high speeds.
4) If matter is energy and travels as waves, these waves can interact and form interference patterns.

Now, let's assume that we have been examining the above model backwards. From relativity, we know that the perception of an event is determined by the relative position of the observer. Just as we can take the view that our good ol' reliable particle suddenly goes berserk and expands out into the universe, we see suddenly that it's just as valid to suggest that the "particle" is the instantaneously berserk condition, and that our reliable order is really this expanded omnipresent condition of all matter and energy. This would certainly be consistent with the illusory nature of the particle we discovered when we looked at the atom and found only oscillating fields!

We have identified a new condition which is:

5) Energy, filling the whole of the universe, suddenly, instantaneously, "collapses" to form a particle.

Now if we combine conditions 4) and 5), we are led to the following incredible postulation:

6) Energy filling all of space, of necessity, forms interference patterns, and out of this all-pervasive condition, matter is instantaneously formed.

We have just described the making of a hologram.

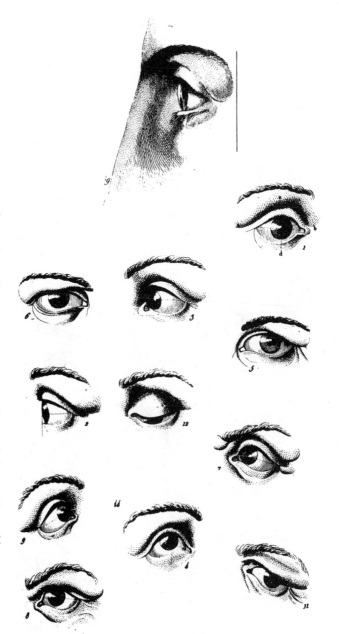

We can now consider the following model: the universe operates holographically. Energy interacts with constructive and destructive interference, to form holograms which we perceive as matter. Just as the optical holograms we made gave us appearances of non-existant 3-D images, energy operating on a much more basic level of perhaps much higher density, forms holograms which we perceive as actual objects. They appear real when viewed as an aggregate of infinitesimal nodes from a standing wave interference pattern.

In order to discuss the validity of a holographic cosmology, certain conditions of hologram construction need to be identified:

1) How standing wave interference patterns are formed, propagated, or manifest themselves.

2) What acts as coherent radiation?

3) If information is stored ubiquitiously throughout the system.

4) Does the hologram store all of the perspectives on the system, i.e., in this case, does it encompass all of the dimensions of spacetime?

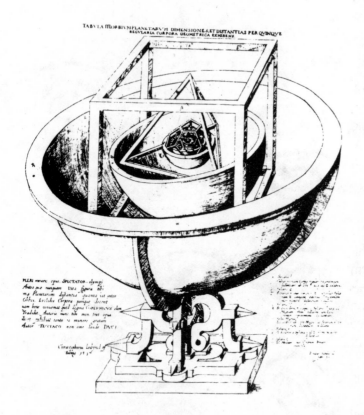

To review briefly, standing waves occur when a wavefront takes on a stationary appearance, while energy continues to pass through the system, each successive wave takes the position of the one before. Standing waves are generated in hologram reconstruction (or in actual object visualization for that matter) since, as the hologram continues to be illuminated over a period of time, the same wavefront continues to be formed.

It is clear that, as with our optical hologram, the standing wave relationships must be maintained throughout the "whole image," or in our case, throughout the universe, to explain with consistency how the whole "ball of wax" can be called a hologram. If this is true, then energy must pass through all particles in such a way as to produce the illusion of being motionless; yet the *energy* must display simple harmonic motion, and be the result of interference from some system of coherent radiation. How can we detect this?

Bentov[22] describes how wave mechanics can provide insight into standing wave consistency. By analogy, when a string vibrates with an integral relationship between its lengths, a standing wave of simple harmonic motion results. This phenomenon can be demonstrated two dimensionally with particles of sand on a metal sheet struck with a violin bow. The same kind of standing waves can also be generated in a solid: the pattern conforms to the kind of lattice found in a crystal.

Since the crystal is vibrating, it can be called an oscillator. If this oscillator is placed near a similar crystal, the two will eventually oscillate in phase, forming a "tuned resonant system." This is also what happens when a struck violin string sets the rest of the instrument vibrating. There is also an obvious amplification of the sound (energy). If still more oscillators are added, they will add to the strength of the resonant system (as does a symphony). Suppose these oscillators are individual atoms. With the total number of atoms in the universe involved, the total amount of energy generated can be astounding! And, "the larger the number of oscillators within such a system the more stable the system, and the more difficult it is to disturb."[23]

Thus, we would expect to see an extremely consistent view of stable particle behavior on the atomic level, as long as the atoms display simple harmonic motion. And they do indeed. Physicists have discovered simple harmonic vibration in all of the basic particles of matter. "The atoms of our bodies are such oscillators, they vibrate at a rate of about 10^{15} Hz. It is very possible that our bodies blink on and off at this very high rate. There is no way of knowing whether this is so, since we currently have no way of registering such rapid phenomena."[24]

We have described a system whereby a set of standing waves may function in phase, thus channeling enormous amounts of energy through a resonant system to maintain the structure of apparently motionless wave fronts. On a macroscopic level, the existence of the structure of moment-to-moment basis , gives rise to the illusion of substance and consistency.

If we search for a source of coherent radiation for this system, we might expect to encounter some problems. We recall in our investigation of ordinary visible light, that it is impossible actually to see the light. All that can be sensed are the effects of the light on our environment. The light itself is invisible; we observe only reflected or modulated wave fronts transformed by the limited optical properties of our eyes. We might expect similar difficulty with this new coherent radiation, with the added handicap of lacking sensory apparatus to process its effects directly. In this case, we might find ourselves one more step removed from directly verifying its existance. We might assume this to be the case, as it would be reasonable to expect that this energy would have to account for all of the manifestations of 4-D space and time, and therefore have to operate outside of them in a 5th or higher dimension.*

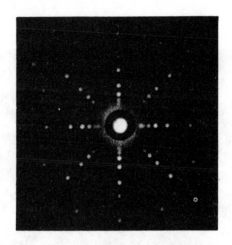

We will describe how information from a dimension escaping detection through our senses or instruments may influence or even shape the universe we do perceive. For an example, let's imagine the following scenario: We begin with a "sea" of coherent energy of an extremely high frequency, one we are unable to perceive directly. To this, we add another energy sea which is slightly out of phase with the first. We can expect the following to happen:

1) Interference will occur separately within each "sea", forming nodes and antinodes.

2) Relative to each other, the two out-of-phase seas will generate "beat" frequencies with periods much lower than the originals. A beat frequency is an illusory secondary wave formed by the *difference* between two primary waves

*Other dimensions are difficult for us to visualize. For an excellent exercise in attempting this, see Sphereland, by Dionys Burger, translated by Cornelie Rheinboldt, Apollo, N.Y.

3) These lower frequencies might fall within the range of our perceivable universe. Now, if this is all we can see, these illusory waves would make up that which we call reality. A universal coherent background wave might be indistinguishable from the void of empty space.

Tiller uses an excellent model to demonstrate the confusion between perceived substance and void:

"There is a well known idea, even in physics, that if you take a crystal which is at absolute zero, it does not scatter electrons. They go through it as if it were empty. And as soon as you raise the temperature and produce inhomogenities, they scatter. Now, if you used those electrons to observe the crystal (e.g., by focusing them with an electron lens to make an image), all you would see would be those inhomogenities and you would say they are what exists and the crystal is what does not exist, right?"[25]

We can observe, in a holographic model, that all of the information about a given particle is present throughout the universe. Earlier we saw, in Bentov's model, how energy might "collapse" to form a particle, and then expand at an infinite velocity to fill the universe. In this way, information about a given particle is transmitted everywhere.

Bohm[26] describes another analogy which very clearly demonstrates the "part contained in all of the whole" concept. It also provides an excellent method of visualizing particle materialization. The model is based on an experiment with a drop of insoluble ink in a vessel filled with glycerine. The device was constructed so that the fluid could be turned slowly without diffusion taking place. When this ws done, the ink drop would be drawn out into a thread that ultimately became invisible. When the turning action was reversed, the drop suddenly became visible again. Bohm calls this the "enfolded or implicate order":

"Ordinarily we think of each point in space and time as distinct and separate and that all relationships are between contiguous points in space and time (but) when we've taken the droplet and enfolded it, it's in the whole thing and every part of the whole thing contributes to that droplet. The implicate order implies a reality immensly beyond what we call matter. Matter is like a small ripple on a tremendous ocean of energy. And the ocean is not primarily in space and time at all."[27]

Bohm elaborates on his model relative to the apparent wave/particle paradox: "We enfolded a droplet by turning the machine a certain number of times, 'n' times. We now put in another droplet in a slightly different place and then enfold that n times. But meanwhile the first one is endolded 2n times, right? Now we have a subtle distinction between a droplet which has been enfolded n times and one which has been enfolded 2n times. They look the same but if we turn one of them n times we get this droplet, turn it n times again we get that one. Now let's do it again with a slightly different position so that it goes n times and the second 2n times and the original 3n times. We keep it up until we put in a lot of droplets. Now we turn the machine backwards and one drop emerges and manifests to our vision, and the next one does, and the next one, so if this is done rapidly, faster than the resolution of the human eye, we will see a particle apparently continuously crossing the field."[28]

We are still faced with the question of why some energy fields appear "stationary" as matter, while others propagate through space and the rest of the electromagnetic spectrum as light. Einstein spent most of his life searching for a "unified field theory" to link matter, energy, and gravity together. We have discussed how the existence of a coherent super-reference energy could be behind the makeup of all; but verification is difficult, and we have no adequate explanation as to *how* this energy becomes matter, and that energy light.

There is one known physical phenomenon which may hold the answer. It is the one body in which all physical constructs merge into a single unified entity. It is, of course, the black hole.

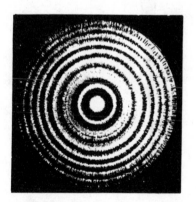

To review quickly, a black hole is an area of collapsed matter which is so dense, that a gravitational field is generated of such magnitude that nothing, not even light, can escape. At this point, both space and time curve inward on each other, and become so distorted that both merge into what is called a singularity - they become a single entity. The definition of a black hole, then, is a body or aggregate of bodies from which no energy or matter can escape, since the escape velocity is greater then the velocity of light itself (this is according to generally accepted theory which places the velocity of light as the *upper* limit in the universe).

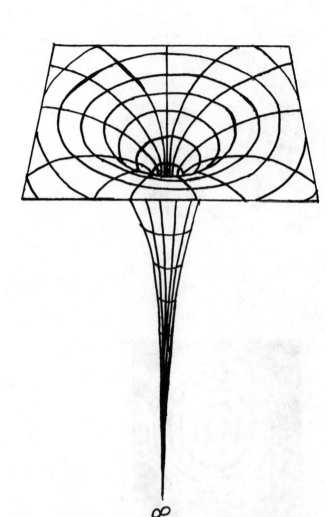

∞

Black holes, incidentally, need not be limited to a certain size. Our observations of what we shall call ordinary black holes (the result of gravitational collapse of massive stars), are always of what we, on the outside, think is going on inside. We may, while still adhering to our definition, describe our entire visible universe as a single black hole. If we imagine ourselves for a moment, stepping outside of this universe, we would observe all of the same phenomena. We would be looking at a system from which no light escapes. Also, by definition, all of the properties of space and time would appear as a single occurrence.

Stephen Hawking[29] has recently done some extraordinary work concerning mini blackholes. He has demonstrated the feasibility of these small holes, perhaps the size of a pea, possibly formed during creation. So, we are faced with the realization that black holes come in a variety of sizes and shapes, suitable for any occasion.

The next step is to imagine what would happen if black holes existed on an atomic level. Actually, what if there were a teeny black hole in the center of every speck of matter. What would happen?

1) Space and time would merge into a singularity, and be indistinguishable. This is precisely what occurs below Plank's distance.

2) Relativity would have to function within each individual black hole. Larger relativistic effects could be described as an aggretgate of these individual monads however.

3) The black hole would be expected to "capture" matter (other holes) under certain circumstances, and annihilate it under others. The atom is capable of both of these functions.

The implications of this model for holocosmology are quite important. It would provide another example of how the whole is also contained in each of the parts. It also may describe another mechanism by which, through tremendous energies generated by resonance, how a consistent picture of atomic particles may be derived. Most importantly, however, it would demonstrate how a unifed universe made up of singularities could account for all or the space-time configurations we have observed, with the distribution of these properties being predominantly holographic in nature.

summary

Certainly, what has been presented here is by no means an exhaustive study of this subject. A great deal of investigative work is needed, and many are currently engaged in this research. We must keep in mind, however, that it is we who are doing the investigating and by doing so, we cannot help but to influence the results. Yet, perhaps some elements of truth can be derived from it all. As Ferguson so eloquently states:

"In a nutshell, the holographic super theory says that our brains mathematically construct 'hard' reality by interpreting frequencies from a dimension transcending time and space. The brain is a hologram, interpreting a holographic universe."[30]

We hope that this book has assisted in the effort to ponder these questions. If, by helping you make some holograms, we have helped to change your view of perceptual reality; the way you see, as well as increasing your understanding of who you are, how you operate, and how the world functions, then our effort here has certainly been worthwhile.

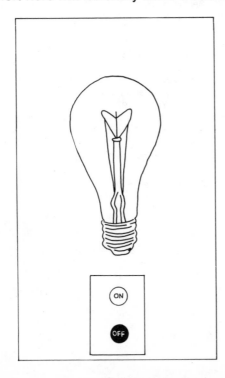

1) A Survey of Holography, #SP-5118, G.P.O. #330-00541, NASA publication, available from: The Superintendent of Documents, U.S.Government Printing Office, Washington, D.C., 20402.

2) Pribram, Karl H., Marr Nuwer and Robert J. Baen. "The Holographic Hypothesis of Memory Structure in Brain Function and Perception", Contemporary Developments in Mathematical Psychology, Vol. 2, edited by R.C. Atkinson, D.H. Krantz, R.C. Luce and P. Suppes, W.H. Freeman & Co., San Francisco, 1974

3) Lashly, K.S. "In Search of the Engram", Psychological Mechanisms in Animal Behavior, Academic Press, New York, 1950

4) Rodieck, Mind, Molecules and Magic, out of print monograph, p.43

5) Lashly, op.cit.

6) Ibid.

7) Pribram, op.cit.

8) Ibid.

9) Rodieck, op.cit.

10) Ibid.

11) Ibid., p.46

12) Pribram, op.cit., p.423

13) Ibid, p.454

14) Ferguson, Marylin. "A New Perspective on Reality", Revision, Vol. 1, #3/4, Summer/fall, 1978.

15) Taylor, Edmund F. and John A. Wheeler, Spacetime Physics, W.H. Freeman & Co., San Francisco, 1966, p. 182

16) Rodieck, op.cit.

17) Bohm, David, "The Enfolding - Unfolding Universe - A Conversation with David Bohm", conducted by Renee Weber, Vol. #1, #3/4, Summer/Fall, Revision, 1978, p. 24-51

18) Bentov, Itzhak. Stalking the Wild Pendulum: On the Mechanics of Consciousness, E.P. DUTTON, New York, 1977, p.26

19) Ibid.

20a) Tiller, William A. "The Positive and Negative Spacetime Frames as Conjugate Systems", Proceedings of the A.R.E. Medical Symposium, Pheonix, Arizona, January 1975

20b) "A Lattice Model of Space and its Relationship to Multidimensional Physics", Proceedings of A.R.E. Medical Symposium, Pheonix, Arizona, January 1977

21) Energy Unlimited - available from E.U., Route 4, Box 288, Los Lunas, New Mexico, 87031, issues Feb. 1979, July 1979

22) Bentov, op.cit.

23) Ibid, p.19

24) Ibid., p.56

25) Tiller, op.cit., p.48

26) Bohm, op.cit., p.25-26

27) Ibid., p.25

28) Ibid., p.26

29a) Milton, S. "Stephen Hawkins", Astronomy, vol. 7, Nov. 1979, p. 28

29b) Ridpath, I. "Awesome Theories in a Wheelchair", Science Digest, vol. 84, Sept. 1978, p.46

29c) Time Magazine, vol. 112, Sept. 4, 1978, p. 56

30) Ferguson, op.cit., p.12

CERTIFICATE of COMPLETION

This is to certify that _____ has successfully

completed a _____ course in holography

this _____ day of _____, 19 ____.

X _____

INSTRUCTOR

Tell you what. Send one of your holograms to the publisher for our collection, and we'll send you back a real, signed and sealed certificate like this one. Send to attention of "Editor, Holography Handbook."

Lloyd Cross in his laboratory. Early stages of stereogram printer construction. Photo by Sharon McCormack.

Filming of ''Dracula and Damsel''. Holographic Sterogram at the Multiplex Co., San Francisco, 1976. Dracula: Dave Schmidt, Makeup: Sharon McCormack, Director: Lee Lacy. Photo © Sharon McCormack.

Lloyd Cross demonstrating optical properties of liquid filled lens. Photo by Sharon McCormack.

appendix

glossary

Abberation: Any imperfection in an optical system which results in an alteration or error in the image.

Absorbtion Hologram: A hologram formed in a material which acquires a certain density in response to exposure. When the hologram is illuminated, part of the light which is not absorbed is diffracted into forming the image.

Achromatic: Free of color. Black and White. In optical systems, the term is used to describe lenses which correct for chromatic aberration.

Acoustical Holography: The making of holograms by using sound waves.

Additive Color Mixing: Means by which two or more frequencies are combined by superimposition to form more colors.

Ambient Light: Light present in the immediate environment. With regards to holographic display, often used to describe background light which is not part of the hologram illumination which may interfere with the viewing of the image.

Amplitude: The maximum value of the displacement of a point on a wave front from its mean value. Graphically, the height or depth of the crest or trough of a wave from its zero point.

Amplitude Hologram: A hologram by which information is stored as variations in transmittance. Another name for absorption hologram.

Antihalation Backing (AH): A dark material placed on the back surface of a plate or film to prevent unwanted light from striking the emulsion. Useful to prevent the formation of *"Newton Rings"* in the hologram. Only to be used with transmission holograms.

Argon Laser: A laser which operates when Argon gas is ionized and controlled by a magnetic field. Produces several blue and green frequencies.

Astigmatism: An abberation caused by the horizontal and vertical aspects of an image forming in different planes.

Attenuation: The loss or reduction of light due to absorption and scattering.

Bandwidth: The range of frequencies over which a given instrument will operate.

Beamsplitter: An optical component which divides a beam into two or more separate beams. A 50:50 beamsplitter produces two beams of approximately equal intensities. A 90:10 beamsplitter transmits approximately 90% of the incident beam and reflects 10% into the second beam.

Beat Frequency: A new frequency formed by the presence of two slightly different frequencies. The beat frequency is equal to the difference in frequency between the original two.

Benton Hologram: Another term for *Rainbow Hologram*. Named for its inventor, Steve Benton. A hologram produced by reducing vertical information in order to correct for image dispersion.

Biconcave: A lens which has both faces curving inward. A type of negative lens.

Biconvex: A lens which has both faces curving outward. A type of positive lens.

Birefringence: The separation of a beam of light into two beams after it penetrates a doubly refracting object. In monochromatic systems, the two beams interfere causing undesirable "rings".

Black Hole: A cosmic phenomenon whereby the mass and density of an object surpasses a critical value so that its escape velocity equals or exceeds the speed of light. Any matter or energy becomes "captured" and cannot escape.

Bleach: In holographic processing, a chemical used to change an absorption hologram into a phase hologram in order to improve efficiency (brightness).

Bragg Diffraction (Bragg's Law): Diffraction which is reinforced by reflection by a series of regularly spaced planes which correspond to a certain wavelength and angular orientation. The angle at which this reinforcement occurs is Bragg's angle.

Brewster Angle: The angle at which transparent material, such as a plane of glass, is placed with respect to incident light such that refracted and reflected rays are mutually perpendicular. The light must be polarized. At this angle, the glass becomes most transparent to the light. Laser tube ends are fitted with "windows" set at this angle (usually 45-55 degrees) for maximum transmissiveness and near zero reflectance.

Brightness: A subjective term describing the amount of light perceived.

BRH: Bureau of Radiological Health. U.S. Government agency responsible for setting laser safety standards.

"C": Commonly used abbreviation for the speed of light, which equals 3×10^8 meters/second or 186,282 miles/second.

Chromatic Aberration: Lens or hologram abberation due to the shifting of image position for each frequency. If severe enough, the image will appear to blur due to the lack of registration of the colors.

Coherence Length: With respect to laser light, the greatest distance between two components of light (i.e. 2 beams) such that interferometric effects will occur. In holography, the coherence length of the laser will determine the depth over which an object can be recorded.

Coherent light: Light which is of the same frequency and vibrating in phase. The laser produces coherent light.

Collage: In holography, a technique used to superimpose various spaces on top of one another by overlapping individual holograms or exposures.

Collimated Light: Light which forms a parallel beam and neither converges nor diverges. Also referred to as collimated beam.

Collimator: A device used to produce collimated light by positioning a light source at the focal point of a lens or parabolic mirror. Such a device is called a collimting lens or collimating mirror respectively.

Color Hologram: A hologram formed by two or more different frequencies of laser light to produce an image which appears in natural color.

Color Spread: The area over which a spectrum is dispersed. In rainbow holography, the amount of vertical viewing area for the image to appear in one color.

Computer Generated Hologram: A synthetic hologram produced using a computer plotter. The binary structure is produced on a large scale and then photographically reduced into a given medium. The technique allows the production of impossible or nonexistant 3-dimensional forms.

Concave Lens: A lens with an inwardly curving surface which causes light to diverge. See also Negative lens.

Concave Mirror: A mirror with an inwardly curving surface which causes light to converge.

Continuous Wave Laser: A laser which emits a beam which is continuous over time.

Convergence: The optical bending of light rays toward each other as by a convex lens or concave mirror.

Convex Lens: A lens with an outwardly curving surface which causes light to converge, usually to a focal point.

Copy Hologram: Another term for image plane hologram, or any second generation hologram produced from a master hologram. A contact copy is produced by placing the plate in contact with the original.

Copy Plate: Another term for copy hologram. Usually refers to the plate before it is exposed.

Cross Hologram: Another name for the type of holographic stereogram which incorporates the advantages of rainbow holography. Named for Lloyd Cross.

Cross Talk: The phenomenon of spurious images formed by color holograms when an interference pattern formed by one color also reconstructs an image in another color.

Cylindrical Mirror/Lens: An optical component which causes light to focus as a slit or line by passing through or reflecting from a surface curved in one dimension.

Denisyuk Hologram: Another name for *single beam reflection hologram*. Named for its inventor, Yu. N. Denisyuk.

Density: The amount of opacity or darkness of a given medium. Technically quantified as the log to the base 10 of the inverse of transmittance. Not the same as absorption.

Depth of field: The area within which satisfactory resolution of an image can be obtained. Also, in holography, used to describe the area within which any image can be formed, due to the constraints of coherence length.

Developer: A chemical solution which changes the latent image of a photographic image or holographic interference pattern (silver salts) into black metallic silver. The term development usually refers to the degree of effect of the developer, or the cause of the amount of density.

Dichromated Gelatin: A light sensitive emulsion made up of a solution of dichromate compound, usually ammonium dichromate, in the presence of a gelatin substrate. Exposure results in the crosslinking of gelatin molecules with those of the dichromate compound.

Diffraction: The change in direction of a wavefront by encountering an object. Usually refers to the case whereby light is bent by passing through a small aperture.

Diffraction efficiency: In a hologram, the percentage of incident illumination light which is diffracted into forming the image. The greater the diffraction efficiency, the brighter the image will appear in a given light.

Diffraction Grating: A holographic diffraction grating is a hologram formed by the interference of two or more beams of pure, undiffused laser light.

Diffuse Reflector: An object which scatters illumination which strikes it. Most objects are diffuse reflectors.

Dispersion: The separation of polychromatic light into component frequencies.

Distances: In holography, usually refers to the matching of beam path lengths in order to maintain coherence.

Divergence: The bending of light rays away from each other, usually by concave lens or convex mirror, so that the light spreads out. Light will also diverge with a convex lens or concave mirror after it passes through the focal point.

Double exposure: The formation of two holograms on the same recording medium. Used to cause either overlapping images or two discrete images to appear under different conditions.

Dye Laser: A laser which employs chemical dyes in order to achieve a tunable range over a wide area of the spectrum. It is usually ''pumped'' or made to function by using the beam from an ion laser to initiate lasing action in the dye.

Electromagnetic Radiation: Radiation emitted from vibrating charged particles all of which travels through space at the speed of light. Visible light is only a small part of the entire electromagnetic spectrum.

Emulsion: The light sensitive material including a base usually of gelatin, which is applied to either a film or glass plate substrate.

Etalon: A very flat optical device with parallel surfaces designed to increase the coherence length of a laser by eliminating modes which are slightly out of phase.

Excited State: A state of an atom which occurs when outside energy is absorbed, causing the atom to contain more energy than normal.

Exposure: The act or time of allowing light to impinge upon the emulsion.

Film plane: The plane at which the recording material is located.

Film speed: The degree of sensitivity of the film to light to cause a satisfactory exposure. Holographic film speeds are usually expressed in ergs/cm².

Fixer: A chemical solution which removes the unexposed silver salts from the emulsion in order to desensitize it and preserve the record stored as metallic silver.

f - number: The ratio of the focal length of a lens or curved mirror to its diameter.

Focal length: The distance from the center of a lens or curved mirror to a position at which light will converge to a point. Using an element with a positive focal length will cause light to actually focus to a point. Optics with a negative focal length only appear to focus at a imaginery point. The *focal point* is the place at which the light focuses. The *focal plane* is a plane through this point.

Focused Image Hologram: A one-step image plane hologram made by using a lens to focus an image directly into the film plane. Also sometimes used as another term for *"open aperture"* hologram.

Fog: The darkening or exposing of film by in advertantly allowing ambient light to strike it. In holography, a fogged plate reduces fringe contrast, resulting in a less efficient image.

Fourier Transform Hologram: A special hologram formed by a plane object situated in the focal plane of a lens. Also called a *"Fraunhofer Hologram"*.

Frequency: The number of crests of waves that pass a fixed point in a given unit of time.

Fresnel Hologram: Another name for the common hologram. Defined as a hologram formed with an object located close to the recording medium.

Fringe: An individual interference band, made up of one cycle of constructive and destructive interference.

Front Surface Mirror: A mirror with the reflecting surface on the front. Conventional mirrors have their reflecting surfaces on the back of a piece of glass and are not useful for holography as the front surface will produce a *"ghost"* reflection.

Gabor Hologram: An *in-line* hologram of the type first invented by Dennis Gabor.

Gas Laser: The most common form of laser, which operates by causing the atoms in a gaseous mixture to undergo a population inversion.

Ghost Image: A duplicate image, usually unwanted, which usually is formed as a result of light reflecting from an undesired surface.

Grating (also Diffraction Grating): A device which bends light due to the effects of diffraction. A hologram is a special type of grating. Gratings can be made by etching, deposition, acoustically, or made holographically.

Ground State: The condition whereby an atom achieves its lowest energy state and greatest stability. Atoms in the ground state are not capable of emitting radiation, but must first be raised to an excited state.

H-1: Another name for master hologram.

H-2: Another name for copy hologram.

Harmonic motion: Motion of a consistant regular period which is induced by a given frequency or integral multiple of it.

Helium-Neon (HeNe): The most common lasing material, which produces a continuous red beam at 632.8 nm.

Hertz (Hz): A unit of frequency equal to one cycle per second.

HOE: Holographic Optical Element. A hologram which may be used to act as a lens, mirror, or some complex optical component.

Holocosmology: A theory which suggests that much of the workings of the cosmos may be explained in holographic terms or by holographic analogy.

Hologram: An interference pattern formed as a result of reference light encountering light scattered by an object and stored as such on a light sensitive emulsion.

Holographic Movie: The animation of a 3-dimensional holographic image by presentation of numerous holograms in rapid sequence in much the same way motion picture film operates. Unlike conventional cinema, it is only with extreme difficulty that the image can be projected, and true hologrphic movies are still very experimental. Term is also often a misnomer for holographic stereograms.

Holographic Snapshot: Term used to denote a hologram which merely replicates an existing object without any creative input on the part of the holographer.

Holographic Stereogram: A hologram made by filming numerous angles of view of a scene and then storing the frames holographically. Each eye views a different frame, displaced so as to result in the illusion of a stereoscopic image. Also called a multiplex or integral hologram.

Hypo-clear: Chemical solution which removes the fixer. Use of hypo-clear usually improves the stability of the hologram by preventing degradation caused by oxidation of residual fixer.

Image Plane Hologram: A second generation hologram formed by positioning a light sensitive plate in the plane of an image formed by a master hologram.

Immersion lens: A lens formed by filling a frame with an index-matching liquid to approximate the function of a solid lens, yet at much less cost and with much greater flexibility.

Incandescent light: Light formed when an electric current passes through a resistant metal wire, which is usually situated in a vacuum bulb.

Incoherent light: Light which is not in phase with itself. Most light is incoherent.

Index-Matching: The process of duplicating the refractive index of a material in order to eliminate effects caused by optical surfaces. A liquid with the same refractive index as the substrate is used to join the two surfaces together. In holography, plates are often indexed to a dark backing to eliminate the effects of Newton's rings.

Index of Refraction: The ratio of the velocity of light in air to the velocity of light in a refractive material for a given wavelength.

Inertial Mass: A large mass which, if at rest, tends to remain at rest.

Infrared: That part of the spectrum which is characterized by wavelengths somewhat longer than those of red light, which are not visible to the eye yet are often perceived as heat. Covers the spectrum from about 750 nm. to 1000 micrometers.

In Line Hologram: A *Gabor hologram*. Made by positioning the object and reference light along the same axis, resulting in a configuration only practical for making holograms of transparencies.

In Phase: The relationship of two waves of the same frequency when they travel through their maximum and minimum values simultaneously and are also polarized identically. Holograms must be made by waves which remain in phase during the course of an exposure.

Intensity: The value of the amplitude squared.

Interference: The combining of two waves so that their amplitudes add at every point. When two coherent waves are so superimposed, the result will be either an increase in amplitude (*constructive interference*) or decrease in amplitude (*destructive interference*). The result is an interference pattern which records the relative phase relationships between the two waves, thus storing the characteristics of the individual waves. This is how a hologram works.

Interferometer: A device that utilizes interference of light to measure changes in systems with extreme accuracy. An interferometer can be used to test the stability of holographic systems.

Ion: An atom which has gained or lost an electron so that it acquires a positive or negative charge.

Ion Laser: A laser within which stimulated emission occurs as a result of energy changes between two levels of an ion. Argon and Krypton are the two most common types of ion laser.

I.P.: Abbreviation for Image Plane.

KISS: "Keep It Sweet and Simple". Also "Keep It Simple, Stupid".

Krypton Laser: An ion laser that produces many frequencies which appear over a large part of the spectrum. The most common lines are blue, green, yellow as well as a very strong red frequency.

Laser: "Light Amplification by Stimulated Emission of Radiation". A laser cavity is filled with a material which, when stimulated with the proper energy, produces a population inversion of excited atoms. The light produced continues to increase in intensity as it oscillates back and forth between two mirrors placed at each end of the cavity. The front mirror is designed to transmit one to two percent of this amplified light, resulting in a beam of very intense, monochromatic, coherent light.

Latent Image: The image or pattern stored in an emulsion before it is developed into a visible image.

Latent Image Decay: A condition that is common to fine grained silver emulsions, including the types used for holography. The decay occurs if the material is not processed soon after exposure, resulting in a lower density.

Leith, Upatnieks Hologram: Another name for *off-axis hologram*, named for its inventors.

Lesliegram (also Leslie Hologram): A particular type of holographic stereogram made by using images projected onto a screen and recording sequential strip holograms by use of a fixed plate and moveable slit mask.

Light Meter: Any device used to sense and measure light. Usually used to sense intensity in order to determine exposure.

Line Spacing: The distance between individual interference fringes in a diffraction grating.

Lippman Hologram: Another name for *reflection hologram*.

Master Hologram: Any hologram which produces an image from which another hologram is made.

Matte Screening: A special type of double exposure whereby two different overlapping yet distinct spaces can be made to appear simultaneously.

Medium: Any substance or material which may be used to convey an idea. In physics, any material through which radiation can travel.

Metal Mounts: Devices to hold optical elements on a vibration isolation table with a metal top.

Metal Table: A vibration isolation system with a metal top.

Mode: The degree to which the beam of a laser is spatially coherent. The tem_{00} mode is characterized by an even spread of light. The tem_{01} mode, or donut mode appears as a ring of light with a dark center.

Moire Pattern: A highly visable type of interference pattern formed when gratings, screens or regularly spaced patterns are superimposed upon one another.

Monochromatic: Light or other radiation with one single frequency or wavelength. Since no light is perfectly monochromatic, the term is loosely used to describe any light of a single color over a very narrow band of wavelengths.

Motion: Term used in holography to describe effects of object or holographic system not remaining rigidly fixed during exposure.

Multichannel Hologram: A hologram formed with two or more separate reference beams or angles.

Multicolor Hologram: Any hologram for which images of more than one pure spectral color may be viewed. Usually used to describe a rainbow hologram made with multiple exposures to achieve color variations.

Multiple Exposure: More than one exposure occuring on the same plate or film.

Multiplex: Another name for *Holographic Stereogram.*

NAH: A holographic plate without an antihalation backing. Also called an unbacked plate.

Negative Lens: A lens characterized by a concave surface which causes light to diverge. A negative lens has a negative focal length.

Newton's Rings: The series of rings or bands which appear due to interference between two nearly parallel surfaces. These rings often form as a result of light interacting between the front and back surfaces of a holographic plate.

Node: The part of a vibrating wave that is not moving - zero point. An *antinode* is a point on the wave of maximum displacement from the zero point.

Noise: The effects of undesired light scattered by an emulsion which interfere with the resolution of the image.

Object Beam: The light from the laser which illuminates the object. Also used to describe the entire beam path from the first beamsplitter to the object and then to the plate. In image-plane holography, it is referred to as the master *illumination beam*.

Objective: The optical element from a microscope which is used to focus a laser beam to a point to pass through a pinhole in a spatial filter.

Off-Axis: The type of hologram invented by Emmet Leith and Juris Upatnieks whereby object and reference beams approach the holographic plate at different angles.

On-Axis: Hologram formed with object and reference beams originating along the same axis. Also called an "*in-line*" or "*Gabor*" hologram.

Open Aperture: A transmission image plane hologram viewable in white light and characterized by both vertical and horizontal parallax and usually a brilliant white image.

Optical Cavity: The space between the two mirrors in a laser. The tube is located within the optical cavity.

Optical Cement: An adhesive used to join optical surfaces. Also useful for laminating a cover glass onto a hologram to protect it. Also known as *Sealant.*

Optical Component: An optical device consisting of the optics (lens, mirror, etc.) and a mount used to affix it to a vibration isolation table.

Optics: Those devices which change or manipulate light including lenses, mirrors, beamsplitters, filters, etc. Also, the science of electromagnetic radiation, its effects, and the phenomenon of vision.

Optics Table: Another name for vibration isolation table. The table upon which holograms are made.

Orthoscopic: An image with correct parallax and front to back orientation.

Oscillator: Any device that converts energy into an alternating electromagentic field, usually of constant period.

Overexposure: Improper exposure resulting from too much light, or light reaching the plate or film for too much time.

Parabolic Mirror: A mirror with a surface curved in the shape of a parabola. Used as a telescope mirror in astronomy or as a collimating mirror in holography.

Parallax: The difference between two different views of an object, obtained by changing viewing position.

PBQ: A bleach usually used in reflection holography containing the chemical p-benzoquinone.

Period: One complete cycle of a wave.

Perspective: The concept of the relationship of various objects in a viewing zone taken from a particular point.

Phase: The position of a wave in space measured at a particular point in time.

Phase Shift: The relationship between the phase of one wavefront which does not match the phase of another. The result of phase shifts of coherent light is interference.

Phase Hologram: A means whereby information is stored as a result of phase shifts between two wavefronts (as are all holograms). A *phase-amplitude* hologram stores information as a result of phase shifts which manifest themselves as variations in amplitude. A *pure phase hologram* stores the phase shift as a result of refractive index variations in the material.

Photochemistry: Any chemistry related to the action of or upon light sensitive materials.

Photo-flo: A chemical wetting-agent used to produce an even coating of water over an emulsion during processing to promote even drying.

Photon: The smallest unit or quantum of electromagnetic energy known today.

Photopolymer: A light-sensitive plastic which is useful for real-time holography due to almost instantaneous processing.

Photoresist: A chemical substance made insoluable by exposure to light (usually ultraviolet). Although most often used to manufacture microcircuits, photo-resist can be used to make holograms.

Pinhole: The small hole used to pass focused light from the objective in a spatial filter.

Plane Waves: Waves which propagate as parallel planes, making up a collimted beam.

Plane Hologram: A hologram for which fringes are large with respect to the thickness of the emulsion, so that interference is mostly stored on the surface of the hologram.

Plano-concave: A lens which has one concave surface and one flat surface.

Plano-convex: A lens which has one convex surface and one flat surface.

Plateholder: Any device which holds a holographic plate or film in place during the exposure.

Platen: Another name for plateholder.

Point Source: A light which appears to come from a point or from a source of relatively small area. The sun as well as an unfrosted light bulb are both point sources.

Polarization: The restriction of light or other radiation to vibration in only one plane.

Population Inversion: A condition whereby more atoms are in the excited state rather than the ground state, resulting in the predominance of stimulated emission.

Positive Lens: A lens with an outwardly curving surface which causes light to converge. Also known as convex lens.

Processing: The entire chemical sequence, from development to final drying of the hologram.

Pseudocolor: The production of colors in a hologram which are not related to the true colors of the original objects. Usually used in connection with multicolor holograms.

Pseudoscopic: The opposite of orthoscopic. An image that is turned inside-out.

Pulsed hologram: A holgram produced with the short pulse of a pulsed laser. May be used to make holograms of live subjects.

Pulsed Laser: A laser which emits radiation in a wave of short bursts and is inactive between bursts.

Quantum: The smallest amount that the energy of a wave may be divided into.

Rainbow Hologram: A transmission image plane hologram made by restricting illumination of the master vertically to produce a horizontal slit, in order to compensate for chromatic dispersion when the copy is viewed in white light. Vertical parallax is sacrificed, however. The hologram is named for the fact that the image is viewable in each of the colors of the rainbow, which change as one's vertical viewing position changes. Also called a *Benton Hologram*, after its inventor.

Ratios: Used to describe the relationship between the intensities of reference to object light as measured at the position of a holographic plate in a given setup.

Real Image: An image formed by light which actually focuses in space.

Real Time Holography: A technique whereby a holographic image is superimposed over a real object in order to observe interference fringes generated by minute changes between the two.

Reconstruction Beam: Light directed at the finished hologram from which the object wavefront will be recreated.

Recording Material: Any substance which may be used to record the interference pattern of the hologram.

Reference Angle: The angle at which the reference beam strikes the plate, usually measured in degrees from the plate surface.

Reference Beam: The unmodulated, pure laser light directed at the plate to interfere with the object light.

Reflection Hologram: A hologram made by allowing reference and object light to impinge on opposite sides of the plate. The finished hologram is viewed by allowing light to reflect from it to the observer.

Refraction: The bending of light which occurs when it passes from a medium of one refractive index to that of another. In a phase hologram, refraction causes a "phase delay" which corresponds to the original phase difference between the two stored wavefronts.

Refractive Index: Same as index of refraction.

Regular Transmission Holograms: Transmission holograms which are viewed in laser light.

Resolution: The ability of a film or an optical system to distinguish between two closely spaced points. Film resolution is usually expressed in terms of how many closely spaced lines per millimeter the film can record. Holographic films must be capable of high resolving capability since the interference fringes are often extremely small and closely spaced.

Resonance: A large amount of vibration in a system which is caused by a small stimulus with approximately the same period as the natural vibration period of the large system.

Resonant Cavity: Another name for optical cavity or laser cavity.

Sand-Based System: A vibration isolation system which uses a sandbox table as its inertial mass. Also, simply called *sandtable*.

Sand Mounts: Optics mounts which may be positioned on sand-based optics tables.

Scatter: Unwanted light which interferes with the making of a good quality hologram.

Setup: The configuration of optical components used to produce a given hologram.

Settling Time: A period of time between the loading of the plate and the exposure in order to allow ambient vibrations time to dampen.

Shadowgram: A hologram made by deliberately moving an object during an exposure, or by using an inherently unstable object, in order to produce a 3-dimensional "hole" or shadow where the object was once located.

Shutter: The device used to block the beam and then allow it to pass unobstructed for the desired exposure time.

Silver Halide: The type of recording material which consists of light sensitive silver particles suspended in gelatin.

Single Beam Hologram: A hologram made with one beam which acts as both reference and object illumination beam.

Slab Table: An optics table which uses a concrete slab as part of its inertial mass.

Slit Optics: Any optical device which causes light to be propagated into a line. Usually formed by light interacting with a cylindrical surface.

Solid State Laser: A laser which uses a solid material, such as ruby, as its lasing medium.

Space: The area between objects. In holography, the area between and including objects.

Spacetime: A four-dimensional model including the three common dimensions of space with the fourth of time.

Spatial Coherence: The degree of correlation between the phases of monochromatic waves emanating at two separate points. Coherence over a given space. Coherence across the diameter of a laser beam.

Spatial collage: Holographic collage characterized by the overlapping of individual dimensional spaces.

Spatial Filtering: The act of "cleaning up" the light of the laser beam by causing it to focus through a tiny aperture. Only the pure light can focus at the desired point, eliminating the effects of dust, optical surface scratches, etc.

Spatial Frequency: Often used with regard to line spacing in diffraction gratings, the spatial frequency is the reciprocal of line spacing, generally expressed in cycles per millimeter. See also resolution.

Speckle: The grainy appearance of an object, or a holographic image, viewed under laser light. It is caused by light reflecting from minute areas of the object and interfering with itself.

Specular Reflection: Any reflection from a smooth polished surface such as a mirror.

Spherical Abberation: The failure of an optical system to form a perfect image due to curvature of the wavefront.

Splitbeam: The act of separating a beam of laser light into two components to separately control the action of reference and object illumination.

Squeegee: A device or action used to remove excess water from the emulsion to facilitate drying.

Stability: The requirement for holographic optical systems to remain motionless during an exposure.

Standing Wave: The combination of two waves of the same frequency and amplitude which are traveling in opposite directions. Also used in a general sense to describe any waves which appear motionless due to each new wave replacing the position of the one before.

Stereogram: An image which creates a 3 dimensional illusion by presenting a different view of an object to each eye.

Stereophotograph: A photographic stereogram made by taking two pictures, at a distance apart corresponding to the average distance between our two eyes. The stereophotograph is viewed in an apparatus which allows only the proper eye to see each view.

Stimulated emission: Radiation produced by incoming radiation of the same phase, amplitude, and frequency.

Stop Bath: The chemical bath immediately following the developer which causes the developer to cease action.

Tachyon (tachion): A theoretical particle which would travel at speeds greater than that of light.

Tem$_{00}$: The lowest mode of a laser, characterized by a beam which is spatially coherent across the diameter of the beam.

Temporal Coherence: Coherence over time. The degree to which waves will remain coherent over time and distance. An elaton improves temporal coherence.

Tension Compression Table: An optics table which is strengthened by tension produced along braces made of tightened, threaded, metal rod. Usually a sand-based system. Invented by Lloyd Cross.

Test Strip: A means of visually determining the correct exposure by making a series of individual exposures of varying times on the same plate. The proper time is determined by selecting the strip which yields the brightest or cleanest image.

Thermoplastic Film: A recording material which works due to the effects of electrostatic forces and heat to produce a deformation corresponding to the interference pattern exposed.

Three Dimensional Viewing: The synthesis of two separate images from two eyes or one eye in motion into a time averaged experience in the brain.

360 Degree Hologram: A hologram made by exposing recording material which completely surrounds an object.

Transfer mirror: A mirror which redirects light from the laser towards the desired working area on the optics table.

Transmission Hologram: Any hologram which is viewed by passing light through it, toward the viewing side. Transmission holograms are made by allowing both object and reference light to impinge on the same side of the plate.

Transmittance: The proportion of light transmitted by a medium to that which is incident upon it.

Triethanolomine: A chemical used to change the thickness of the emulsion to produce different color playback, usually with reflection holograms.

Ultraviolet: An invisable part of the spectrum characterized by wavelengths somewhat shorter than violet (approx. 100-400 nm.).

Unbacked Plate (NAH): A holographic plate without an antihalation backing. Essential for reflection holograms.

Variable Beamsplitter (VBS): A beamsplitter whereby the ratio of transmitted to reflected beam changes as the beam intercepts the component at different points.

Vibration Isolation: The pratice of removing a system from the effects of ambient vibrations which may induce changes, particularly in optical systems. Vibration isolation must be used in making a hologram to prevent the movement of interference fringes during an exposure.

Virtual Image: An image generated by light which does not actually focus in space, yet it appears to do so. The image can be seen, yet neither projected nor imaged onto an emulsion.

Volume Hologram: A hologram made up of fringes which are small with respect to the emulsion thickness. The fringes are thus stacked up through the volume of the emulsion. Volume holograms must be illuminated at the original reference angle due to the effects of *Bragg diffraction*.

Wave Form: The characteristic shape taken on by a wavefront.

Wavefront: The surface of a propagating wave which represents all points equidistant from the light source, and is characterized by one period of the wave.

Wavelength: The physical distance over which the complete cycle of one wave occurs. Wavelength is inversely proportional to *frequency*.

White Light Transmission Hologram: Any transmission hologram which can be displayed using ordinary white light.

Xylene: The most common used chemical for index-matching. Xylene has a refractive index very close to that of glass.

YAG Laser: A solid state laser using Yttrium Aluminum Garnet as the lasing material.

Zone Plate: A pattern consisting of a central spot surrounded by concentric zones, alternatingly opaque and transparent, the total area of each zone being equal. It represents the real image of a point produced by diffraction.

bibliography[*]

Benrey, Ronald. *How to Build a Low Cost Laser*, Hayden Book Co., Rochelle Park, New Jersey, 1974.

Plans for building a 1 mw laser (a little weak for most holography). Power supply diagram is somewhat dated yet assistance can be had by having it interpreted by an electronics supplier.

Berner, Jeff. *The Holography Book*, Avon Books, New York, 1980.

A disappointing book with mis-information. Definitely does not live up to its promise as a complete manual. Nice to read quotes from some holographic artists, however.

Brown, Sam. *Popular Optics*, Edmund Scientific Co., Barrington, New Jersey, 1974.

A basic introduction to optics for the layman. Non-Technical. Valuable for those interested in optics who find themselves without much of a math background.

Camatini, E. (editor), *Optical and Acoustical Holography*. Plenum Press, New York, 1972.

Very general first look at Holography.

Caulfield, H.J. *Handbook of Optical Holography*, Academic Press, New York, 1979.

Some excellent up to date information. Collection of articles by some of the top researchers in the field. The book, however, is grossly overpriced for what you get (approx. $60.).

Collier, Robert J., Christoph B. Burckhardt and Lawarence H. Lin. *Optical Holography*, Academic Press, New York, 1971.

One of the most complete coverages of optical holography we have seen. The book is, in some minor ways, beginning to become a little dated, yet it has the best technical treatment of the field we've seen. No serious holographer should be without this book. An excellent classroom text.

Dowbenko, George. *Homegrown Holography*, American Photographic Book Publishing Co., 1978.

A disappointing book with numerous errors and misleading drawings. Could have been a good practical guide had more attention been paid to detail accuracy. Perhaps another edition, George?

Francon, M. (translated from French by Grace Marner Spruch). *Holography*, Academic Press, New York, 1974.

Very technical. More up to date than many others of its kind yet often difficult to follow for the layman.

[*]Note: Most of these books, and a few others not listed, are available from the Museum of Holography Bookstore, 11 Mercer Street, New York, New York 10013

Freeman, W. H. *Lasers & Light*, W.H. Freeman & Co., San Francisco, 1969.

Selected readings from Scientific American. Good, non-technical discussions of how lasers and other light sources work. Somewhat dated information, however.

Gregory, R. L. *The Intelligent Eye*, McGraw Hill, New York, 1970.

Good coverage of topics related to visual perception.

Hewitt, Paul. *Conceptual Physics: A New Introduction to Your Environment*, Little & Brown, Boston, 1977.

One of the best general physics texts we've ever seen. Designed for the non-scientist, it has rather complete and fascinating discussions of most physical phenomena. Excellent sections on optics. Brief treatment of holography. Excellent for acquiring a background sense of how holography relates to other fields.

Jeong, T. H. *A Study Guide on Holography*, American Association for the Advancement of Science, 1975.

Although limited in scope, a fairly good and basic practical guide.

Klein, Niler V. *Optics*, John Wiley & Sons Inc., New York, 1969.

Technical text about science of optics. Good for math-inclined holographers with an interest in scientific applications. Helps develop a background in which holography may be applied.

Lehmann, Matt. *Holography: Theory and Practice*, Focal Press, London, 1970.

An excellent book. Matt keeps the math bearable and the book is written in an easy to read style. However, much practical information is now dated.

Outwater, Chris and Eric Van Hamersueld. *A Guide to Pratical Holography*, Pentangle Press, 1974.

Very good down to earth practical approach although some of the information is now dated. Somewhat limited in scope. Good descriptions for the layman. A few errors, however.

Rainwater, Clarence. *Light and Color*, Western Publishing Co. & Golden Press, New York, 1971.

An excellent little book dealing with the nature of light and how we see. One of the best we've seen that covers this subject conceptually.

Saxby, Graham. *Holograms*, Focal Press, London, 1980.

A book we like very much. Good, up-to-date information which complements our own. Easy to read and straightforward approach with math left to an appendix. Only criticism is a little weakness in setup descriptions, and layout of material. Best description of building liquid-filled lenses we've yet seen.

Smith, Howard M. *Principles of Holography*, John Wiley & Sons, New York, 1975.

Fairly good introduction with a separation of heavy math theory and descriptive material. Good information on the state of recording materials and processing information as of 1975, although information is somewhat dated as of now.

Stroke, George W. *An Introduction to Coherent Optics and Holography*, Academic Press, New York, 1969 Good.

Very technical approach. If you're not into math, better pass this one up. Although an older book, the mathematics, of course, still holds and is excellent.

(TERC), Jones, John. *Holographic Techniques and Equipment*, Laser Electro-Optics, Module #7-4, Technical Education Research Center (TERC), 4800 Lakewood Drive, Waco, Texas 76710 (48 pages).

Better than module #5-9 for quick review of technique, yet still has much incomplete information.

(TERC), Pedrotti, Leno S. *Holography*, Laser Electro-Optics, Module #5-9, Technical Education Research Center (TERC), 4800 Lakewood Drive, Waco, Texas 76710 (31 pages), 1980.

Very basic. Quick overview for use as teaching aid as part of another course (e.g. Physics Lab I). Lacking much information, however.

Wenyon, Michael. *Understanding Holography*, David & Charles, North Pomfret, Vermont, 1978.

Good general description of lasers, holography and possible applications.

some magazines and magazine articles on holography:

Applied Optics Magazine
American Institute of Physics
335 E. 45th Street
New York, New York
10017
(212) 644-8025

Technical journal which often contains papers regarding holography. Available in most technical libraries.

Holosphere
Museum of Holography
11 Mercer Street
New York, New York 10013
(212) 925-0581

The best digest of current information on holography available. Published monthly. Back issues available.

Laser Focus
1001 Watertown St.
Newton, Mass.
02165
(617) 244-2939
Good monthly which keeps up to date with laser applications - occasional articles on holography.

Optical Spectra
P.O. Box 1146
Pittsfield, Mass.
01202
(413) 499-0514

Good monthly which keeps up to date with numerous optical applications - occasional articles on holography.

a few good magazine articles on holography:

Popular Photography
Sept. 1977 - Vol. 81, P. 100-103
"Holography - Gimmick or New Visual Art" by P. Sealfon

General discussion of value of display holography in art. Somewhat biased, we feel.

Optical Engineering
P.O. Box 10
Bellingham, Washington 98225
Sept./Oct., 1980 - Vol. 19 #5

Excellent collection of recent papers on holographic applications.

Science Digest
July 1981 - Vol. 89 #6, P. 44-51
"Holography: Laser Pictures That Live" by Tekulsky and Asinof

Average article with some nice photos of recent holographic work (also a photo of one of the authors of this book and quotes from another).

Proceedings of the Society of Photo-Optical Instrumentation Engineers (SPIE)

P.O. Box 10
Bellingham, Washington
98225
(206) 676-3290

Proceedings of technical optics symposiums. Several devoted to holography. See especially:

Vol. 215 "Recent Advances in Holography", Feb. 4-5, 1980 - Los Angeles, Ca.:

Vol. 212 "Optics and Photonics Applied to Three Dimensional Imagery", Nov. 26-30 - 1980, Strasbourg, France.

Vol. 120 "Three Dimensional Imaging", August 25-26, 1977 - San Diego, Ca.

There are a number of interesting articles from Scientific American. Here are some:

1) October, 1976 - Vol. 235 #4, P. 80-95. "White Light Holography" by Emmet Leith (Excellent discussion of white light reflection and rainbow holography. Excellent pictures and diagrams.)

2) April, 1977 - Vol. 236 #4, P. 116-127. "The Theory of the Rainbow" by H. Moyses Nussenzveig

3) June 1974 - Vol. 230 #6, P. 122-127 (The Amateur Scientist) "An Unusual Kind of Gas Laser that Puts out Pulses in the Ultraviolet" by C.L. Stong

4) December 1977 - Vol. 237 #6, P. 108-129 "The Retinex Theory of Color Vision" by Edwin H. Land

5) December 1977 - Vol. 239 #6, P. 182-188 (The Amateur Scientist) "Moire Effects, the Kaleidoscope and other Victorian Diversions" by Jearl Walker

6) June 1979 - Vol. 240 #6, P. 189-200 (The Amateur Scientist) "Experiments with Edwin Land's method of getting color out of black and white."

7) July 1979 - Vol. 241 #1, P. 136-151 "The Visual Perception of Motion in Depth" by David Regan, Kenneth Beverly and Max Cynader.

8) August 1980 - Vol. 243 #2, P. 158-167 (The Amateur Scientist) "Dazzling Laser Displays That Shed Light on Light" by Jearl Walker

9) October 1980 - Vol 243 #4, P. 204-209 (The Amateur Scientist) "A Homemade Mercury-Vapor Ion Laser That Emits Both Green and Red-Orange" by Jearl Walker

places to see holograms

The Museum of Holography
11 Mercer Street
New York, New York 10013
Tel. (212) 925-0581

See-3 Holograms Ltd.
13 Bovingdon Road
London SW6 2AP
England Tel. (01) 736-0076

The Anhalt/Barnes Gallery
750 North La Cienega Blvd.
Los Angeles, Ca. 90069
Tel. (213) 657-4038

Gallery 1134
Fine Arts Research and Holographic Center
1134 W. Washington Blvd.
Chicago, Ill. 60607
Tel. (312) 226-1007

The Green Cow
Via Piascane, 16
20129 Milan, Italy

Holos Gallery
1792 Haight Street
San Francisco, Ca. 94117
Tel.(415) 668-4656

Lasergroup/Holovision
Sandhamnsgatan 25
s-11528 Stockholm
Sweden

House of Holograms
29291 Southfield Rd.
Southfield, MI 48075

Let There Be Neon
451 West Broadway
New York, N.Y. 10013
Tel. (212) 473-8630

Odyssey Image Center
8853 Sunset Blvd.
Hollywood, Ca. 90069
Tel. (213) 652-0983

Optical Laboratory
3 Rue de Universite
67000 Strasbourg, France

Sapan Holographics
240 East 26th Street
New York, N.Y. 10010
Tel. (212) 286-9397

Edmund Scientific
101 E. Gloucester Pike
Barrington, N.J.06007
Tel. (609) 547-8900

3488

Sirens of Light
Goodies Warehouse
200 Second Ave., 2nd Floor
Nashville, TN.
Tel. (615) 254-9697

Musee Francais de l'Holographie
4 Rue Beaubourg
75004 Paris, France
Tel. 277-15-12

classes in holography

Holografix
1420 45th Street #35
Emeryville, Ca. 94608
Tel. (415) 658-3200

School of Holography
550 Shotwell Street
San Francisco, Ca.
Tel. (415) 824-3769

Rick Silberman
148 Penn Street
Providence, RI 02909
Tel. (401) 351-8985

New York Holographic Lab
34 W. 13th Street
New York, N.Y. 10011
Tel. (212) 242-9774

Holography Workshop
Goldsmith's College
The Millard Building
Cormont Road London SE5 9RG
England

Foundation Ideecentrum
P.O.Box 222
5600 MK, Eindhoven
The Netherlands

Fine Arts Research and Holographic Center
1134 W. Washington Blvd.
Chicago, Ill. 60607
Tel. (312) 226-1007

Holography Workshops
Lake Forest College
Lake Forest, Ill. 60045
Tel. (312) 234-3100

societies to belong to

Museum of Holography
11 Mercer Street
New York, N.Y. 10013

Holographic Display Artists and Engineers
Club
℅ Fujio Iwata
Toppan Printing Co.
1-5 Taito, Taito-Ku
Tokyo 110, Japan

Laser Arts Society for Education & Research
(L.A.S.E.R.)
P.O. Box 42083
San Francisco, Calif. 94101

Society of Photo-Optical Instrumentation
Engineers (S.P.I.E.)
P.O. Box 10
Bellingham, Wa. 98225

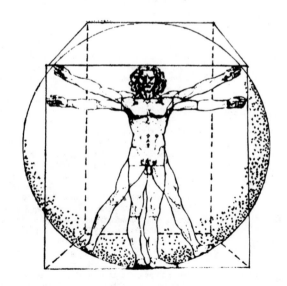

acknowledgements

Photo by Brad Cantos.

Photo by Sharon McCormack.

Carol, Norman & John Schlesinger, In memory of Edith Unterseher, Fedor Unterseher, The whole Unterseher family, Fern Figgie, Mrs. Straight, Jack Barker, Dale Clark, Paramahansa Yogananda, Krishnamurdi, Kenneth Nack, Margaret Moody, Mrs. Shaw, Sharon Henkie, Eddie Pickett, Pim Giebles, Allen Beach, Lindsey Deurer, Katie Ziegler, Jerry Werner, Mike Smith, Alan Erlbaum, Steve Danaher, Molly ?, Staff of Life Bakery, Gini Dean, Alice Roszczewski, Alan Smithline, Lloyd Zimet, Wanda Snyder, Charlie Davies, Mike Williams, Abbey Salit, George Egan, Annie Hayes, Barbara Schwartz, Erix Metz, Randy Levitz, Linda Woods, Julie Thompson, Chris Bodie, Michelle Lovelace, Nils Abramson, Peter Nicholson, Rick Long, Richard Rallison, Jim Feroe, Kent Caroll, Roland Rice, Stephanie Gilliland, Kim Peterson, Gary Zellerbach, Jeff Milton, Nancy Gorglione, Bruce Goldberg, Posy Jackson, Ed Bush, Lee Zemann, Museum of Holography, Brad Cantos, Sharon McCormack, JR, LR, GR, CR, Ricker, Kit, DS, Sceubs,m RS, JS, JS, Szeles, WS, Fran & Jan, Snow, BS, JS, KIS, TT, Beverly, Princess Cruises & RT, RW, SW, MW, MZ, Dennis, Eva & Zsa Zsa Gabor, Ernie Kovaks, Brandy, Greenpeace, Ike & Ren i e Schlesinger, Helen & Dave Lieberman, Barbara & Dick Lieberman & family, Rod Wiley, Richard, Bobby, Gene and Lois Buday, Jeff Trenk, Liz, Sierra Club, Misty, George, Kimmy, Sandy, Jamie Farrel, Gary Deter, Michael Lewis, Lenny Fieber, Mrs. Clark, Mr. Lewlevidge, Ralph Celebre, Tim Butler, Rick Flaster, Lynn Koffler, the Cowens, the Lenoxes Hawkeye, Jon Rottenberg, John Rosenberg, Amy Goldberger, Karen Heller, Alison Lubman, Jon Ben-Asher, Bill Burnett, Paul Erlbaum, Gene Schwartz, Peggy Bikales, Eslee Samberg, Neal Rorke, Cynthia Wong, Aaron Berman, Amy Schnapper, Marianne Bloniarz, Duncan Harp, Kim Millet, Jeff Salz, Mike Cohen, Ruthie Denholtz, Lynn Rolnick, Jill Kessler, Dave English, Lloyd Cross, Emmet Leith, Juris Upatnieks, Jerry Fox, Michael Kan, Dave Schmidt, Dennis Gabor, Yu. N. Denisyuk, Dan Schweitzer, Rick Silberman, Margaret Benyon, Ruben Nunez, John Kaufman, Steve Slater, Vic Rice, Vince DiBiase, Peter Claudius, Bill Molteni, Robert Kurtz, Scott Nemtzow, Paul Barefoot, Robert Pole, Matt Lehmann, Graham Saxbe, Nigel Abraham, Nick Phillips, Wright Huntley, Bob Parker, Kasper Kinosian, Francie King, Kathy Getty, Scott Donahue, Paul Herzoff, Rick Grafton, Rick Magistrali, Francie Oman, Anya Horvath, Norman Spivok, Donn Manzeske, Mark Rennie, Mark Scheiss, Fred Hahne, Chris Outwater, Steve Benton, Milton Chang, B.J. Chang, Michael Foster, Andrejs Graube, Steve Anderson, Ralph Wuerker, Hans Bjelkhagen, Tung H. Jeong, John Caulfield, Anait Stevens, Ken Johnson, Richard Katz, Peter Miller, Sherwin Chew, Joe & Laura Hall, Barbara Weisman, Frank & Judy Foreman, Dr. Carp, Bruce Nauman, Don Day, Rick, Gary Adams, Chris Benton, Curtis Schrarer, Jennifer Merrill, Soupy Sales, Bob Yazell, Steve McGrew, Paul Robinson, Arlie Conners, Ed Lowe, Stephen Hawking, Edwin Taylor, John Weller, Elaine Yano, Robert Erwin, John & Agge Franske, Pruner & Augie, Franz Ross, Dr. Steve Canson, Stuart Heller, All the people at Adeline St., The makers of charts cathedral , All the members of the Laser Arts Society, All the left handed people in the world, All the people with learning disabilities, Linda Charlop, Roy Bonnie, Dr. Edwin Land, Steve Provenoe, Karl Pribram, K.S.Lashly, Marylin Ferguson, David Bohm, Itzhak Bentov, William Tiller, Hattie Cohen, Don Martin, Dr. Shapiro, The Force, The Moon, The Sun, Sue Behrstock, Chris & Marianne Thalken, Bob Lane, Steve Axlerod, Michael Krupnick, T.H. Maiman, Joan Burke, Captain Copy - Multi, Steve & Deni Anderson, Don Broadbent, Woody Archer, Kathy French, Gloria Di Biase, Seth Lefferts, Dr. Hla Tin, Dr. Ken Hanes, Roc Fleishman, Susan DeGrande, Vic Rice, Arlan Hunt, Dave Evans, Rufus Friedman, Leonard Gassensway, Jaye Oliver, Ben, Lili, Susie, Reedproductions, Ecky & Villian, 18%, x-ray, Jim Alinder, Gato, Hugh, Verla, Boy Wonder & Laurie, Carol Rea, The Drake Women/Persons, The Galardi Boys, Doug Baird, Dr. Z, Clyde and his Alley Scoop, Kodak, Reflectasol, Sunpack, Pentax, Erwin & Peter, Meg, Tom at Alien, Los Conilio, Boother, Moss-l-Photo, Peters, Danny Soto, Vickie Simms, Helvim, Bubba, Holy Sobel, Jan Mason, Vince & Charlene, Uncle Van, Kristen, Bentz, Roots & Mohr Aka McQuillan, Coffman, Mohr-Miller, City and County of Los Angeles, Gotham, Petrow, S of Cedd, Food Villa, Wechsler, Tom, Jeremy Kramer, Kent, other productions, the Pinks, Randy & Dale, Buck Bonham, Cranna & Elk, Femprov (Carol, Terry, Pat, Susan & Myself), AWB, JB, RB, SB, GB, WB, JC(4), RC, GC, AC, WD, TDP, DD, ED, RD, JD, RE, JE, Dr. Gonzo, PG, DH, KH, AH, BJ, JWJ, SK, DK, PK, RK, RL, ML, RM, KM, JM, HM, PM, Leah, Cousin Kelley, Sharry & Lisa, Wendy Blair, Larry Menkin, Jane Dornacker, Billy Farley, Rex & David, O Donna O, Hanna & Bruce, Lisa E.G., Screaming Memes, Laura Numeroff, Norton & Cecil & Ben, Deb & Durst, Ed & Kathy, Phil &Linda, Fleminger, SP, DAR, CP, Photographers Supply, JP, MP, & Mary Jo & Kita, Pitta, BP, Susan Houde, Will Walter, Larry Lieberman, Lee Lacy, Don Gillespie, Harriet Casdin-Silver, Sam Morree, John Brown, Randy James, Dorianne Steinberg, Paula Kasler, Nancy Finley, Lon Moore, Allison Studdiford, Pam Brazier, Abe Rezny, Jerry Pethick, Albert Einstein, Robert Collier, Christoph Burckhardt, Lawrence Lin, CB Gaines, Gordon Gould, Isaac Newton, Thomas Young, James Maxwell, Max Planck, Neils Bohr, Robert Hooke, Christian Huygens, Fourier, Fraunhofer, Herman Helmholz, Vander Lugt, Fresnel, W.L. Bragg, D.J.De Bitetto, H.M.A. El-Sum, Brent Puder, Glen Sincerbox, Poohsan Tamura, Everett Hafner, G.L.Rodgers, G.W.Stroke, Nile Hartman, G.Lippmann, George Eastman, C.S. Ih, K.S. Pennnington, B.L. Booth, P. Hariharan, T.A. Shankoff, L.Joly, C.H.Townes, A. Schawlaw, Lorraine Bahrick, Lyrinda Snyderman, Weiser, Marx Brothers, EDF, Nature Conservancy, all people working towards making this a humane planet in the face of incredible odds, Amnesty International, all those who oppose the dangerous rise to power of extreme fundamentalists Dave Meitas, Rosie the Rivetor, Simon Weisenthal, Raoul Wallenberg, all victims of human cruelty and all oppressed peoples, NASA, Ann Holtzmann, Space, Time, you and me.

Photo by Peter D. Miller

index

This is the first book to come complete with its own hologram, which may be viewed using ordinary light. The hologram is a white light transmission hologram which has been converted into a hologram viewed by reflection, by embossing it into a mirror-like plastic sheet. It was made as follows:

First, an open aperture hologram of the type discussed on pages 277-281 was made at Holografix, Emeryville, Ca. Next, the hologram was copied into a material with surface relief by a process perfected by Stephen McGrew. The surface-relief hologram was then plated with metal to protect it, and used as a stamping master to imprint the characteristics of the hologram into the enclosed plastic sheet, much in the same way phonograph records are pressed out. This replicated hologram stores the same interference characteristics as the original, and thus produces the same image.

VIEWING:

To view the image, light from a "point source" should approach the hologram from above. This can be direct sunlight, light from a clear unfrosted light bulb (such as a high intensity lamp), a flashlight or candle in a darkened room, or any light which originates from a small area. The image will not appear as well in florescent lights or in the shade, and multiple lights will form multiple, slightly overlaping, images which will appear "fuzzy". In the case of viewing in the presence of several different types of light, the image quality will tend to be characterized most by that which is brightest. Experiment with different lighting conditions to determine the effects you like best!

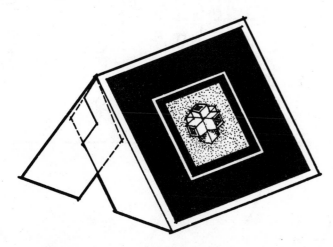

ASSEMBLED DESK-TOP DISPLAY

HOW TO ASSEMBLE DESK-TOP DISPLAY:

1. CUT OUT ALONG HEAVY SOLID LINES.

2. SCORE WITH KNIFE & FOLD OR JUST BEND CAREFULLY ALONG DOTTED LINES.

3. GLUE FLAPS TOGETHER OR TAPE TOGETHER FROM BEHIND OR CAREFULLY CUT SLOTS & TABS & INSERT TABS INTO SLOTS WHERE LETTERS MATCH. ALIGN TOP OF FRONT FLAP AGAINST LIGHT SOLID LINE ON REAR FLAP.

REAR FLAP

FRONT FLAP

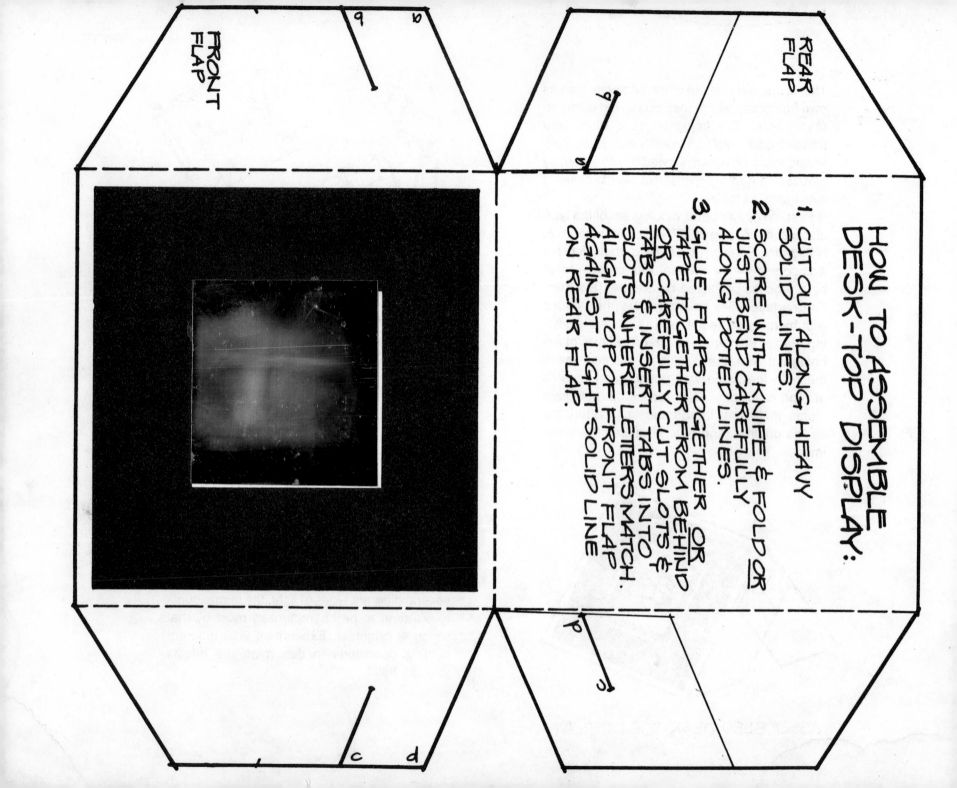